IntechOpen

Supporting open minds since 2005

Bioethics in Medicine and Society
http://dx.doi.org/10.5772/intechopen.87361
Edited by Thomas F. Heston and Sujoy Ray

Contributors
Paulo Santos, Alberto Hespanhol, Pedro Teixeira, Helena Beça, Suresh S. S Suryawanshi, Prajakta P. P. Kamble, Viawas A. Bapat, Jyoti P. Jadhav, Omur Sayligil, Patrick Rech Rech Ramos, Doina-Carmen Manciuc, Georgiana Alexandra Alexandra Lacatusu, Cristina Vasilescu, Maria Alexandra Largu, Reg Green, Jairo Eduardo Eduardo Márquez Díaz, Hiromichi Fumoto, Augustine Pampalny, Jonathan Comyn Comyn De Rothewelle, William Novick, Rodrigo Goncalves, Maria Carolina Formigoni, Edmund Chada Baracat, José Roberto Filassi, José Maria Soares, J. Filipe Monteiro, Liliana Lorettu, Jocelyn Aubut, Rosagemma Ciliberti, Ungureanu Mihail Adeodatus, Astărăstoae Vasile, Daniel Menkes, Jose Marcos Da Silva, Tudor-Ştefan Rotaru, Andrea Montalto, Francesco Musumeci, Edward Collins Vacek S.J., Julio César Ballesteros Del Olmo, Dabota Yvonne Yvonne Buowari, Kehinde Kazeem Kanmodi, Pawan Kumar Kare, Neerja Aggarwal, Sudip Kumar Datta

First published in London, United Kingdom, 2021 by IntechOpen
IntechOpen is the global imprint of INTECHOPEN LIMITED, registered in England and Wales, registration number: 11086078, 5 Princes Gate Court, London, SW7 2QJ, United Kingdom
Printed in Croatia

British Library Cataloguing-in-Publication Data
A catalogue record for this book is available from the British Library

Additional hard and PDF copies can be obtained from orders@intechopen.com

Bioethics in Medicine and Society
Edited by Thomas F. Heston and Sujoy Ray
p. cm.
Print ISBN 978-1-83881-177-8
Online ISBN 978-1-83881-178-5
eBook (PDF) ISBN 978-1-83881-179-2

For EU product safety concerns:
IN TECH d.o.o., Prolaz Marije Krucifikse Kozulić 3, 51000 Rijeka, Croatia, info@intechopen.com or visit our website at intechopen.com.

Meet the editors

Thomas F. Heston, MD, is a clinical associate professor at the Elson S. Floyd College of Medicine, Washington State University, Spokane, Washington, USA. After graduating from St. Louis University Medical School, St. Louis, Missouri, USA, Dr. Heston went on to do his internship at Duke University, Durham, North Carolina, USA, then completed residencies in nuclear medicine and family medicine. He subsequently completed a fellowship in molecular imaging at Johns Hopkins University, Baltimore, Maryland, USA, and served on the Hopkins faculty as an adjunct assistant professor. He has served as the chair of the Ethics Committee at Deaconess Medical Center, Spokane, Washington.

Dr. Sujoy Ray holds an MBBS and MD in Psychiatry. He is also a member of the Royal College of Psychiatrists, London, England. He is currently an assistant professor at IQ City Medical College, India. Dr. Ray has published many articles and reviews. He has been involved in screening trials and providing clinical answers based on evidence about psychiatry. He has regularly participated in medical education programs and various workshops related to psychiatry and research methodology. Dr. Ray held presentations at national and international conferences during his undergraduate and residency days. His topics of interest include mood disorders, schizophrenia, delirium, and obsessive-compulsive disorder.

Contents

Preface

Bioethics in Medicine and Society provides readers with an overview of several topics in the broad field of ethics, ranging from the very personal topics of end-of-life decision making and cancer screening to the larger global practices of geoengineering, the Internet of Things, and medical missions to low- and middle-income countries. Several chapters also provide a broad overview of the field of medical ethics to lay a solid foundation for the more specialized topics. Few readers will be familiar with the broad range and applications of bioethics. Bioethics not only applies to direct patient care but also to laboratory medicine, to our response to climate change, and even to the evolution of religious marriage. Although no single textbook can comprehensively address the entire field of bioethics, this book offers in-depth discussions of a broad range of several important topics in the field.

Thomas F. Heston, MD, MS, FAAFP, FASNC, FACNM
Clinical Associate Professor,
Elson S Floyd College of Medicine,
Washington State University,
Spokane, Washington, USA

Sujoy Ray
IQ City Medical College,
Durgapur, West Bengal, India

Section 1

Clinical

Chapter 1

The Quality of Life in the Light of Immanence and Its Sacrality in the Light of Transcendence

Ungureanu Mihail Adeodatus and Vasile Astărăstoae

Abstract

One of the biggest current issues of European society is the dramatic decrease of the natality rate in most countries. Comparing these low natality rates and the increased mortality rates, we can conclude that in a not too distant future, the European demographic fund will differ greatly from the current one. In this context, the biotechnological industry is looking for a way to extend life for as long as possible. The paper analyses the way in which the two categories – people who want to extend their life and people who want to bring it to an end – report to the transcendental relationship. The questions for which an answer is sought are: Can we, in the light of immanence, reach the truth? Which are the answers of medicine, bioethics, and theology?

Keywords: quality of life, immanent, transcendent, biotechnology

1. Introduction

The concept of "quality of life" has several traits that have been analysed by researchers to assess human life in society [1]. Without the intention of analysing this phrase exhaustively, we want to show how this concept should be understood in the light of the notions of *immanent* and *transcendent*.

Immanence is a term which, referring to a thing or a being, states that existence and reality exist and evolve by themselves, undetermined by an outside cause and it is characteristic to the nature of the object or being.

The term **transcendence** refers to what lies beyond any given domain, beyond the material world. In Immanuel Kant's philosophy, the term designates what lies beyond the limits of experimental knowledge, inaccessible to experience-based knowledge, which exceeds the limits of reality. In religious philosophies, it refers to that instance of divinity that lies above its creations.

The quality of life (in a comprehensive interpretation) represents the totality of conditions that ensure the integrity of biological life, the satisfaction of socio-economic requirements related to the level of material and spiritual living that allows balance, the formation and assertion of human personality. The paradigm that characterises the research related to the quality of life refers to bridging the de facto conditions (the conditions of existence) with the people's perceptions and evaluations, with their moods of satisfaction/dissatisfaction, happiness/frustration.

The concept of "quality of life" leads us to the crucial question "What is life?" whose answer, from a scientific-rational perspective, is ambiguous, since there are live beside God forever while other to live separated from God, for the punishment of their sins committed in their lives. The Bible tells us that only God the Father, His Son, Jesus Christ, and the Holy Spirit (the three Entities of Divinity) possess natural immortality, in Itself, and that they are the source of life and immortality for the created beings. The angels themselves created by God received immortality, provided that they obey God's law. In the Christian faith, the stake of love for the others lived in this life is nothing else but eternal life: "whoever lives in love lives in God, and God in them" (John, 4,16).

The right way regarding the interpretation of immortality can only be found when we will be concerned more about what we live behind, in the collective memory, more about the spirit and less obsessed by physical immortality. When the great values and virtues of mankind will be acknowledged and acquired, then our role on Earth will have been found and we will know immortality.

3. The perspective of bioethics

With the extraordinary development of state-of-the art technology, medicine has turned into a "public interest enterprise that creates a stringent necessity of moral orientation, which the existing medical deontology – preoccupied more with the relationships among the medical personnel – cannot satisfy". [5]. It is in this context that **bioethics** has emerged, with the mission of "accounting not only for the individual, confused conscience, but also for the public, undivided one" [5].

Nonetheless, to be able to use bioethics, we should know the moral it is based on. For instance, for Christian bioethics, abortion is forbidden (with rare exceptions), while for secular bioethics, the ban on abortion is a "forced imposition of the pregnancy, which violates a woman's fundamental rights" [6]. These are fundamental notions related to life and death, to good and bad and to right and wrong.

At the moment, a social marginalisation of Christian moral theology is in place, as well as a development of a "global secular cosmopolite culture, and this theology is presented as belonging to the past, to the Middle Ages, being no longer able to cope with the requirements of the modern individual [6]." The current problem is that secular bioethics cannot have a vision that exceeds the immanent and leads towards the transcendent. Being stuck in their immanence, individuals cannot go "beyond" the limit of rational knowledge and that is why there arise laws in "His image and likeness", of a being that considers itself autonomous in relation to the transcendent. The tragedy of the immanent human being is that they consider themselves too proud to establish a contact with the transcendent, as they are "the measure of all things" and therefore need no help.

Therefore, the following question arises: **do we or do we not collaborate with the transcendent**? Do we want to listen to what comes from "there" or do we live our life according to our own laws created by the human mind? If we want to see what is "beyond", the Christian theology states that we should approach a single way, that of seeking the "information" that comes from the transcendent, and this "information" is, in fact, the Divine Revelation, namely the "voice" of God descending from heavens to tell us that He is the "Life, the Wy and the Truth" (John 14,6) and that we should not look for other laws and ways, since there is only one "Life in Christ". Deviation from the Divine Revelation and the neglection of God's word coming from heavens will only lead us to the darkness of ignorance, since knowledge without God is darkness. Even if technologization will be taken to an extreme, we will end up stating what Socrates did: "I know that I know nothing". Which is true, without God, are nothing, know nothing and will know nothing.

The full connection between the immanent and the transcendent can be made by a Christian theology, which asserts that it has kept unaltered the divine truth; that is why it is the only way that can lead to the transcendent. Sufferance, diseases, infirmity, and death are allowed by God for a well determined purpose: to save humans. "By placing the experience of sufferance and death inside a life that aims at transcendence, the immanent preoccupations, and concerns, including those related to medical assistance, are radically relativized. As a consequence, the traditional Christian approach to these problems – such as the continuation or discontinuation of the treatment, the acceptance of medically assisted suicide or euthanasia – acquires another significance that powerfully contrasts with their meaning presented as certain and safe inside a secular moral" [6].

We need, therefore, to enter the "realm" of religion. It is necessary to see the relationship between bioethics and religion (if any) or whether it should be considered. If the relation of bioethics – and of medicine, in general – with religion is not considered at all, then the data change. As Romanians, with an orthodox tradition of over 2000 years and currently reporting to this religion, we are bound to analyse the relationship between medicine and religion, between bioethics and religion. From this perspective, when an individual lives under the immanence of their action, without reporting to the transcendent, it is impossible for them to see that life on Earth does not end with death but that it lasts forever. Without thinking about immortality and living with the eyes fixed only on the finitude of telluric life, individuals act and relate to this desideratum: live your life as well as you can on Earth, by any means, because death is waiting and there is nothing else after it. In this context, it is understandable that humans, severed from the transcendent, attempt to fulfil their dreams and pursue happiness only in relation to their immanent thinking.

This thinking is not novel, but a 1200 years old, when the European human rationality split from the thinking of the Saint Fathers and all sorts of rationalistic concepts appeared, looking for the truth in immanence, and refusing to relate to the transcendent. A new theology was defined as a theology that evolves and adapts to the times, leading to a rethinking of the cult unit, prayer and Christian life in the eyes of human rationality, which in turn led to a multitude of opinions that tore Christianity to pieces. A divided Christianity became incapable of holding a unitary vision regarding moral guidance and a new global moral came to birth in this void of morality, the secular moral. This secular moral separated itself from theology, aspiring to the status of global moral that should bind people as they weaken their ties with the ethnical roots, cultural traditions, and religious constraints.

The 17th century hails the *Enlightenment* which, paradoxically, aimed at being similar to Christian thinking but refused to relate to traditional Christianity. Enlightenment creates a reinvented Christianity in immanent rational terms.

The current thinking supports loudly the "upgrade" (aggiornamento) of Christianity to cosmopolite liberal culture. Edward Schillebeekx confirms it: "*in the form it was taught to us, Christian revelation no longer provides a valid answer to the questions about God formulated today by most people.* It seems to no longer significantly contribute to the modern individual's understanding of the self, in this world and in the history of humanity. It is obvious that more and more people are dissatisfied and disappointed by the traditional Christian answers to their questions" [6].

This reinvented Christianity has led to the secularisation of Christianity. The 20th century makes millions of victims in the name of a new secular future, allegedly better: that of justice, equity, and human rights. Human life has been channelled to attempts of bringing "heaven on Earth" (for instance, national-socialism and communism) but instead, they brought hell instead of heaven [7].

Orthodoxy states that the divine teaching came to us through individuals inspired by God, who prepared themselves in fasting and prayer before conceiving laws for people. Due to human pride and vanity, the European started making laws for humans that adapt to the times, without considering the laws given by God to humanity through *chosen* and *inspired* people. That is why we have come to the *multitude of opinions* regarding how the world can be led to reach happiness and wellbeing. This desire for happiness was not related to heaven, though, being directed only towards Earth, with its finitude here. Thus, a new type of moral appeared from the benches of rational academies, which was applied in the human society, at the beginning in as many "morals" as opinions there were, later unified in a single moral, the *secular* one.

3.1 Christian or secular bioethics

Christian bioethics argues that life begins at conception and that starting with the zygote, there is a human being in the making, which needs to be treated as such. Reporting human life to eternity, Christian bioethics defends the embryo, acknowledging its soul. Christians do not "raise the problem of the embryo just for fun but reach it starting from the resurrection" [8] but still, they are different in thought. Tristam Engelhard was wondering, "which Christianity is better for a Christian bioethics?", answering that "we should see Christianity from a historical perspective, in the sense that it is at the basis of the historical roots of contemporary Christianities, or that we should see it not like something from the past, old and obsolete, but as something present, animated and alive" [6]. In the West, theology went towards aggiornamento, meaning that the accent transferred on the "accommodation" with the world, while in the East, orthodox Christianity kept the tradition, the accent being placed on sanctification of the world. The aggiornamento of theology has decisively influenced bioethics, which in turn became "accommodated" with the worldly, immanent interests to the detriment of the transcendent. There arises thus a dispute between bioethics built on the grounds of the transcendent Divine Revelation and bioethics built on the grounds of immanent human rationality.

In the context of secular moral there also appears the *principle of permissiveness* which is a procedural one. This principle "will justify, support and explain moral practices based on procedures, such as the rights and contracts, to give up on what we are entitled to (forbearance rights and contracts), including contracts for health care services. The principle of permissiveness will be central not because it would be valued, but because the people's permission is the only accessible source of secular authority. In the absence of a canonical ethic, the bioethics of such a society will prioritize certain practices such as informed consent, the right to refuse the treatment, the development of contracts for health services and the right to do to oneself and to the others who consent as it was mutually agreed upon (for instance, a doctor assisted suicide or euthanasia)" [6]. Kierkegaard states that "Christian bioethics should never become a matter of academic erudition" [6]. As it is impossible to talk about love if one does not love, about the good if one does no good or about forgiveness if one does not forgive. The good that God wants for us differs from the good we want, as humans. Transcendent good differs from the immanent good. In transcendental terms, the "good" of a medically assisted suicide becomes "evil" and in immanent terms, the "evil" of human sufferance becomes a "good" performed through euthanasia. In the absence of transcendental communication, sufferance acquires extreme dimensions in the immanent world, which is why it is put to an end by medically assisted suicide. "In transcendental Christianity, the accent falls on experiencing God, which implies the content of moral life (including what is related to health care (...) therefore, Christian bioethics should be more of a lifestyle than a collection of principles, rules, ideas, or conclusions to arguments" [6] since "nowhere are the questions regarding the meaning of life more troublesome than in healthcare. The hospitals are the arena where sickness, infirmity, and death come to play on people" [6].

4. The quality of life from a theological perspective

Sufferance leads to a crucial question: is there life after death? Does sufferance have any meaning in relation to the transcendent? Even if medicine argues that sufferance can be controlled and death postponed, eventually death wins. Supporting euthanasia these days is like Seneca's pagan stoicism expressed in his letters on suicide: "One death involves pain, another is simple and easy, why not take the easy one?" [9]. It is known that Seneca, to avoid being captured and tortured by Nero, commits suicide together with his wife. In the current secularised society, which encourages self-determination in an exacerbated way – with individuals no longer observing any moral principle – the intention is that religion should encourage active voluntary euthanasia and medically assisted suicide. In other words, Christians should be encouraged to avoid sufferance and choose death as great dignity. This is how the commercials that encourage these suicidal acts have appeared: "The Death Club: leave this life with the same spark you have lived it", "The last journey: experiment death that you have always wanted! Leave in dignity, pleasure and style!" or "Executive death: for those who have always been in control" [6].

Through biotechnology, the life expectancy has increased but the 21st century's individual "lives a spiritual crisis, namely a crisis of significance because they want to maximize happiness in this earthly life, seeking to transform the biological in the search for perfection" [10]. The Church expressed its concern that genetic research is not closely monitored and regulated to ensure the protection of the community [10]. The Romanian Orthodox Church has initiated an action, by setting up in 2001 certain Bioethics Committees that debate litigious themes where the faith in life's Sacrality is expressed, as well as in human dignity and its individuality [11]. There needs to be an open dialogue between researchers (laic people) and the representatives of the Church regarding the implications of genetic technologies: "The determination of a relationship between science and religion should seek common points in time to observe the gap between metaphysics and epistemology. Hence, the importance of initiating a dialogue between religion and science by means of Bioethics" [11].

The influence of evil on those in pain is often remembered in theology. Thus, in the case of St. Martyrs Timotheus and Mavra who were in the ordeals of death, the devil showed to them under an angel's face to allegedly save them from sufferance; however, his presence was meant to take from them the crown of martyrdom [12]. Although individuals suffer with their body, the temptation addresses the soul, because the "soul is the vital principle of the body – the one that gives it life, structures, puts it into motion and keeps it alive [13]. Unfaithful people are afraid of death while the faithful see death as a "gate to eternal life" [14], death "not being evil in itself, what would be bad is to die badly" [15] as St. John Chrysostom says, namely unprepared for eternity. "The world, in its entirety, as well as each thing created by God, has a rationality, they were created for a reason and towards a certain finality" [16]. "The physical world, aware or not of God's love, is created as a means of His love for humans" [16]. Christian theology speaks about the *necessity of sufferance*. Saint Isaac the Syrian argues that sufferance "is necessarily useful to people. For the sinful for humbleness and return from sin while for the more spiritually advanced to strengthen them and help them move forward towards

5. Changing the way organ donation is thought of

From the very beginning the global explosion of interest showed us that we had an opportunity to make a permanent change in the way organ donation was thought of. Everywhere we went the media came too.

Building on this, one or other of us has given speeches in countries as diverse as Venezuela and New Zealand, Russia and South Korea, written dozens of articles for or been interviewed by publications ranging from The Times of India to Oprah Winfrey, from Buddhist television in Taiwan to Vatican radio, spoken in evangelist churches, synagogues and cathedrals, to Muslim and atheist groups and to every age from nursing homes to primary schools.

I have written two books on organ donation. One of them, "The Nicholas Effect" [3], was the basis for the made-for-television movie, "Nicholas' Gift" [4], starring Jamie Lee Curtis and Alan Bates which has been seen by 100 million people worldwide. The other, "The Gift that Heals," has been used by hospitals across the United States as the easiest and most comprehensive introduction available to the human side of organ donation.

One result is that people all over the world feel close to Nicholas and have responded with an enthusiasm that has multiplied the message. To give just one example: a school in Sicily put two clocks in its hallway, one set for local time, the other set to 'Bodega Bay time,' the village in California where we and Nicholas lived, so that every day students are reminded that as they go through life there is always something they can do for others.

The message traveled up as well as across. Pope John Paul II was so moved by Nicholas' story that he authorized the casting of a magnificent bell for a tower designed and built by a San Francisco sculptor, Bruce Hasson, in Bodega Bay, which is dedicated to all children who have died [5].

I always felt, however, there was one cause that needed to be taken up in Italy where the law effectively prevents the two sides of a transplant from making any sort of contact, even by anonymous letters. For more than twenty years I stayed clear of the issue not wanting to be seen as interfering in the laws of a foreign country. But in 2016, at the age of 87 I thought time was running out, so with the help of just one friend, Andrea Scarabelli of Rome, I started a lonely campaign—we called ourselves Don Quixote and Sancho Panza—bombarding the media with information, writing articles, being interviewed on national television. Key elements of the media responded so enthusiastically that we reached tens of millions of people and everywhere in Italy those people began to ask, "If two families want to meet, why not?" Marco Galbiati, the bereaved father of a much-loved 15-year old boy, for example, collected 50,000 signatures in favor of changing the law.

Faced by these two prongs, the Italian National Transplant Center referred the issue to the National Bioethics Committee, which almost everyone saw as a delaying tactic. We bombarded the committee with evidence, most tellingly of how in the great majority of cases in the United States, when families have contacted each other, the health and happiness of both have improved. To general surprise, having scrutinized all the information presented to it, the committee decided in favor of allowing contact—including face-to-face meetings—under the usual conditions and if both sides express a wish for it. The Italian Department of Health endorsed the decision [6] and legislation has been introduced into the Italian Parliament.

6. Contacts between families are strongly encouraged in the United States

Unlike most countries, communication between the two sides is strongly encouraged in the United States, and it works with great success. Naturally, how families respond to that possibility is different in every case. Some do not want any part of it. Others embrace it eagerly and for them the steps are carefully chosen and watched over by their doctors all the way. The normal procedure is for both sides to be asked if they are willing to receive an anonymous letter from the other. If either of them says no, the process stops cold and is abandoned however persistent the other family is. If both agree to communicate, however, one side writes an unsigned letter.

If it is the recipients who make the first move their letters typically start by saying they do not know how to express their gratitude: but then they dig deeply into themselves and even the ones for whom writing is difficult find a power they did not know they had. On reflection, this should not be surprising: they are speaking their most intense thoughts. They then say the transplant has worked well (it generally has) and that they can now do things that were impossible while they were sick.

That part of the letter astonishes many donor families: they had not fully understood until then that before the transplant many of these people had stayed indoors permanently and were afraid to be alone at any time; or that their hopes of marrying, having a baby, getting a degree, traveling, playing games or having a career had been put out of their minds as impossible fancies; or that every night when they went to bed they wondered if they would wake up in the morning. Accompanying that understanding comes a surge of pride in the donor family that even in death their loved one is still bringing peace of mind to complete strangers.

The process of communication is handled by one of the organ procurement organizations (OPOs) designated by the US Department of Health to look after transplant families and staffed by specially trained health professionals. Before the letter is sent it is read at the OPO for signs that it might cause difficulties: does it suggest the writer is looking for a closer relationship than normal? Does it indicate extreme views that might not be compatible with the other family's lifestyle? If something seems wrong the letter-writer may be warned to go easy or the letter may be stopped entirely. If all looks well, however, as it generally does, it is forwarded, unsigned.

Reading it, the donor families might weep a little at the memories it stirs, but the predominant feeling is one of excitement and fulfillment. Having read the letter they have complete freedom of action, anything from putting it away and never doing anything more or, if they wish, replying with an anonymous letter of their own, which will also be scrutinized by the OPO, and if it sends up no cautionary signs (as it rarely does) will be passed on to the first family. Alternatively, it might be the donor family who makes the first move but the process is the same.

If all goes well (as it normally does) the two families can start sending letters that they sign and tell more about themselves. In time (with their medical advisers' approval) they may want to telephone, send photos or email each other—each stage an exciting step that reveals more and more about themselves—and, ultimately, if both agree, they can meet, like families who happen to come into contact with each other under any other circumstance. Why not? Here the bond is far more meaningful than two families who meet because they have a common interest in yoga or their kids' baseball team.

7. Contacts that turn negative are rare

If the people who want to know more about the other side were just a handful, and unrepresentative of the typical family, this would be a side issue. But it is not. This is what Alexandra Glazier, CEO of New England Donor Services, an OPO responsible for organ donation in six states with a population of 14 million, says: "A recent review of our data indicates, that about 52 percent of donor families will connect with a recipient, either by receiving a communication from or sending a

Author details

Reg Green
The Nicholas Green Foundation, USA

*Address all correspondence to: rfdgreen@gmail.com

References

[1] Annual Report on Italian Organ Donation, to December 31 2018; Transplant Information System, Italian National Transplant Center. Available from: https://www.cnt@iss.it

[2] The Gift that Heals by Reg Green copublished by the United Network for Organ Donation. 2007. p. 76. Available from: www.authorhouse.com

[3] The Nicholas Effect by Reg Green. 2009. Available from: www.authorhouse.com

[4] Nicholas' Gift Movie. 1998. Available from: https://www.imdb.com/title/tt0143767/

[5] Santa Rosa (California) Press Democrat; 27 October 1996. [daily newspaper]

[6] "Organ donor families should be free to meet their recipients under controlled conditions if both sides wish, Italian National Committee for Bioethics says". Annali Istituto Superiore di Sanità (ISS) 2019;55(1). Available from: https://www.annali-iss.eu/index.php/anna/article/view/789/525

difficult) wherein, the withholding and withdrawing life-sustaining treatments are the daily bread of intensive care units. Euthanasia, legalized in few countries is a subject of discussion, while ethical concerns about topics like drugs double effects and induced coma also deserve a reflection.

The discussion about the ethical dilemmas about end of life care in terminal diseases have been a subject of concern in all the civilizations, although written documents about the entanglement of the opinion makers, the philosophers and thinkers of the societies, are more easily traced in Western civilizations. Later, with the involvement of the church, the priests had a say regarding the end-of-life and finally with the evolution of medical knowledge the clinicians, had progressively a scientific ascendant regarding the dilemmas about treating terminal illness.

The delaying of the process of death with lengthening of the suffering is, nowa-days at the center of end-of-life ethical debates: the non-acceptance of suffering which can windup with the treatment limitation, at the request of the patient or as a decision of medical team, or as a request of euthanasia, also known as a merciful death, at the request of patient.

In the democratic societies, the decision itself has been subject of discussion. Who should be responsible for decision? The epistemic authority of those who have the knowledge. Or the moral authority of the patient, the family, the surrogate, or a judge in the name of state?

In a nearby future this is a debate that will continue to focus the attention of the modern societies.

4. Withdrawing and withholding life-sustaining treatments

As described previously, in a medical procedure, the treatment is a consequence of diagnosis and should also take into consideration the expected prognosis. Moreover, the treatment strategy is not linear, that is, it can suffer abrupt changes mainly in intensive medicine where life sustaining treatments are involved: they may shift from maintenance of vital functions to palliative care.

As far as life sustaining treatments are concerned, there is a study in the USA, that revealed that in a five years interval of time, deaths in two intensive care units in a period of one year that resulted after withholding or withdrawing these treatments increased, in the same period, from 51 to 90% [3].

A French study involving 43 ICU's revealed that 52% of patients died after they had their treatment withdrawn or withheld [4].

Despite various meetings to standardize the criteria regarding the withdrawal or withholding of end-of-life treatments, cultural and religious barriers have made it difficult to have a uniform code of conduct. However, there is a consensus regarding the guidelines relevant to general principles of treatment renouncement, which can be summarized as: [5].

- The treatment renouncement should result when the treatments have no longer any medical indication or do not offer any well-being to the patient

- The withholding of future treatment is morally and legally equivalent to the withdrawal of treatment

- A mindset, whose only aim is to hasten death, is morally and legally condemnable.

- Any treatment can be withdrawn or withheld

- The withdrawal or withholding of life sustaining treatments is a medical procedure

- In case of any deterrence to withhold life sustaining treatment, the withdrawal of the treatment already prescribed with the same objective should be reconsidered.

The decision of treatment renouncement deserves some reflections and considerations.

These decisions are seldom an urgent decision and, as such, it should not be a hasty and sudden verdict. As far as possible, it should be a consequence of a broad consensus. Any doubt, from any staff member, should be respected and the reason for the apprehension analyzed in minutia.

Ethically, the withdrawal and withholding of treatment are identical attitudes, although, for some clinicians, it is more admissible to withhold than to withdraw treatment.

The treatment renouncement is considered, by some, as passive euthanasia. It is extremely important to realize that the intent of treatment renouncement is to withdraw or withhold an undesired treatment that can lead to the death of the patient, but not to induce the death of the patient. The distinction between dysthanasia and euthanasia is that in the latter there is an intention to administer a drug or a poison with the sole purpose to hasten or cause death.

On ethical reasoning, in intentions and acts, there is a clear divergence between treatment renouncement and an attitude whose main and sole purpose is to cause death.

The treatment renouncement decision has been a seat of disagreeing between the Anglo- Saxon and Central and Southern European countries.

For the former countries (particularly the United States) the decisions, after the due explanations, rests entirely on the patient, while in European countries, particularly those in Southern Europe, the physicians are accountable for the decision. In ethical rationale, so far as the authorship of decision is concerned there is a confrontation between the two principles: autonomy from the Anglo-Saxon and beneficence from the Mediterranean Europe side.

In my opinion, considering that the treatment renouncement is a medical procedure, the responsibility should be on the physician, after all the necessary information is provided to the family.

Is it morally acceptable, that the epistemic knowledge being on the physician side, the decision should rest on the patient or family part? Moreover, when any one of them (patient or family) are extremely fragile, weak and exhausted?

It is be retained and emphasized that treatment limitation is not synonymous with ceasing of any form of treatment. It is a shift from cure to care as the primary goal of providing health services.

5. Dysthanasia and euthanasia

In medicine, end-of-life care is made up of two constituents: the process of death and the moment of death.

The process of death is a stage wherein an individual has been diagnosed with an infirmity, that by the existing biomarkers death will be a natural outcome in a rather short interval of time.

The physicians, with the technological equipment's and procedures at their disposition, can lengthen or hasten the process of death.

On the other hand, it is impossible to portray the moment of death. It is a moment of irreversibility that belongs to the sphere of the unknown.

5.1 Dysthanasia

In this context, the word dysthanasia that emanates from Greek – *dys,* in medical terms, painful and *Thanatos* meaning death – in common language means to retard as much as possible the process of death.

Although conceptually slightly different, treatment stubbornness, therapeutic doggedness, or medical futility have been used as synonymous. In dysthanasia, the attention is focused on the process of death, while in its synonymous, the point of convergence on persistency of cure-oriented treatment decisions, whose consequence may drag out the process of death.

In a context of a medical act, dysthanasia should be perceived as an approach where there is an excessive treatment in relation to the clinical condition and its expected outcome. From the perspective of a medical procedure and in the light of deontological precepts, treatment should consider the expected prognosis, as highlighted previously.

A basic rationale for dysthanasia can be a treatment that presents no beneficial odd for the patient.

For some time, dysthanasia was considered, in a broadest sense, futile care that does not benefits the patient. However, the term futile raised some polemic, since, futile, refers to anything that is unable or ineffective of bringing forth any useful result. Nonetheless, there are treatments that can cause some effect on patients' biological parameters without any beneficial good. This evidence highlights the argument that the effects and the benefits are different facts. The prolongation of life without any cognitive capacity and confined to an intensive care bed cannot, in my opinion, be considered the aim of Medicine. I am fully aware that this is a value judgment, and, as such, it is intrinsically difficult to reach a consensus.

The cause effect correlation, to be unequivocal, should be clearly defined and reproducible. The dispute around treatment stubbornness has been focused around difficulty in deciding what should be considered a medical futility and who should be responsible for this decision.

Regarding the definition, there is a distinction between quantitative and qualitative futility. The previous (quantitative) futility is based on statistical premises – a treatment is futile when the last 100 cases of a certain medical treatment for a distinct medical situation have been unsuccessful. On the other side, qualitative futility is related to a treatment that maintains a patient unconscious or does not withdraw his total dependency in relation to intensive care measures.

Summarizing, should the definition be mathematical or clinical?

Mathematics is a science of certainties while medicine (clinical) is a science of probabilities.

Can there be a minimum common denominator among this epistemic ambivalence?

Another worrisome dispute is related to the decision-making: who should have the ultimate say about the futility of a treatment: someone who has the scientific knowledge about the treatment and its effect (epistemic authority) or one, or his surrogate, who is the subject of treatment (moral authority)?

In this dispute I sign up the point of view of Theodore Brown. In his statement:

"*Moral authority is the capacity to convince others of how the world should be. This distinguishes it from expert or epistemic authority, which could be defined as the capacity to convince the others of how the world is*" [6].

From all the consideration, previously exposed, it seems obvious that an act of dysthanasia or treatment stubbornness can be considered as an act of medical malpractice. How can this demeanor by the physicians be explained?

In my opinion, a feasible and rational justification for dysthanasia can be met, on one side, at the light of philosophical underpinnings and, on the other side, as a safeguard against a complaint of substandard medical practice [7].

Philosophically, dysthanasia can be explained, among others, by the phenomenology of knowledge. Edgar Morin, the French sociologist, noted that "*The great contribution of knowledge left by the twentieth century was the knowledge of limits of knowledge. The major certainty is that uncertainties are unable to be eliminated not only in action but in knowledge*" [8].

On the other side, the good or bad application of technique can be understood by the dialectic. Ethically, every man of science, in this case the physicians, serves two gods: the first god is that of ethics of knowledge – everything should be sacrificed to safe the thirstiness for knowledge. The second god is that of civic and human ethics. In dysthanasia prevails the first one.

Axiology, the philosophical study of value, can also be of relevance in explaining the treatment stubbornness. Since the Hippocratic oath, life is considered as a supreme value. By opposition, death has no value or is a non-value. If this rationale is righteous and undistorted, then treatment stubbornness can be justified.

Finally, a foundation for treatment stubbornness can be explained at the light of hope and escape. For Ernst Bloch, the German philosopher, hope is the most human of all emotions and the denial of anguish [9].

The physicians, particularly those dealing with severe cases, know from their experience, that there is, however small, a probability that the process of death may not be irreversible. Dysthanasia can find an underlying rationale in this hope or in other existential attitudes like escape or absurd rebellion.

As pointed out previously, an argument for treatment stubbornness can rest in a reaction to an accusation of medical malpractice - defensive medicine. Currently, doctors are afraid of malpractice lawsuits; a physician response, entirely or to a certain extent, is based on medical procedures to evade any blame rather than to help the patient in his illness.

Defensive medicine can be positive or negative. In the first setting unnecessary procedures are carried out by the doctors to safeguard himself against any complaint. In the second case, he abstains, from procedures and patients, to protect himself from the same recrimination.

In brief, in defensive medicine, the procedures result not from his innate values and beliefs, but from self-protection against accusation of misconduct in the advent of a detrimental outcome [7].

5.2 Euthanasia

Perhaps, the most disputed end-of-life dilemma in the Western contemporary societies, is around euthanasia. However, its debate can be traced to the Renaissance and Age of Enlightenment, as mentioned earlier.

There are multifold descriptions of Euthanasia. In a medical understanding, it is an intentional act to end a life, to relief pain and suffering. The death is brought about by a doctor, family member or friend through a lethal injection and is at the request of the patient who suffers an incurable disease manifested through unrelievable psychic or physical pain.

The word comes Greek "eu" (goodly or well) and "Thanatos" (death). It has referred as "assisted death" or "friendly death".

In this definition, in my opinion, the most inclusive, there are two premises – unrelievable pain and incurable disease – that need an added analysis and clarification.

Besides dissecting this definition, many a times the expression passive euthanasia is used to describe withholding of some medical procedures or treatments, which was already addressed in a previous section (Withdrawing and withholding life-sustaining treatments) that will need further consideration.

Unrelievable pain.

The state-of-the art in pharmacology has presented the medical science with drugs that can control entirely the pain. The main obstacle with the use of one or more pain killers lies with its side effects when there is a need to increase progressively the dose or upgrade the drugs. The most frightening side effect is the respiratory arrest.

In short, the drug outcome can result in a double effect. Reliving the pain but with a significant odd of causing death. Is it morally acceptable?

Besides the relief of pain, the terminal sedation has also been questioned.

This reflection and discussion will be done in next section.

Incurable diseases.

In Medicine, in a classic definition "incurable" implies an illness without cure that will lead, in a short span of time, to death. In natural history, some diseases, when untreated, end up with the failure of the organ and, ultimately, in the death of the patient. The organ failure can be a consequence of an acute condition or an end-stage chronic situation. With modern technological achievements, many organs can be temporarily or permanently substituted by devices or transplants.

In my point of view, illnesses resulting from an end-stage chronic organ failure cannot be strictly defined as incurable in the sense that the outcome will be, unquestionably, the death of the patient, since the devices that substitute the failing organs can do their function.

The question in debate is whether there is any limitation to the use of these devices.

The permanent use of mechanical devices should consider the prognosis, the quality of life from the perspective of the patient and, first and foremost, the patient autonomy.

As previously stated, the treatment should be proportional to the expected prognosis.

It is accepted by some medical associations and by the Catholic Church, that "the use of extraordinary means to maintain life should be discontinued in an unrecoverable situation of a nearby, certainly fatal, prognosis and when the persistence of such treatments will not bring any benefit to the patient" [10].

Let us consider a situation of a patient with a chronic end stage disease but in full possession of his cognitive abilities who refuses any mechanical device to maintain his life. Should his will be denied because the withdrawal of the mechanical device will be considered by the physician or society an act of euthanasia? If, the alternative to a failing organ was its transplant, could the patient be forced to accept it?

Can a society or a physician impose their will? Is the informed consent a mere rhetoric?

By refusing the mechanical device the patient is rejecting to live permanently with a mechanical device. He is not asking to be killed, although he knows that the consequence of his wish will be his death.

In my opinion, the removal of a device that does nor suppress the evolution of the illness, instead prolongs the process of death cannot be considered an act of euthanasia.

To recognize the restrain of science and technology is an act of matureness. To comply with the patient request is to respect the principle of autonomy, which, according to Kant is the only principle of morality.

It is my view, that many disputes and polemics in relation to end-of-life ethical dilemmas have its outset in the premise that the refusal of dysthanasia is an act of passive euthanasia. The objection of treatment stubbornness is act of good medical praxis and meets the *leges artis.* It cannot and should not be labeled as act of euthanasia, unless in bad faith.

6. Drugs double effects

All medicines used in a treatment may cause unwanted symptoms. They are also called "adverse effects" or "adverse reactions". Side effects happen when a treatment causes a problem because it does more than treat the target issue. Side effects can range from mild to life-threatening conditions.

In end stage diseases, when symptoms like pain or breathlessness are a source of great suffering of the patient, the physician is compelled to prescribe powerful analgesics or sedatives and these medicines may cause an undesired double effect.

In case of an analgesic, besides alleviating the pain, they may depress the respiratory center and cause a respiratory arrest and ultimately the death of the patient.

A major doubt, at the light of ethical principles, is whether the double effect of a drug is acceptable or not?

The principle or doctrine of double effect, often abbreviated as DDE, is a set of ethical criteria which Christian philosophers, like Thomas Aquinas' in his work *Summa Theologica*, have advocated for evaluating the permissibility of acting when one's otherwise legitimate act (for example, relieving a terminally ill patient's pain) may also cause an effect that he would, otherwise, be obliged to avoid (sedation and a slightly shortened life) [11].

In his assessment, this set of criteria is justifiable if the following are true:

The nature of the act is itself good, or at least morally neutral.

The agent intends the good effect and does not intend the bad effect either to do good or as the end in itself.

The good effect outweighs the bad effect, in circumstances sufficiently grave to justify causing the bad effect, and the agent exercises due diligence to minimize the harm.

Resuming, the DDE is based on the idea that there is, morally, a pertinent difference between an "intended" outcome of an act and one that is foreseen by the actor but not deliberately planned to achieve his motive.

This doctrine has been criticized by the consequentialist, like John Stuart Mill, advocate of the utilitarian version of consequentialism. He argues that our moral analysis should ignore matters of motivation, which appeals to a distinction between intended and unintended consequences. In his opinion the scrutiny of motives will reveal a man's character, but utilitarianism does not judge character, only the rightness or wrongness of actions [12]. Thus, he concludes that the DDE should be rejected.

Analyzing and reflecting the DDE at the light of ethical principles, namely of beneficence and nonmaleficence, it is clear that there is a clash between the duty to suppress harm or suffering (do good) and, perhaps, the oldest of codes of conduct that reminds the physician that his main attitude towards the patient should be not to harm him (*primum non nocere*).

In my opinion, even in common jurisprudence there a clear distinction between intention and motive. The intention is the basic element for making a person liable for the crime, which is commonly contrasted with motive. While intention means the purpose of doing something, motive determines the reason for committing an act.

If, there is no other way to suppress the suffering of the patient, than the prescription of an analgesic, should take in consideration the minimal dose to achieve the effect.

In this situation, it is my point of view that the procedure is morally acceptable even knowing that it might be a cause of respiratory arrest and death.

6.1 Terminal sedation

Terminal sedation, sometimes considered as an act of euthanasia, is a procedure wherein a patient is prescribed with a drug to induce sleep or unconsciousness until death occurs as result of the primary disease, while maintaining all other palliative medications. A typical example is a respiratory failure in end stage pulmonary fibrosis. In this stage, the breathlessness induces a suffering that a patient cannot tolerate. The only procedure is to sedate with a minimum effective dose that will induce the patient to lose his cognitive capacities but will preserve his organic functions. The drug prescribed has a short biological half-life. If, for any reason, the medication is brought to a standstill, the patient recovers in no time from his unconsciousness. Thus, this is not an irreversible procedure. Can this practice be considered as euthanasia? In my judgment it does not seems judicious to consider terminal sedation as an act of euthanasia.

7. Religious perspectives on end-of-life dilemmas

Is there any special reason to include religious perspectives in a document on end-of-life ethical dilemmas? In other words, is there any space for religious overview in a field based in a scientific knowledge?

From my point of view there is every reason to entail religious perspectives in a reflection and discussion of end-of-life ethical dilemmas.

First and foremost, I will enumerate the arguments to entail religion in this discussion and in in a second section how the major religions overview these dilemmas.

It is evident that a human being has biological and cognitive functions. In an instance of biological suffering the response should come entirely from medicine, a science based on knowledge; however, in cognitive discomfort, the psyche also has a say.

According to Pew Research Center, 2015 in 2020, 98,1% of world population will be adherent of a religion, with Christianity with 31,2 and Islam with 24,1% occupying the first and second places, respectively (**Figure 1**).

These numbers display, in my opinion, that the religiosity of people cannot and should not be forsaken when analyzing the end-of-life ethical questions. On the other side, even unbelievers, atheists and agnostics can have spiritual concerns, a need in the human psyche to understand the ultimate meaning of our existence and values in life. Spirituality is intricately linked to religion. It is difficult to imagine someone professing a religion and not being spiritual, while the inverse is possible; it is not imperious for spirituality to be coupled to religion.

In mid-nineties, a new term, spiritual intelligence, was introduced by some philosophers, psychologists, and developmental theorists [13]. Spiritual intelligence relies on the concept of spirituality as being distinct from religiosity - existential intelligence. It is, therefore, reasonable to accept that in human suffering religion

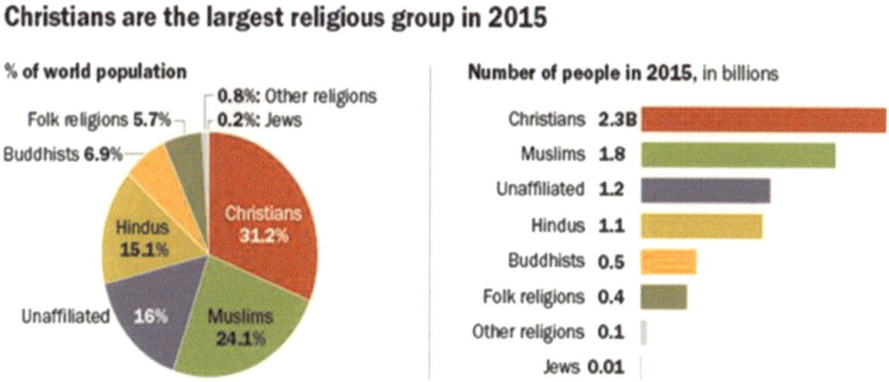

Figure 1.
World population 2015 by religion. Source: Pew research center.

or spirituality and medicine are bound to intersect. To understand the concepts of ethics linked to the end-of-life dilemmas it is fundamental to question "Why?" Get to the bottom to understand the things.

Another good reason to include religious perspective is the historical contribution that Catholic Church thinkers have given to the analysis and discussions of important topics on end-of-life ethical dilemmas.

In the previous section, we had a brief reference to saint Thomas Aquinas on his *Suma Theologica* when he advocated the intention, and not the result, in the doctrine of double effects.

Almost seven centuries later, in 1957, Pope Pious XII in a speech addressed to anesthesiologists, accepted the proposition of double effects of drugs based on principle of liceity of prescription [14].

Another important doctrine in end-of-life ethics, about the difference between ordinary and extraordinary means, was developed by three Spanish Dominican friars (Francisco de Vitória, Domingo de Soto and Domingo Báñez) in the seventeenth century [15].

Other thinking's of Catholic Church, namely regarding treatment stubbornness, were expressed in catholic Catechism and encyclicals (*Evangelium Vitae* by Pope John Paul II) [16].

In the subset of this theme I will make a reference to the most practiced religions and their stand regarding end-of-life ethical dilemmas - treatment stubbornness, euthanasia, drugs double effects, and nutrition and hydration.

7.1 Christian perspective

7.1.1 The Roman Catholic Church

According to the Catholic Catechism *"Discontinuing medical procedures that are burdensome, dangerous, extraordinary, or disproportionate to the expected outcome, can be legitimate; it is the refusal of "over-zealous" treatment"* (treatment stubbornness) [17].

Regarding euthanasia the catechism says that an act or omission which, of itself or by intention, causes death in order to eliminate suffering constitutes a murder gravely contrary to the dignity of the human person and to the respect due to the living God, his Creator.

The position of Catholic Church in relation to nutrition and hydration through artificial means was clarified by Pope John Paul II when he stated that its administration, even when provided by artificial means represents a natural means of preserving life, not a medical act. According to him, its use should be considered ordinary and proportionate. However, he added" *insofar as and until it is seen to have attained its proper finality, which in case of a vegetative state consists in providing nourishment to the patient and alleviating his suffering*" [18]. This conditioning has raised some doubts in theologians and clinicians.

7.1.2 The Greek orthodox church (GOC)

The GOC rejects death resulting from human decisions and condemns as unethical any medical procedure that does not commit to the prolongation of life.

According to the bioethics committee of the Greek Church, withholding or withdrawing of treatment including artificial nutrition is not allowed since there is a possibility of a medical mistake, an unforeseen outcome or even a miracle. Euthanasia is not allowed, and pain relief medication prescription is allowed only in doses that are certain not to depress the respiratory center [19].

7.1.3 Protestant churches

In Protestant churches euthanasia is accepted.

7.2 Judaism

According to the Jewish law, euthanasia is not allowed. A significant divergence regarding Western medical and Jewish ethics, resides in withdrawing and withholding treatments, since in Jewish law, treatments may be withheld while withdrawal is not allowed, considering that this deed may be a factual cause of patient death.

Artificial nutrition and hydration are considered as a form of primary care, and, as such, must be provided.

Treatment for the palliation of pain can be prescribed without fear of an eventual respiratory compromise [19].

7.3 Islam

In Islamic principle, life-sustaining treatments can be withheld or withdrawn in terminally ill patients, while euthanasia is proscribed. The withdrawal of nutrition is considered an unlawful act; however, no reference is made in case the nourishment is through artificial methods. Mitigation of pain is admitted even if death is hastened, provided it was not the physician intention [19].

7.4 Buddhism

From a Buddhist ethics perspective, there is no moral obligation to preserve life at all costs (rejection of dysthanasia). The respect for life forbids killing of living things (euthanasia is outlawed). Artificial nutrition and hydration are not imperative, since they may avert the individual from securing the next stage of his life, the rebirth.

The use of pain killers and the principle of double effect is accepted [19].

7.5 Hinduism

In Hinduism there is no single central authority to enforce submission to Hinduism. When making concrete end-of-life decisions, their attitudes are flexible;

this includes the individual circumstantial background to which religious analysis and arguments can be added.

7.6 Confucian and Taoist perspective

Most Chinese do not consider Confucianism a religion but rather a philosophical system. Unlike the West, in China, cultural and social relations sustain the basis for moral judgment. Thus, it is the family who is responsible for decision making.

There is a difference in religious and philosophical Taoism in regard to the end-of-life: in the former case (religious Taoism), one should accede longevity and immortality, while for the latter (philosophical Taoism) death should be perceived with peace of mind and detachment, and, as such, artificial measures that confront the natural course should not be attempted [20].

8. The living will in the end-of-life

The controversies about the end-of-life ethical dilemmas can be traced to ancient Greece and Rome. In that distant past, it was mainly focused in the treatment stubbornness. However, it is with the development of knowledge and research in lifesaving drugs and technologies, that the debates medialize on various fields, the main point of convergence being euthanasia and dysthanasia.

In a nearby future, with an increase in average life expectancy and innovations in medical technoscience it seems little probable that there will be a decrease in these disputes. Anyhow, a realistic hope in the reduction of the controversies should have its mainstay based in prevention. Hence, an objective of all those who have a leading position in the society should be to curtail the disposition towards this confrontation.

How to downsize this problem?

The struggle to curtail has evolved through the years and there has been no consensus for an acceptable agreement. Starting with the denomination of the dilemmas, passing to the definition itself and ending with the parties involved, an acceptable accordance is far from being a reality.

For example, so far as the denomination of futility is concerned there have been various suggestions to shift for a different terminology like non-beneficial treatment, medically inappropriate, medically inadvisable or not medically indicated, among others, but each one with some drawbacks.

In any dispute of end-of-life dilemmas, there is an involvement of three parties, the patient, the family, and the team of physicians and the institution that provides the health care.

In my opinion, the solution to ease this challenging problem will be met, at least partially, by the unveiling of the living will. Henceforth, in this demanding issue, every effort should be directed to engage all the involved parties in supporting and diffusing the living will.

In the western societies there have been a progressive acceptance and legislation of the living will. The main reason for its broad recognition and approval is the affirmation of the principle of autonomy, through the informed consent, in the Anglo-Saxon countries.

Other arguments for its recommendations are religious creeds (no acceptance of blood transfusions by Adventists cult) and those who reject resuscitation maneuvers fearing a bed quality of life that could result from this procedure.

On the other side, in the eastern countries, or societies where there is a predominance of principle of beneficence and family-oriented decision-making, the living will have hardly made any inroad in this matter.

8.1 The living will

The living will or advanced directive specifies what types of medical treatment are desired by a person in circumstances in which he is no longer able to express informed consent.

A living will can be very explicit and precise or very general. The most frequent statement in a living will, appeals that if the patient suffers an incurable, irreversible illness or condition, and the attending physician decides that the condition is terminal, life-sustaining measures, that would serve only to prolong dying, be withheld or discontinued. More explicit living wills may include details regarding an individual's desire for measures such as pain relief, antibiotics, hydration, feeding, and the use of ventilators, blood products, or cardiopulmonary resuscitation.

The intrinsic objective of living will has two main intents: give the concerned person a control regarding his health in an end-of-life setting and take a burden and distress off the shoulders of his family and thus avoid the self-condemnation complex that sometimes curtails a painful decision. Moreover, the living will elude the discords which may arise among family members about the prescription or withholding of specific treatments.

In this circumstance, it is expected that the author of the will is mature enough to interiorize his illness, the natural process of the disease and his own death.

The decision to draft a living will is not a sprint against time; it must be a follow-up of various steps that entails the acceptance of the illness by the patient, the treatment limitation and the evolution of the disease till the death. For all these discernments required, the living will should not, a priori, be addressed in acute settings.

It should be retained that the living will can be revoked or changed as often as the person wishes. However, he should notify all parties who were informed of the living will.

It is desirable, but not indispensable, for the patient to discuss with his physician his apprehensions, fears, and values. No one, better than his physician, to explain him the natural history of his illness, the prognoses, the technical means, and their limits.

Not under any condition, can the physician use his knowledge to shape or imprint the decision in a negative way.

In brief, the living will have, typically, two parts:

On the one hand is named the person who will be answerable to fulfill with treatment orientations and care in the end-of-life. The attorney should be someone in proximity and trustworthy to the person and with awareness of his line of rationale. The attorney, one or more, can be a family member or a close friend. In case of more than one attorney it should be overtly established if the resolution should be collective or individual and how to decide in a case of a stalemate.

On the other hand, there should be clearly stated what diagnostic methods and treatments should be authorized and those that are to be refused.

The living will is not an alchemy for all the dilemmas related to end-of-life ethics, but it is, beyond any doubt, a good means to obviate many scenarios of anguishing treatment decision making and provide the patient a dignified death.

8.2 The physician role in the living will

What can be the role of the physician in the patient's living will?

The living will that is considered in this text is the one dealing with chronic diseases in their advanced stages. In this setting the physician should consider the timing of the discussion, nature of the illness, quality of life, the end-of-life care, and prognosis.

The dialog should take place not in a specific visit of the patient but through the various follow-up assignments. The physician should explain in a clear and accessible conversation the natural history of the illness with its effect on his quality of life, the end-of-life care that he may eventually need and the treatments options that he may require in the acute exacerbations.

This conversation can be handled by the primary care physician or the specialist consultant who has been following the patient. In my opinion, the consultant physician, with all his experience, will be in a better position to clear the doubts and eagerness of the vulnerable patient.

8.3 The family part in the living will

The feelings and attitudes towards the end-of-life depend on the sociocultural environments of the societies.

In some settings, the family can refuse home nursing the household or allow the treatment limitation and bring back home the family member. This posture can take place for various reasons: spiritual and psychological unpreparedness for the death of their beloved, not knowing how many days would ensue till the patient's death or the physical, familiar and financial concerns that would imply to take care of the patient at home.

8.4 The hospital involvement

The role of the hospital should be focused mainly in preventive measures that should be aimed to remind the physicians, through codes of conduct, to clarify the patient and the family of the evolution of the illness, and, at the proper moment to consider, not enforce, the living will.

In case of disagreement between the family and the physician, the back-office team involving a representative of ethics committee, a psychologist, and an eventual patient religious representative, can have good chances of solving the struggle.

9. Concluding remarks

In Medicine, bioethics is a field of study concerned with the ethics and philosophical implications of certain medical procedures, technologies, and treatments; in this case, the end-of-life ethical dilemmas, are directly or indirectly related to those presumptions.

The main interrogation is to know why these procedures raise so many doubts and uncertainties?

As previously outlined, in the words of French sociologist Edgar Morin, the great bestowal of knowledge left by the twentieth century was the awareness of the limits of knowledge. And he endowed that the major conviction is that uncertainties are unable to be dismissed not only in action but in knowledge.

For this scholar, the knowledge is imbued by three principles of uncertainties:

The brain, the psychic, and the epistemological uncertainties.

In the pursuit of medicine, despite of countless progress in the fields of physiopathology and technical advances that evaluates and modifies the natural history of numerous clinical ailments, the skepticism and the unpredictability can overshadow the result.

In short, the science and its execution entail the uncertainty and sometimes the conflict. This unpredictability leads, from time to time, to question the procedures, the consequences, and the results.

Hence the dilemmas.

Conflict of interest

The author declares that there is no conflict of interest.

Notes/Thanks/Other declarations

Thanks to Pew Research Center for no objection in using their source of investigation.

Author details

J. Filipe Monteiro
Pneumology Department, Santa Maria University Hospital, Lisbon, Portugal

*Address all correspondence to: jfpmonteiro@yahoo.com

References

[1] Monteiro J. Filipe. Deviation from medical Procedures and Ethical Principles. In: Dysthanasia: Delaying the Process of Death through Treatment Stubbornness. Cambridge Scholars Publishing 2018. P. 19-30.

[2] L T Cowley, E Young and T A Raffin, 'Care of the Dying: An Ethical and Historical Perspective.', Critical Care Medicine, 20.10 (1992), 1473-82 <http://www.ncbi.nlm.nih.gov/pubmed/1395671> (accessed on 23/10/2017).

[3] Prendergast TJ, Luce JM. Increasing incidence of Withholding and Withdrawing of Life Support from Critically Ill. Am. J. Respir. Crit Care Med. 1997:155: 15-20.

[4] Lesieur O, Leloup M, Gonzalez F, Mamzer M-F and EPILAT study group. Withholding or withdrawal of treatment under French rules: a study performed in 43 intensive care units. Annals of Intensive Care 20155:15 DOI: 10.1186/s13613-015-0056

[5] Rubenfeld GD, Crawford SW. Principles and Practice of Withdrawing Life-Sustaining Treatment in ICU. In: Managing Death in the Intensive Care Unit. J Randall Curtis and Gordon D Rubenfeld. Oxford University Press 2001. p 127.

[6] Brown Theodore. n.d. "What Is Moral Authority?" http//www.bigthink.com/articles/what-is-moral-authority/.

[7] Monteiro J. Filipe. The Philosophical Underpinnings of Dysthanasia. In: Dysthanasia: Delaying the Process of Death through Treatment Stubbornness. Cambridge Scholars Publishing 2018. P. 45-59.

[8] Morin E. Enfrentar a Incerteza. In Reformar o Pensamento. A Cabeça bem Feita (Tr. La Tête bien Faite). Ed. Instituto Piaget. Lisboa, 1999. P. 61.

[9] Block E. Le Principe Esperance. Paris: Gallimard Editions. 1982. P. 52-55.

[10] Ordem dos Médicos. Novo Código Deontológico. Revista Ordem dos Médicos. Ano 25; N° 97 Janeiro 2009. P.58.

[11] Doctrine of Double Effect. Stanford Encyclopedia of Philosophy. https://plato.stanford.edu/entries/double-effect/

[12] Consequentialism. Stanford Encyclopedia of Philosophy. https://plato.stanford.edu/entries/conse

[13] Zohar D, Marshall I. Spiritual Intelligence: The Ultimate Intelligence. Bloomsbury Publishing 2000.

[14] Pope Pius XII. 1957, "Address to an International Congress of Anesthesiologists." http://www.lifeissues.net/writers/doc/doc31 resuscitation.html_

[15] Gracia D. The old and the new in the doctrine of the ordinary and extraordinary means. In: Wildes KW (Eds.). Critical choice and critical care: catholic perspective on allocating resources in intensive care medicine. Kluwer Academic Publishers. Boston 1995. P. 119-125.

[16] Pope John Paul II. "Evangelium Vitae". https://www.w2.vatican.va/.../john-paul-ii/...hf_jp-ii_spe

[17] Catechism of Catholic Church. Part 3-Section 2, Chapter 2; Article 5-2278. www.vatican.va/archive/ccc_css/archive/catechism/ccc_toc.htm

[18] Pope John Paul II. Address to the participants in the International Congress on "Life-sustaining Treatments and Vegetative state: Scientific Advances and Ethical dilemmas. March 2004. https://

w2.vatican.va/.../john-paul-ii/.../
hf_jp-ii_spe_20040320.

[19] Bülow H-H, Sprung CL, Reinhart K *et al*. The world's major religions' points of view on end-of-life decisions in the intensive care unit. Intensive Care Med (2008) 34:423-430. DOI 10.1007/s00134-007-0973-8

[20] Bowman KW, HUI EC. Bioethics for Clinicians 2000: Chinese bioethics. Can Med Assoc J 163: P. 1481-1485.

Chapter 4

The Psychological Impact on Families of Departed Patients with Infectious Diseases

Doina-Carmen Manciuc, Georgiana Alexandra Lacatusu, Cristina Vasilescu and Maria Alexandra Largu

Abstract

Looking back into history, infectious diseases played an important role in human history being responsible, in terms of pathologies, for more deaths than any other disease. Considering that infectious diseases have a high rate of transmissibility, with an acute debut and sometimes with a fast evolution to exitus, the impact of the news on families of the departed patient diagnosed with an infectious disease can come as a shock. Processing the unexpected death of a family member needs not only the implication of the physician but also the counseling of a specialized psychologist which can help the families through all stages of loss and grief.

Keywords: infectious disease, death, shock, loss, grief, mourning

1. Introduction

Mourning is the emotional response to loss. The mental suffering caused by mourning has emotional, physical, behavioral, social and spiritual consequences [1]. Suffering may manifest differently depending on each person's personality, family values, culture and religious belief [2].

In medical practice, 2 types of mourning are often present which include the anticipatory mourning, occurring when a dear person is suffering from an incurable disease and the mourning associated with personal experience.

Mourning is influenced by age and gender and it is a known fact that women are more vulnerable to loss, due to the level of anxiety and insecurity of attachment ties. Also, the elderly are the most vulnerable in experiencing the mourning to the extent that it reminds them of an expected denouement. In association with age and gender, personality traits such as the level of emotional control and regulation, frustration resistance, traits of optimism and resilience, are important mediation factors of mourning [3].

Identifying the meaning of loss and finding a new meaning in life remains the main challenge for all mourning people. The coping mechanism of the person facing a loss is the recovery process that will allow the continuation of a normal life in the absence of the loved one [4].

2. Stages of processing a loss

Even if its course and manifestations are different, the researchers have shown that the processing of loss has several sequential and predictable stages:

- *Shock and disbelief* which appears immediately after a traumatic loss or event, one cannot believe that the event happened or even denies it.

- *Sadness* represents a healthy and normal feeling of being in a situation of loss. The feelings of inner emptiness, despair, desire for loneliness may also occur in situations of major loss. If there is no improvement from one day to the next, depression has settled in.

- *Guilt* - another common feeling is regret or guilt over things that have remained unspoken or unfulfilled and also can be associated with guilt over feelings of relief after dealing with a long and painful illness of the loved one.

- *Anger/annoyance* - in many cases, loss can seem unfair and make us feel angry or upset. For many people, the therapy for overcoming loss is an exercise in forgiveness or anger release.

- *Fear* - a significant loss can trigger feelings of anxiety, helplessness and insecurity and even lead to panic attacks.

- *Physical symptoms* - due to the intense levels of stress associated with psychological suffering caused by loss, our body also responds physically, not only emotionally by symptoms that may include fatigue, nausea, dizziness weight loss or gain, pain and insomnia [5–7].

All life problems are seen as social problems, and as with any other expressions of life, the suffering must be seen holistically (biopsychosocially) and in relation with the personality as a whole.

3. The role of the therapist

The therapist provides the emotional support needed to express, manage and respond to pain and supports the patient in mourning and establishes the defense mechanisms and strategies that the person uses to cope with the loss through psychological counseling, crisis intervention and supportive therapy. The active, empathetic and thoughtful engagement of the professionals coming in contact with persons living in grief leads to the adaptation of the way the subject constructs his reality and the meaning of loss, the first step towards a humanistic and efficient intervention [8].

Loss is a syndrome with psychological symptoms and somatic signs that may evolve normally to remission or abnormally to a mental disorder that must be addressed like any other mental disorder [9].

The researchers have shown that the loss processing has several sequential and predictable stages (**Table 1**):

- denial of loss;

- anger;

- bargaining;
- depression;
- acceptance [10, 11].

For the family of the deceased patients in hospitals, the shock of loss followed by potential depression of mourning is a reality that the psychologist and the attending physician face every day and try to prevent [12].

Freud	Bowlby & Parkes	Kubler-Ross	Worden	Stroebe &Schut
The process of hyper cathexis (pain, thoughts and memories related to the missing person)	The phase characterized by shock, perplexity and denial that keeps the subject in a state of unreality	The denial phase - the individual does not accept that the loved one has died	The phase of accepting the reality of loss	The phase of loss of orientation that is given by the effort to overcome the loss, such as intrusive thoughts about the missing person, remembrance of the past, deep sadness
The decathexis process (relaxation of the emotional connection with the deceased person)	The phase characterized by deep grief towards the missing person and by waves of crying, pain and anxiety and sometimes by the feeling of the missing person's presence	The rage/anger against oneself for not doing the right thing, the tendency to blame others.	The phase of grief and loss of regret	Orientation restoration phase which includes avoiding feelings of mourning by focusing on the future and solving secondary transition problems
The cathexis process (recovery of the emotional energy invested in the missing person)	Disorganization and detachment phase characterized by strong sadness and feeling of hopelessness	The negotiation phase - the attempt to negotiate in the imaginary or with a supernatural force the return of the lost person.	Phase of adaptation to the environment in which the deceased person no longer exists	
	The reorganization phase, involving the missing relaxation and the return to the future	The sad phase - in which the subject is inhibited, withdrawn, disillusioned, hopeless.	Phase of emotional relocation of the missing and return to the current life	
		The acceptance phase - the individual realizes and accepts the loss through contemplation, reflection, finding meaning in loss and continuation of life		

Table 1.
Models of the stages of mourning.

Challenges of loss in medical practice:

- anticipatory mourning which occurs when a dear person is suffering from an incurable disease;
- the mourning associated with personal experience.

There are 3 basic theories on mourning counseling - some might call them beliefs. The first one suggests the idea of mourning counseling offered to all individuals, especially to families where a parent or child has died. While this belief is understandable, the costs and other factors make impossible the offer to help on such a large scale. Moreover, it may not be necessary for everyone. The second belief is that some people will need help to overcome their guilt, but they will wait until they have difficulties, they will recognize their own need for help and will seek for assistance. This belief is perhaps more economical than the first, but it requires that individuals to go through some discomfort before help is offered. A third belief is based on a preventive approach in the form of an early intervention for avoiding an unresolved/complicated mourning reaction [13–15].

The role of the therapist can also be to introduce the person to the resources offered by the community (support groups, therapists, doctors). Also, is important to mention that mourning counseling can be done in the context of a group. This thing is not only very effective, but it can also be a way to provide emotional and social support during the interaction with people who are experiencing loss and adaptation to mourning. Mourning groups usually exist for more reasons which include emotional support or educational or social purposes. The groups formed for emotional support can continue with the same people for a period of time and embrace more social goals, even if the emotional support continues to be offered [16, 17].

4. Treatment

There are numerous efficient cognitive-behavioral protocols for treating depression

In case of complicated pain, the most effective treatment is Complicated Grief Treatment (CGT), a form of psychotherapy derived from cognitive-behavioral therapy and attachment theory, which is focused on addressing factors that impair adaptation: dysfunctional thoughts, maladaptive behaviors, inefficient emotional regulation strategies. Sessions include the idea of loss, healing, as well as elements of exposure to loss which finally leads to accepting the loss and its consequences by reviewing the relationship with the deceased person and the ability to imagine a future in which the possibility of happiness can be seen, despite the fact that the next person is missing [18]. Further research is needed to understand the role of attachment relationships in emotional and physiological regulation that seems to be deficient in persistent complicated mourning [19–23].

Anger can be a symptom of suffering. Families can be angry at doctors and nurses because they think they have not looked after the deceased well enough. Some are angry even on the deceased, considering that they had neglected their own health. Sometimes, the feeling of anger towards the missing person is associated with all the burdens that the deceased leave on the shoulders of the surviving partner.

Some people feel guilty because of the anger they experience. In other words, they feel condemned because of this anger. The family of patients may feel guilty because they have not offered better medical care, haven't agreed to an operation, haven't consulted a doctor earlier or didn't choose the right hospital. Whatever the reasons are, most of this guilt is irrational and focuses on the circumstances of death [24, 25].

The particularities of death also influence the adaptation to grief and pain, as in the case of the sudden death of a child when mourning is prolonged by more than one year compared to the pain of the death of a sick person. In cases of a child's death, there is always a feeling of guilt. Parents whose children died are highly vulnerable to feelings of guilt that focus on being unable to prevent the child's suffering and death The role of a parent is to protect their child, to take care of their happiness and health, so whatever the circumstances are, mothers feel that they have failed. The death of an only child adds to the trauma of death the loss of the role, identity and social status of the parent [26].

There are different factors that are influencing parents grief and include a) the characteristics of the bereaved person (personality, age, physical and mental health), previous experiences of loss, social, economic, cultural, ethnic and religious background; b) the family typology and the relationship with the extended family; c) age and personality of the child at the time of death; circumstances in which the death occurred (sudden death, anticipated death) [27].

From a social point of view, the impact on the loss of a child is major not only for the parents but also for the grandparents. From our experience, children who are lost at a young age have young grandparents who also can become socially deprived due to their restraint from the social activities they used to participate in and to the inability to focus in a workplace with prolonged downtime. The professional activities also have a low yield, the subject being overwhelmed by memories, feelings of guilt, uselessness, with their somatization perceived as physical and mental asthenia with the impossibility of concentrating on professional and social duties. These can lead in extreme cases to the denial of social status, absenteeism in society and isolation, until the physical aspect of the person concerned is neglected, with the perception of uselessness of one's own.

The final determinant that can give rise to complicated reactions of mourning is generated by social factors. Pain is a process and it is good to be experienced in a social context in which people can support and comfort each other. Lazare has found three social aspects that can inhibit or facilitate mourning. The first is when one cannot talk about loss socially, a second social factor that complicates mourning reactions occurs when the loss is socially denied (when the person and those around him act as if the loss never happened) and third social dimension that can cause complications is the absence of a network of social support, a support matrix formed by the people who have known the deceased [28].

The social impact is often smaller than in the case of the loss of a child, with a lower amplitude of the psychosocial phenomena mentioned above, because each of us considers this to be the natural course of life.

5. Conclusions

In conclusion, the infectious diseases physician and the clinical psychologist find the relationship with the caregivers and the family of the deceased patients difficult to manage. There are social consequences for the family that extend over a long period of time associated with decreased professional efficiency, absenteeism, lower productivity. This leads to emotional issues such as the tendency to blame, with the need for psychological and psychiatric intervention.

For the family of deceased patients in hospitals, mourning and depression are a reality that the psychologist and the attending physician face every day. The team of psychologists and medical doctors are facing cases of severe shock and depression in parents, varying with the age of the child and of the young adult, in cases with an acute or severe disease leading to death.

The psychologist and the physician are in close contact, so that the psychologist succeeds in rebalancing the psychoemotional state in an extremely difficult moment and in profiling the depression caused by mourning with the whole range of feelings of guilt, sadness and isolation.

Author details

Doina-Carmen Manciuc[1]*, Georgiana Alexandra Lacatusu[2], Cristina Vasilescu[2] and Maria Alexandra Largu[3]

1 University of Medicine and Pharmacy "Grigore T. Popa", Iasi, Romania

2 "Sf. Parascheva" Clinical Hospital of Infectious Diseases, Iasi, Romania

3 COTE Foundation, Edina, MN, USA

*Address all correspondence to: dmanciuc@yahoo.com

References

[1] Nicole J. LeBlanc, Naomi M. Simon, Charles F. Reynolds, M. Katherine Shear, Natalia Skritskaya, Sidney Zisook, Chapter 20 - Relationship Between Complicated Grief and Depression: Relevance, Etiological Mechanisms, and Implications, Neurobiology of Depression 2019,: (20)231-239

[2] Bryant RA. Grief as a psychiatric disorder. British Journal of Psychiatry. 2010;**201**:9-10

[3] P. Lazare, A. (1979). Unresolved Grief. In A. Lazare (Ed.), Outpatient psychiatry: Diagnosis and treatment (pp. 498-512). Baltimore: Williams & Wilkens.

[4] Altmaier EM. Best practice in counseling grief and loss: Finding benefit from trauma. Journal of Mental Health Counseling. 2011;**33**(1):33-45

[5] Granek L. Grief as pathology: The Evolution of Grief Theory in Psychology From Freud to the Present. History of Psychology. 2010;**13**(1):46-73

[6] Shear MK, Mulhare E. Complicated grief. Psychiatric Annals. 2008;**39**:662-670

[7] Wright RM, Hogan NS. Grief theories and models: Applications to hospice nursing practice. Journal of Hospice and Palliative Nursing. 2008;**10**(6):350-356

[8] Corruble E, Chouinard VA, Letierce A, Gorwood PA, Chouinard G. Is DSM-IV bereavement exclusion for major depressive episode relevant to severity and pattern of symptoms? A case-control, cross-sectional study. Journal of Clinical Psychiatry. 2009;**70**:1091-1097

[9] Lieberman, P. B., & Jacobs, S.C- Berewment and it's complications in medical patients: A guide for consultation-liaison psychiatrists. International Journal of Psychiatric Medicine. (1987)17, 23-39.

[10] Worden JW, Kubler-Ross E. Attitudes and experiences of death workshop attendees. (1977-78). Omega;**8**:91-106

[11] Worden JWP d a. Englewood Cliffs. In: NY. 1976

[12] Mallon B. Dying, Death and Grief: Working with Adult Bereavement. London: Sage; 2008

[13] Rothaupt JW, Becker K. A literature review of western bereavement theory: From decathecting to continuing bonds. The Family Journal: Counseling and Therapy for Couples and Families. 2007;**15**(1):6-15

[14] Fred D. Davis -Perceived Usefulness, Perceived Ease of Use, and User Acceptance of Information Technology MIS Quarterly Vol. 13, No. 3 (Sep., 1989), pp. 319-340

[15] Welshons J- Awakening from grief, (2002); Open Heart Publication, NJ.

[16] Reilly DM. Death propensity, dying and bereavement: A family system prospective. Family Therapy. 1978;**5**:35-55

[17] Schleifer SJ et al. Suppression of lymphocyte stimulation following bereavement. Journal of American Medical Association. 1983;**250**:374-377

[18] Zisook S, Shuchter SR. Treatment of the depressions of bereavement, American Behavioral Scientist, Special Issue: New Directions in Bereavement Research and. Theory. 2001;**44**:782-792

[19] Prigerson HG, Horowitz MJ, Jacobs SC, Parkes CM, et al. Prolonged

elements include a detailed explanation of the nature of the disease and the treatment options. Risks, benefits, potential consequences of the recommended therapy, and possible alternatives must be elucidated. The likelihood of success, the prognosis without treatment are other aspects to explain. The elements relating to consent conclude the informed consent process. The decision made by the patient reflects which therapeutic option among those proposed to him he considers the best based on the information obtained. The consent is signed by both parties, like a real contract [3].

As proposed by the Italian Society of Cardiac Surgery, there are five informed consents that most commonly concern the cardiac surgeon's activity:

1. Consent to coronary examination, coronary angioplasty, and interventional procedures (in collaboration with the cardiologist);

2. Consent to the processing of personal data for scientific purposes

3. Consent to cardiac surgery

4. Consent to transfusion of blood products

5. Consent of the patient's family members (in particular situations)

Before signing the consent, the patient must make autonomous decisions, without external pressure (from the doctor, the family), understand the severity of the disease, and the risk/benefit ratio of the proposed interventions.

However, the surgeon remains a point of reference for the patient, who, however, informed and aware, entrusts his or her life in the hands of another person; for this reason, the surgeon must direct and advice based on their knowledge and scientific updating. This situation, for example, emerges in cardiac surgery when the patient, who is going to undergo a valve replacement, must choose between a biological and mechanical prosthesis. It is also possible that the surgeon finds himself in a condition of "paternalism, "which means he must make the decision "for the patient" who intends to leave him the choice of the valve prosthesis to be implanted. In this case, this must be reported in the written consent. Furthermore, the surgeon has to stratify the risk for each patient based on his experience and use the most recent risk calculators (EUROScore I and II - European System for Cardiac Operative Risk Evaluation, and STS score - to help him). Society of Thoracic Surgeons score) useful for risk stratification and calculating the risk/benefit ratio [4].

2. Appropriateness of care

The theory of evaluative dynamism considers each therapeutic choice as proportionate/disproportionate; ordinary/extraordinary. The use of a means of life support is considered "proportionate" to the extent that it proves adequate to achieve a specific health goal. The proportionality of the therapeutic option is given by availability, technical possibility, and efficacy. Side effects and risks, possible therapeutic alternatives must be evaluated, and required technical and economic resources should be quantified. The extraordinary judgment correlates with a reasonably high probability of severe risks to the patient's life and health concerning his clinical condition and a low overall efficacy rate. When the effort to find the therapeutic option for life support is very high, or if the choice is linked to severe physical pain and the economic costs are too burdensome for the patient or his relatives, a judgment of extraordinary nature and, therefore, sometimes of disproportionality is configured [5]. In clinical practice, it is mandatory to use this ethical axiom: it is the duty of the doctor to use any proportionate and ordinary means for the treatment of the disease, the use of disproportionate and ordinary methods or proportionate but extraordinary measures may be optional; while it is not ethical to use means that are disproportionate and extraordinary at the same time.

3. Professional self regulation

Professional self-regulation is a personal or collective process of supervision that reflects the ethical concept of responsibility for patient care and includes accountability for that care's scientific and clinical quality. The self-regulation process is best exemplified by introducing the clinical practice of critical reviews through routine meetings on mortality and morbidity [6].

4. Ethics in LVAD as DT

Left ventricular assist systems (LVAD) are widely used to treat patients with heart failure refractory to medical therapy as a bridge to heart transplantation [7]. In subjects who have severe comorbidities such as compromising their inclusion in the transplant list, LVADs can be used as definitive therapy (Destination Therapy - DT). This use implies a series of ethical aspects that include the correct selection of the patient, the patient's precise information, and the definition of a possible protocol for deactivating the LVAD [8]. There are three ethical considerations regarding the selection of patients: the inclusion and exclusion criteria must be clearly explained, justified by rational diagnostic-therapeutic paths, and frequently discussed with the patient during the evaluation phase: The medical criteria must be supported by the most current medical, scientific evidence; The assessment must be conducted based on an explicit weighting of the costs, benefits, and risks associated with the procedure to be applied. The physician is responsible for providing clear information to the patient and family members, clarifying the risks and benefits of different treatment options, including medical treatment, palliative care, and VAD as destination therapy. The patient must be informed about the treatment's goal as an innovative surgical procedure aimed at improving the quality of life. An advanced treatment plan must be presented to the patient, clarifying the patient's preferences in case of complications. It is advisable to propose different scenarios that may occur during daily life through an interview with other subjects with LVAD.

Another ethical aspect concerns the possible interruption of therapy with mechanical support [9]. According to the principle of self-determination, the patient is fully entitled to stop medical treatment so that natural death can occur. A natural death can occur as a consequence of the deactivation of an LVAD due to a dysfunction of the device, due to coexisting comorbidities, or as a consequence of the progression of the disease. There are two ways of deactivating the LVAD. A team can help the patient disconnect the device from the power supply to cause a stop of the pump. The device's interruption does not result in euthanasia since the ventricular function was already severely depressed before implanting the LVAD; the patient's death occurs from the underlying heart disease. The second option involves surgical removal, but this alternative is somewhat complicated because it could cause the patient's death and then. After all, it would contravene the goal of ensuring patient comfort [10].

5. Innovation and research, conflict of interests

Scientific research is an essential part of a heart surgeon's activity. Research must be understood as a systematic investigation aimed at a reflection, followed by scientific evaluation and analysis of the results. The scientific research results are the final product of the work and the correct analysis of the collected data. The faithful exposure of the latter is necessary to respond "ethically" to the need for health [11].

An ethical violation in the field of scientific research is present in case of falsification/fabrication of the results (failure to publish complications related to the intervention, operative mortality, causes of death during a clinical follow-up), or when it is present excessive influence of industry. In randomized clinical trials, in which patients are generally divided into two or more groups to receive a different treatment, absolute impartiality of judgment (not influenced by industry, personal interests) is required to analyze the results. For this reason, many of the studies are conducted in a "double-blind" manner, in which neither the doctor nor the patient is aware of the type of treatment received. Conflict of interest occurs when there are divergences between individual interests and the surgeon's profession's obligations [12]. Conflict of interest can arise due to problems related to the relationship between surgeon and patient and between surgeon and industry, or finally, between public and private activities. Conflict of interest is frequent and even wholly unavoidable. Therefore, the moral purpose is not to cancel it altogether - it would be impossible - but to manage it, always managing to guarantee the prevalence of the primary interest (the patient's health) over secondary interests (e.g., personal earnings and gratification, remuneration). The conflict of interest that may occur between surgeon and industry may be, for example, the choice of a prosthesis that does not give the same guarantees as other more tested prostheses (in this case, personal interest prevails).

6. Ethics in pandemic

In the event of a pandemic, the crucial role of the government, the community, and the health system is to intervene to isolate the disease and slow the spread of the infection. The health system should prepare for an epidemic peak by implementing care capabilities. If the epidemic peak exceeds the available resources necessary to cope with it, a careful resource allocation policy is indispensable. An appropriate resource allocation policy requires first identifying the decision-makers in charge of setting priorities in the distribution. Precise identification of the resources that are limited in time and require a distribution plan is essential. It is necessary to identify the subjects who can benefit most from the use of that particular resource. Finally, it is useful to determine if there is an ethical justification for giving certain groups a priority over others.

Author details

Andrea Montalto* and Francesco Musumeci
Department of Cardiac Surgery and Heart Trasplantation, San Camillo Hospital, Rome, Italy

*Address all correspondence to: andrea.montalto@libero.it

References

[1] Plato. Translated into English with Introduction, Analysis, Marginal Analysis and Index. 3rd Ed.. Oxford University Press at the Clarendon Press; London: 1888. The Republic of Plato.

[2] Kon AA. The shared decision-making continuum. JAMA. 2010; 304(8): 903-904. [Pub Med: 20736477] * Kon provides an excellent discussion of what shared decision making is and the many forms it can take.

[3] Boyle, RJ. The process of informed consent. In: Fletcher, JC.; Lombardo, PA.; Marshall, MF.; Miller, FG., editors. Introduction to Clinical Ethics. 2nd Ed.. University Publishing Group; Frederick, MD: 1997. p. 89-109.* Fletcher's Introduction to Clinical Ethics has several excellent discussion of the foundation of bedside ethics that are still current and pertinent.

[4] Windecker S., Kolh P., Alfonso F., *et al*. (2014). The Task Force on Myocardial Revascularization of the European Society of Cardiology (ESC) and the European Association for Cardio-Thoracic Surgery (EACTS) Developed with the special contribution of the European Association of Percutaneous Cardiovascular Interventions (EAPCI). *European Heart Journal*, 35, 37: 2541. DOI: 10.1093/eurheartj/ehu278.

[5] Calipari M. (2006). *Curarsi e farsi curare: tra abbandono del paziente e accanimento terapeutico. Etica dell'uso dei mezzi terapeutici e di sostegno vitale*. Milano: Edizioni San Paolo.

[6] Cruess S.R., Cruess R.L. (2005). The medical profession and self-regulation: a current challenge. *Virtual Mentor*, 7, 4. DOI: 10.1001/virtualmentor.2005.7.4.oped1-0504.

[7] Bond AE, Nelson K, Germany CL, Smart AN. The left ventricular assist device. Am J Nurs 2003;103(1):32-40.

[8] Rose EA, Moskowitz AJ, Packer M, et al. The REMATCH trial: rationale, design, and end points. Ann Thorac Surg 1999;67(3):723-30.

[9] Bramstedt KA, Wenger NS, McGregor M, et al. When withdrawal of life-sustaining care does more than allow death to take its course: the dilemma of left ventricular assist devices. J Heart Lung Transplant 2001;20:544-8.

[10] Paola F, Walker R. Deactivating the implantable cardioverter- defibrillator: a biofixture analysis. South Med J 2000;93:20-3.

[11] Morreim H, Mack MJ, Sade RM. Surgical innovation: too risky to remain unregulated? Ann Thorac Surg. 2006; 82(6):1957-65. [PubMed: 17133664] * Innovation is differentiated from research and a minimally intrusive method of overseeing innovation is described.

[12] Lo, B.; Field, MJ., editors. Institute of Medicine. Committee on Conflict of Interest in Medical Research, Education, and Practice. National Academies Press; Washington, DC: 2009. What is a conflict of interest? Conflict of Interest in Medical Research, Education, and Practice; p. 45-48.

Chapter 6

Ethical Concerns Regarding Breast Cancer Screening

Rodrigo Goncalves, Maria Carolina Formigoni, José Maria Soares, Edmund Chada Baracat and José Roberto Filassi

Abstract

The incidence and mortality of breast cancer are rising in the whole world in the past few decades, adding up to a total of around two million new cases and 620,000 deaths in 2018. Unlike what occurs in developed countries, most of the cases diagnosed in the developing world are already in advanced stages and also in women younger than 50 years old. As most screening programs suggest annual mammograms starting at the age of 50, we can infer that a considerable portion of the new breast cancer cases is missed with this strategy. Here, we will propose the adoption of an alternative hierarchical patient flow, with the creation of a diagnostic fast track with referral to timely treatment, promoting better resources reallocation favoring the least advantaged strata of the population, which is not only ethically acceptable but also a way of promoting social justice.

Keywords: breast cancer, screening, mammogram, public health, ethics

1. Introduction

According to data from the World Health Organization (WHO), the number of deaths due to cancer will increase up to 45% between 2008 and 2030 and 70% of those deaths will occur in developing countries [1]. To try and change this scenario, the WHO recommends the implementation of cancer control programs that must include cost-effective measures on healthy life style, vaccination programs and screening programs [2]. A screening program consists in a set of coordinated actions with the objective of reducing cancer mortality through early stage diagnosis in an asymptomatic population, with adequate referral to diagnostic and treatment facilities. These programs have four main components: the definition and recruitment of the target population, adequate offer of diagnostic tests with quality assurance, guaranteed offer of follow up exams and biopsies to confirm findings from the initial diagnostic tests, and referral to treatment facilities and timely navigation through the health system [3]. Although screening programs present the potential benefit of reducing mortality, they are not risk-free. The main risks of such a program are the false-positive and false-negative results, and also the occurrence of over diagnosis. All these can lead to clinical and psychological repercussions and, also, to the increase in the health care system expenditure. To address this issue, the Public Health Agency of Canada performed a study to estimate the harms

IntechOpen

3. Ethical implications of mammographic screening

Carefully considering the recommendations of these three different countries with very diverse populations, we can conclude that although mammographic screening in women between 40 and 49 provide a modest benefit in reducing breast cancer mortality, the occurrence of adverse effects is more pronounced.

We can also note that the BCR and BSM adopt a paternalistic approach, reflecting the principle of beneficence. In the meantime, the USPSTF and the CPSTF advocate that the screening decision should be shared with the patient. This way, patients that are more risk averse could opt out of the screening program and patients that value more the potential benefits could opt in, following the principles of non-maleficence and autonomy. However, what we must ask is whether it is possible to convey important information regarding the risks associated with a screening program in a clear and, more important, neutral manner. In this sense, it is of utmost importance that the autonomy principle is respected and that patients are not manipulated to undergo tests or treatments which they do not agree with, due to the use of biased information.

Addressing this issue, Biddle introduced the concept of epistemic risk, defined as the risk of error that comes up at any moment in the process of knowledge production [32]. These errors can happen because of biases during the data collection step and also because of decisions made in scenarios of uncertainty. These decisions reflect the set of values of the involved researchers and have consequences to human health and to the definition of public policies. Rudner agrees with this argument and suggests that it's impossible to prove any hypothesis with full certainty, as there is always a possibility of error. This way, researchers must judge what is the necessary amount of data to accept or reject a hypothesis and this judgment depends on the set of values of the researcher and on the importance of the consequences an error can lead to [33]. Pramesh et al. discuss such a conflict in depth when they justify the necessity of a randomized clinical trial to prove a hypothesis raised by a cross-sectional study, as they believe the data gathered in the latter is not sufficient to support the decision-making process [34].

The reasoning to support a mammographic screening program for asymptomatic women below the age of 50 is not free of the risk of epistemic risks. One kind of epistemic risk associated with to mammographic screening is the inductive risk, defined as the risk of incorrectly accept or reject a hypothesis based on the available evidence [35]. Breast surgeons must accept or reject that a patient has a disease, frequently a ductal carcinoma *in situ* (DCIS), that will evolve causing symptoms and death based on evidence that does not guarantee the veracity of this hypothesis. That happens mostly because of the lack of evidence to predict which cases of DCIS will evolve to become invasive carcinomas. Another epistemic risk, the one in the gathering of data of breast biopsies, occurs in the evaluation of the differential diagnosis between atypical hyperplasia and DCIS. While the first ones are treated with a small surgical procedure, the latter requires surgical excision followed by radiation therapy and, in some cases, endocrine therapy for 5 years. This way, an error committed by the pathologist might lead to an enormous impact in the treatment of the patients. As pathologists have different formations and different experience backgrounds, and as the biopsy evaluation is a subjective process, this is an epistemic risk that is hard to be assessed. In order to try and decrease the odds of such an error is the development of image analyses software. Mercan et al. evaluated 240 breast biopsies comparing the performance of three experienced pathologists and an automated image analysis method. In this study, the automated method performed better than the pathologists in differentiating atypical hyperplasia and DCIS, becoming a promising alternative for the near future [36]. As we saw in these two examples, the information conveyed to the patients eligible to screening are not obtained in the absence of the researchers personal judgment and values. Thus, more than just respecting the autonomy principle in the shared decision-making process, healthcare workers must convey not only the necessary information but also their values and personal beliefs used by them to define their diagnostic and therapeutic decisions. As long as there is ambiguity in the results of mammographic screening studies in asymptomatic women below the age of 50, the priority should be debating the advantages and disadvantages of this strategy, instead of discrediting their opponents [37].

4. Current situation of mammographic screening

In Brazil, as in many developing countries, there is no public policy to the implementation of an organized mammographic screening program. As mentioned previously, there is a recommendation from the Brazilian National Cancer Institute (INCA) for mammogram every two years for women between 50 and 69 years old [7] and the main medical societies recommend an annual mammogram for women starting at 40 years of age [30].

This difference in recommendations happens due to complex interactions between the country's decision makers' interests, beliefs, perspectives and personal values. In the present scenario, with this disparity of recommendations, patients present late stage diagnosis, worse than the ones observed in Norway before the implementation of the local mammographic screening program (**Table 2**) [38].

To evaluate the necessity to expand the INCA's recommendation to other age ranges, Brito et al. analyzed all breast cancer cases, all DCIS cases and all breast cancer related deaths in the city of Aracaju between 1998 and 2014, dividing pages according to age groups [39]. The breast cancer incidence trends remained stable over the studied period across all age groups. Both incidence and cancer-specific mortality in that municipality were similar to the ones observed in countries with the same human development index. The authors concluded that, as these rates remained stable in all age groups, including the ones in which screening is recommended, the investment of public resources to screen women below the age of 50 or over the age of 69 is not justifiable.

A broader study by Rodrigues et al. evaluated retrospective data regarding mammograms between 2008 and 2016 in the public health system [40]. Around nineteen million mammograms were performed in this period with an increase in coverage of 14.5% annually between 2008 and 2012 followed by a stable period

Stage	Brazil (n = 22,527)	Norway (n = 26,883)
I	21.3%	48.5%
II	35.2%	38.5%
III	25.2%	5.3%
IV	8.9%	6.5%
Unknown	1.6%	—

Adapted from Tiezzi et al. [38].

Table 2.
Prevalence of breast cancer according to stage in the state of São Paulo between 2000 and 2017, and in Norway before the implementation of mammographic screening.

between 2012 and 2017. The population coverage of mammogram varied in the period from 14.4% to 24.2% of the target population. This number is far from the 70% coverage recommended by the WHO, necessary to effectively reduce breast cancer-specific mortality [41]. Rodrigues et al. also evaluated the number of mammogram machines available in the country, their geographical distribution and the total number of exams performed in 2016 [42]. In this study, it was demonstrated that Brazil has 4628 machines with a capacity of 14,279,654 exams per year. In 2016, however, only 4,073,079 exams were performed, 29% of the total capacity, displaying a clear under-use of the available infrastructure. The low coverage of the target population with the stable trend in the past few years associated with the under-use of the available infrastructure raises the hypothesis of the existence of barriers to access to the healthcare system.

The Barretos Cancer Hospital adopted an alternative to improve the coverage of the screening program with the use of mobile mammogram machines in trucks reaching 108 municipalities in the northeastern region of São Paulo, targeting women between 40 and 69 years old. Greenwald et al. evaluated the efficacy of this initiative from 2010 to 2015 [43] and, in this period, 122,634 women were evaluated with a coverage of 54.8% of the target population, referral of 12.25% of these women were referred for additional exams with a cancer detection rate of 3,63/1000 women. 92.51% of the referrals to treatment centers were successfully accepted. The results obtained by this program are very promising, showing the potential to be expanded to other regions and other countries.

5. Barriers to access the healthcare system

Brazil is a developing country with a population of 209.3 million inhabitants with enormous social and economical disparities between its 5 regions [44]. Moreover, there are also inequalities in the distribution of human resources and health infrastructure with a significant variation in the number of hospital beds and physicians dedicated to oncological patients leading to significant differences in health outcomes [8]. Another source of outcome variability is the duality of access to the healthcare system. Every Brazilian citizen has unrestricted access to the public health system (PHS) and the richer portion of the population also has access to private healthcare providers through out-of-pocket direct expending or through healthcare insurance companies. This duality of the system is perverse in a way that it perpetuates the idea that a small portion of the population has access to state-of-the-art diagnostic and treatment facilities while the majority of the population, around 71%, depends exclusively on the PHS with all its limitations. When comparing this two scenarios, we observe a striking difference in the initial stage of the breast cancer patients; the majority of the patients seen in the private setting is diagnosed with early stage tumors, whereas the majority of patients that depend on the PHS is diagnosed with locally advanced tumors [45], a clear indication that difficulties to access the healthcare system are the main obstacles to early detection. 37% of the breast cancer cases diagnosed in the PHS are stage III or IV while in the private sector this number falls to 16.2% [46]. These data are corroborated by national studies that showed intervals of 75 to 185 days between initial symptom presentation and initial biopsy [45] and a median interval of 113.4 days between indication and initiation of radiation therapy [47]. For comparison purposes, patients seen in the private setting can have diagnostic tests and start treatment in less than 30 days.

To identify the main barriers to access to the PHS, Vieira et al. conducted a systematic review of the literature identifying 30 publications on this topic [48].

In a general analysis, it has been identified an underuse of mammogram machines on the north and northeast regions of the country and a mammogram coverage of only 35% of the Brazilian women, most of them in the private setting. The main issues related with not having a mammogram performed are non-white ethnicity, low educational level, low familiar income and not having health insurance. Another interesting finding of this study is that normally the treatment of breast cancer is performed in big cities and patients end up traveling more than 100 miles from their residences to the hospitals [49]. Even before the start of treatment, patients have to face delays of more than 60 days in 36.9% of the cases, because of inefficient referral and navigation. The main issues related with delays in the initiation of treatment were non-white ethnicity, not having a partner, low educational level, early stage breast cancer and dependence of the PHS [50]. Exclusive dependence on the PHS and non-white ethnicity were also associated with higher breast cancer-specific mortality [51].

6. Recommendation and discussion of the ethical dilemma

Considering the inefficacy of the screening programs in developing countries and the lack of solid evidence supporting screening of asymptomatic women below the age of 50 years old, we recommend resources reallocation to improve access to the healthcare system and the implementation of a fast track between diagnostic and treatment facilities to symptomatic patients, based on the hierarchical flow proposed by Migowski et al. (**Figure 1**) [52]. This algorithm proposes three different actions: educational activities in primary care facilities to raise awareness regarding breast cancer and also the potential benefits and harms of mammographic screening; to offer the option of screening mammogram to asymptomatic women aged 50 to 69 during their visit to the primary healthcare provider; and to promote priority access to symptomatic patients, without the need of prior scheduling, in which the ones with suspicious lesions will be referred to diagnostic facilities. This recommendation is supported by Rawls' two principles of justice [9]. The first principle governs that all persons have equal rights and freedoms. The second principle governs that the adoption of policies that generate social or economical inequalities is only acceptable if it favors the least advantaged portion of society. The promotion of educational activities proposed by Migowski et al. [52] is supported by Rawls' first principle since it standardizes the access to a basic right, education. The second part of the recommendation is justified by Rawls' second principle of justice. The adoption of a fast track to symptomatic patients, removing the need of a prior appointment or referral, promotes the reallocation of public resources to remove barriers in the access to care, reducing delays in diagnosis and treatment and, therefore, reducing inequalities in favor of the least advantaged part of the population that relies solely in the public health system. Although the recommendation favors a part of the population, it does not violate individual rights, as asymptomatic patients will still have access to screening mammogram in their routine visits to their primary healthcare providers. Moreover, the proposed recommendation promotes equal access to breast cancer diagnosis and treatment as it removes the age boundaries, starting to provide care to women below the age of 50 years old, an age range responsible for a large amount of new cases in developing countries and that were not previously included in the past recommendation [6].

Let us consider for a moment a hypothetical scenario in which the healthcare system works perfectly without any access barriers. Even in this setting, mammographic screening as it is currently suggested would not be ideal in developing countries. The current evidence that recommends mammographic screening is not

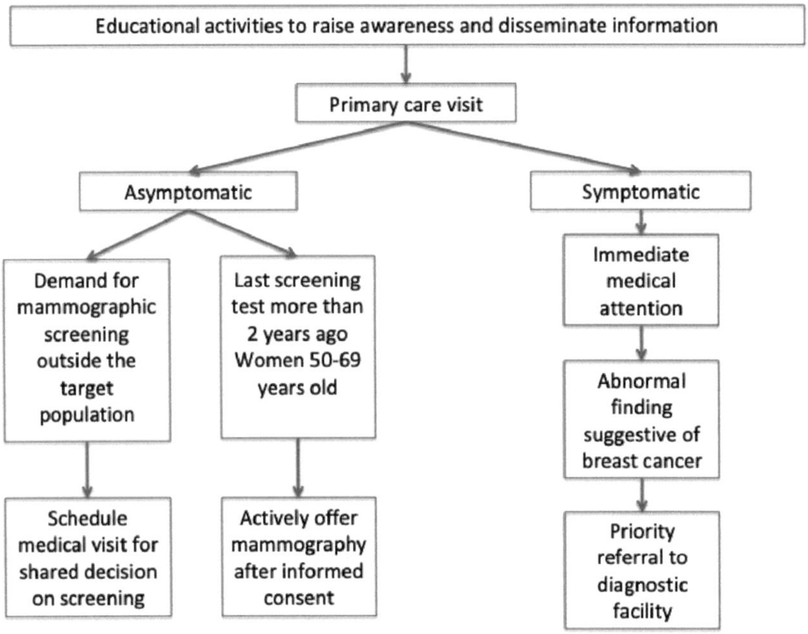

Figure 1.
Hierarchical flow to promote early diagnosis and treatment of breast cancer. Adapted from Migowski et al. [52].

unanimous and large randomized clinical trials did not show a robust mortality reduction attributable to it [13, 53]. Moreover, even if these studies showed a significant mortality reduction attributable to screening mammogram, their results would hardly be applicable to the developing countries' realities. Those studies were conducted in countries with high human development index (HDI) and in the context of organized screening. Brazil and most developing countries have lower HDIs and promote opportunistic screening due to the weak organizational structure of the healthcare system. This way, the international studies that assessed the effectiveness of mammographic screening lack the necessary external validity to be applied in developing countries. A recent article published by Vale et al. suggested, that the opportunistic screening program employed in the state of São Paulo, Brazil, promoted an increase in early stage diagnosis without, however, presenting data regarding mortality reduction [54]. Without data showing mortality reduction attributable to the screening program it is impossible to conclude whether this model is effective or not. Adding up to that data we have some concerning facts associated with screening women between the age of 40 and 49; we observe that less than one death from breast cancer is avoided for every one thousand screening mammograms performed; two hundred and ninety-four false-positive results (**Table 1**) generate additional diagnostic procedures leading to economical impact to the health system and also physical and psychological impacts to the patients. Based on everything that was exposed in this paragraph, we can conclude is not adequate from Bentham's and Mill's utilitarianism perspective [10, 11].

In this context, with the shortage of resources to invest in an organized mammographic screening program and without solid data to justify its implementation, can we accept a sub-optimal program? On the one hand, the inexistence of a screening program can lead to the increase in the number of cases diagnosed in late stages, for which the treatment options might be inaccessible and, sometimes, ineffective. On the other hand, developing countries, such as Brazil, sometimes lack the necessary infrastructure to perform timely screening mammograms to the whole eligible population and the consequent breast biopsies of lesions identified through screening [55]. In order to consider a screening program adequate, it must be acceptable, accessible, and sustainable, it must promote equity and it must be economically efficient to the target population [56]. As it has been demonstrated in this chapter, the Brazilian mammographic screening program is not accessible, since the coverage does not reach 30% of the target population [42]. This program is not sustainable either since there is a huge delay between the identification of an abnormal mammogram and the necessary biopsy to confirm the diagnostic. Due to the incapacity to adequately follow-up and refer patients with abnormal findings, the risk of a false-positive result must be considered clinically and ethically relevant. Finally, it's been demonstrated that mammographic screening in developing countries is not cost-effective when compared to the alternative of treating patients with palpable initial lesions [57, 58]. Contemplating all these issues, Sedhom et al. argued that clinical examination of the breast with fast referral to avoid delays in diagnosis and treatment, although not a screening program, must be considered a more pragmatic and adequate choice than screening mammogram in developing countries [59].

7. Conclusion

When weighing the benefits and harms of a mammographic screening program in a developing country, in a context where breast cancer-specific mortality has been increasing in the past few decades, it is extremely hard to justify increasing the age range to women aged 40 to 49 years old from an utilitarian perspective, since the amount of resources to establish and make the system work adequately is prohibitive. An alternative strategy that promotes easy access and fast referral of symptomatic patients, relegating a secondary role to mammographic screening, favors a larger and more vulnerable part of the population that depends solely on the PHS. This reallocation of resources to favor the least advantaged members of society is not only ethically justifiable but also a way of promoting social justice.

Conflict of interest

Rodrigo Goncalves has received consultation fees from EMS Pharmaceuticals in 2019 and 2020 and from Novartis in 2019, not related to the topics of this chapter.

The remaining authors do not have any conflicts of interest to disclose.

Author details

Rodrigo Goncalves[1*], Maria Carolina Formigoni[1], José Maria Soares[2], Edmund Chada Baracat[2] and José Roberto Filassi[1]

1 Breast Surgery Sector, Discipline of Gynecology, Department of Obstetrics and Gynecology, Hospital das Clinicas of the University of Sao Paulo Medical School, São Paulo, Brazil

2 Discipline of Gynecology, Department of Obstetrics and Gynecology, Hospital das Clinicas of the University of Sao Paulo Medical School, São Paulo, Brazil

*Address all correspondence to: rodrigo.g@hc.fm.usp.br

References

[1] WHO. Key statistics on cancer. 29/02/2020]; Available from: https://www.who.int/cancer/resources/keyfacts/en/.

[2] WHO. National Cancer Control Programmes. Available from: https://www.who.int/cancer/nccp/en/.

[3] INCA, Rastreamento organizado do câncer de mama 2011.

[4] 1000 person tool. 2020 [cited 2020 09/11/2020]; Available from: https://canadiantaskforce.ca/tools-resources/breast-cancer-update/1000-person-tool/.

[5] INCA. Estatísticas de câncer. 2020 29/02/2020]; Available from: https://www.inca.gov.br/numeros-de-cancer.

[6] Simon, S.D., et al., Characteristics and prognosis of stage I-III breast cancer subtypes in Brazil: The AMAZONA retrospective cohort study. Breast, 2019. 44: p. 113-119.

[7] INCA, Diretrizes para a detecção precoce do câncer de mama no Brasil. 2015.

[8] da Mota Almeida Peroni, F., et al., Realizing the right to health in Brazil's Unified Health System through the lens of breast and cervical cancer. Int J Equity Health, 2019. 18(1): p. 39.

[9] Rawls, J., A theory of justice. Rev. ed. 1999, Cambridge, Mass.: Belknap Press of Harvard University Press. xxii, 538 p.

[10] Bentham, J., An Introduction to the Principles of Morals and Legislation.

[11] Mill, J.S., Utilitarianism.

[12] Siu, A.L. and U.S.P.S.T. Force, Screening for Breast Cancer: U.S. Preventive Services Task Force Recommendation Statement. Ann Intern Med, 2016. 164(4): p. 279-96.

[13] Nelson, H.D., et al., Effectiveness of Breast Cancer Screening: Systematic Review and Meta-analysis to Update the 2009 U.S. Preventive Services Task Force Recommendation. Ann Intern Med, 2016. 164(4): p. 244-55.

[14] Nelson, H.D., et al., Screening for breast cancer: an update for the U.S. Preventive Services Task Force. Ann Intern Med, 2009. 151(10): p. 727-37, W237-42.

[15] Marmot, M.G., et al., The benefits and harms of breast cancer screening: an independent review. Br J Cancer, 2013. 108(11): p. 2205-40.

[16] Medical Advisory, S., Screening mammography for women aged 40 to 49 years at average risk for breast cancer: an evidence-based analysis. Ont Health Technol Assess Ser, 2007. 7(1): p. 1-32.

[17] Humphrey, L., et al., in Screening for Breast Cancer. 2002: Rockville (MD).

[18] Ringash, J. and C. Canadian Task Force on Preventive Health, Preventive health care, 2001 update: screening mammography among women aged 40-49 years at average risk of breast cancer. CMAJ, 2001. 164(4): p. 469-76.

[19] Olsen, O. and P.C. Gotzsche, Screening for breast cancer with mammography. Cochrane Database Syst Rev, 2001(4): p. CD001877.

[20] Alexander, F.E., The Edinburgh Randomized Trial of Breast Cancer Screening. J Natl Cancer Inst Monogr, 1997(22): p. 31-5.

[21] Andersson, I. and L. Janzon, Reduced breast cancer mortality in women under age 50: updated results

from the Malmo Mammographic Screening Program. J Natl Cancer Inst Monogr, 1997(22): p. 63-7.

[22] Bjurstam, N., et al., The Gothenburg breast screening trial: first results on mortality, incidence, and mode of detection for women ages 39-49 years at randomization. Cancer, 1997. 80(11): p. 2091-9.

[23] Frisell, J. and E. Lidbrink, The Stockholm Mammographic Screening Trial: Risks and benefits in age group 40-49 years. J Natl Cancer Inst Monogr, 1997(22): p. 49-51.

[24] Miller, A.B., et al., The Canadian National Breast Screening Study-1: breast cancer mortality after 11 to 16 years of follow-up. A randomized screening trial of mammography in women age 40 to 49 years. Ann Intern Med, 2002. 137(5 Part 1): p. 305-12.

[25] Moss, S.M., et al., Effect of mammographic screening from age 40 years on breast cancer mortality at 10 years' follow-up: a randomised controlled trial. Lancet, 2006. 368(9552): p. 2053-60.

[26] Shapiro, S., Periodic screening for breast cancer: the HIP Randomized Controlled Trial. Health Insurance Plan. J Natl Cancer Inst Monogr, 1997(22): p. 27-30.

[27] Tabar, L., et al., The Swedish Two-County Trial twenty years later. Updated mortality results and new insights from long-term follow-up. Radiol Clin North Am, 2000. 38(4): p. 625-51.

[28] Hellquist, B.N., et al., Effectiveness of population-based service screening with mammography for women ages 40 to 49 years: evaluation of the Swedish Mammography Screening in Young Women (SCRY) cohort. Cancer, 2011. 117(4): p. 714-22.

[29] Jonsson, H., et al., Service screening with mammography in Northern Sweden: effects on breast cancer mortality - an update. J Med Screen, 2007. 14(2): p. 87-93.

[30] Urban, L., et al., Breast cancer screening: updated recommendations of the Brazilian College of Radiology and Diagnostic Imaging, Brazilian Breast Disease Society, and Brazilian Federation of Gynecological and Obstetrical Associations. Radiol Bras, 2017. 50(4): p. 244-249.

[31] Decision-making for cancer care: A review of current practice and opportunities for improvement in Latin America. 2019, The Economist Intelligence Unit.

[32] Biddle, J.B., Epistemic risks in cancer screening: Implications for ethics and policy. Stud Hist Philos Biol Biomed Sci, 2019: p. 101200.

[33] Rudner, R., The Scientist Qua Scientist Makes Value Judgments.

[34] Pramesh, C.S., et al., Ethics of "standard care" in randomised trials of screening for cervical cancer should not ignore scientific evidence and ground realities. Indian J Med Ethics, 2013. 10(4): p. 250-1.

[35] Douglas, H., Inductive risk and values in science.

[36] Mercan, E., et al., Assessment of Machine Learning of Breast Pathology Structures for Automated Differentiation of Breast Cancer and High-Risk Proliferative Lesions. JAMA Netw Open, 2019. 2(8): p. e198777.

[37] Ennis, R. and A. Jotkowitz, Good ethics begins with sound medicine: prostate cancer screening and chemoprevention. Am J Bioeth, 2011. 11(12): p. 26-7.

[38] Tiezzi, D.G., et al., Current Breast Cancer Screening Scenario in Brazil. Revista Brasileira de Ginecologia e Obstetrícia, 2019. 41: p. 633-635.

[39] Brito, E.A.C., et al., Assessing trends of breast cancer and carcinoma in situ to monitor screening policies in developing settings. Sci Rep, 2019. 9(1): p. 14144.

[40] Rodrigues, D.C.N., et al., Temporal changes in breast cancer screening coverage provided under the Brazilian National Health Service between 2008 and 2017. BMC Public Health, 2019. 19(1): p. 959.

[41] WHO, Cancer control: knowledge into action. WHO guide for effective programmes: prevention. 2007, Geneva: WHO.

[42] Rodrigues, D.C.N., et al., Difficult Access and Poor Productivity: Mammography Screening in Brazil. Asian Pac J Cancer Prev, 2019. 20(6): p. 1857-1864.

[43] Greenwald, Z.R., et al., The performance of mobile screening units in a breast cancer screening program in Brazil. Cancer Causes Control, 2018. 29(2): p. 233-241.

[44] Bank, W., Brazil - systematic country diagnostic. 2016.

[45] Werutsky, G., P. Nunes, and C. Barrios, Locally advanced breast cancer in Brazil: current status and future perspectives. Ecancermedicalscience, 2019. 13: p. 895.

[46] Lee, B.L., et al., Breast cancer in Brazil: present status and future goals. Lancet Oncol, 2012. 13(3): p. e95-e102.

[47] TCU, Relatório de Auditoria Operacional: Política Nacional de Atenção Oncológica. , S.d.F.e.A.d.P.d. Governo, Editor. 2011.

[48] Vieira, R., A. Formenton, and S.R. Bertolini, Breast cancer screening in Brazil. Barriers related to the health system. Rev Assoc Med Bras (1992), 2017. 63(5): p. 466-474.

[49] Oliveira, E.X., et al., [Access to cancer care: mapping hospital admissions and high-complexity outpatient care flows. The case of breast cancer]. Cad Saude Publica, 2011. 27(2): p. 317-26.

[50] Medeiros, G.C., et al., [Determinants of the time between breast cancer diagnosis and initiation of treatment in Brazilian women]. Cad Saude Publica, 2015. 31(6): p. 1269-82.

[51] Guerra, M.R., et al., Breast cancer survival and health iniquities. Cad Saude Publica, 2015. 31(8): p. 1673-84.

[52] Migowski, A., et al., Guidelines for early detection of breast cancer in Brazil. III - Challenges for implementation. Cad Saude Publica, 2018. 34(6): p. e00046317.

[53] Miller, A.B., et al., Twenty five year follow-up for breast cancer incidence and mortality of the Canadian National Breast Screening Study: randomised screening trial. BMJ, 2014. 348: p. g366.

[54] Vale, D.B., et al., Downstaging in opportunistic breast cancer screening in Brazil: a temporal trend analysis. BMC Cancer, 2019. 19(1): p. 432.

[55] Vizcaino, I., et al., Short-term follow-up results in 795 nonpalpable probably benign lesions detected at screening mammography. Radiology, 2001. 219(2): p. 475-83.

[56] Shah, S.C., et al., Cancer Control in Low- and Middle-Income Countries: Is It Time to Consider Screening? J Glob Oncol, 2019. 5: p. 1-8.

[57] Gelband, H., et al., Costs, affordability, and feasibility of an essential package of cancer control interventions in low-income and middle-income countries: key messages from Disease Control Priorities, 3rd edition. Lancet, 2016. 387(10033): p. 2133-2144.

[58] Groot, M.T., et al., Costs and health effects of breast cancer interventions in epidemiologically different regions of Africa, North America, and Asia. Breast J, 2006. 12 Suppl 1: p. S81-90.

[59] Sedhom, R. and B. Gyawali, When Is a Suboptimal Approach to Cancer Screening Better Than None? AMA J Ethics, 2020. 22(2): p. E93-101.

Chapter 7

Ethical Evaluation of Clinical Research on Complementary and Alternative Medicine

Omur Sayligil

Abstract

Traditional medicine (TM) as well as complementary and alternative medicine (CAM) practices have been used more frequently; since modern medicine has gravitated toward a dehumanistic situation due to the extensive workload of healthcare professionals and thus lack of time given to each patient and mistrust of patients due to some side effects of latest treatment options, in addition to TM and CAM practices having been more affordable, accessible, most often noninvasive, and seen as a hope during terminal periods of some diseases. In order to ensure TM and CAM complying with the standards as other healthcare services, it is necessary to address and evaluate scientific and ethical issues for these clinical researches as well. On the other hand, so far, the ethical side of TM and CAM has not been discussed in detail. Issues such as misleading information, informed consent, publications, patient-physician relationship, and confidentiality should be discussed within the framework of ethics. Ethical issues on CAM and TM research can be sorted as patient's autonomy and consent, principle of justice, patient-physician relationship, use of public resources, and health insurances. This chapter aims at evaluating CAM and TM research according to fundamental ethics principles, as well as discussing legislations on CAM and TM research in Turkey.

Keywords: traditional medicine, complementary alternative medicine, research ethics, voluntariness, beneficence

1. Introduction

According to the World Health Organization (WHO), TM consists of knowledge, skills, and practices used for protection from physical and mental illnesses, diagnosis, improvement, and treatment of these diseases, as well as maintaining the health well, which is based on theories, beliefs, and experiences indigenous to different cultures, whether explicable or not. TM practices have a long history [1]. In general terms, the use of supplementary methods in addition to modern medicine in the patient treatment process is called "complementary medicine," while the use of other methods instead of modern medicine is called "alternative medicine." Similar terms had also been used in Turkey for a long time. However, as a result of the evaluations done based on the definition of WHO, it has been concluded that medicine cannot have an alternative, only the treatment can have one. Accordingly, the definition, "traditional and complementary medicine," has come to the fore [2].

People resort more to TM and CAM applications because these applications have low cost, are easy to access, are often free of invasive procedures, are considered promising in some chronic and terminal stage diseases, and some side effects of new treatment options can cause distrust. Also, in countries where TM and CAM applications are covered by public insurance, the inclination toward these applications will inevitably increase [3]. Friends talk about the solution to certain problems and articles about TM and CAM are frequently seen in the printed media. Many hospitals include these treatment methods in their care plans. Courses on TM and CAM are placed in the curriculum of medicine, pharmacy, and nursing faculties [4].

When the literature is evaluated, TM and CAM applications are observed to become widespread all over the world, and especially the increase in the use of TM and CAM in the western world in the last decade has increased the need for field-specific evidence-based research [5]. Although CAM has an effect, there are many unknown points about these treatments regarding scientific research that will convincingly reveal the value of individual treatments.

1.1 Scientific research and clinical research on TM and CAM

Scientific knowledge reveals the cause and effect relationship logically based on experiments. It aims to clarify the cause and effect link between events investigated as a result of accepted, proven, and reproducible experiments. Scientific research should be planned in a way to show useful results and healing in patient care. Clinical research consists of scientific studies that are carried out with the participation of volunteer people and aim to obtain medical information. For new drugs, medical devices, and other treatment methods to be beneficial to humans, scientific information should be provided on topics such as showing their safety and effectiveness, developing new treatment methods, and determining the side effects of medications and medical devices employed in treatment before they are used widely. After all treatments have been shown to be successful in the laboratory and on laboratory animals, the effects/side effects on human volunteers are investigated. Answers to questions such as the effectiveness of treatment, suitable doses for treatment, and the possible side effects can be found with clinical research results.

Medical research is carried out systematically using scientific methods based on a plan to obtain scientific information. Medical research that is conducted on humans is carried out systematically with scientific methods based on a protocol by staff (physicians) who are authorized to perform their profession in the field of medicine on patients or healthy volunteers by doing physical, chemical, or psychological medical interventions, or using biological samples taken from these people, or utilizing other personal data of these people. Medical research on humans primarily aims to obtain new and generalizable scientific data. On the other hand, its effects on the human body may not be foreseen precisely [6].

Researchers should know and apply common methods, measures, and standards so that necessary evidence can be produced and interpreted to make their decision on the use of TM and CAM treatments. Scientific thinking must remain within the ethical framework while focusing on the solution of problems. Scientific research that can be carried on human volunteers should be designed with great scientific and ethical sensitivity. Each stage of the research should be applied in a way that will protect human rights and dignity and will not cause permanent harm for future generations.

1.2 Traditional medicine and complementary alternative medicine health surveys

Research carried out in the field of TM and CAM can be handled under two headings.

1.2.1 Clinical research into traditional and complementary medicine practices

These are studies carried out on human beings to reveal or verify the clinical or pharmacological effects of one or more products and/or methods falling within the field of traditional and complementary medicine applications, to identify their adverse events or reactions, and to investigate their safety and effectiveness.

1.2.2 Noninvasive clinical studies

These can be defined as traditional and complementary medicine applications that will be carried out using methods that will not require direct attention or intervention of a physician or dentist provided that observational studies, survey studies, retrospective studies, and traditional and complementary medicine applications are administered under conditions that are specified within the legal regulations of countries.

In studies to be conducted in the field of TM and CAM, natural, synthetic, or biotechnology-rooted active substance or combination of substances administered to human beings to prevent, diagnose, or treat diseases, correct, regulate, or change a physiological function can be defined as a drug or human medicinal product.

To ensure that studies are conducted in international scientific and ethical standards, good clinical practice titles consist of rules that should be obeyed by parties participating in the research and cover regulations about topics such as designing, conducting, following, budgeting, evaluating and reporting the research, protecting all rights of the volunteer and body integrity, providing the reliability of research data, and maintaining their confidentiality.

Randomized clinical trials, which are shown as a source of evidence for the applicability of a particular intervention, provide important data for making practical decisions. They were accepted as the gold standard by A. Hill. Although the level of evidence is evaluated according to the type of the research, each study should also be evaluated in terms of strengths and weaknesses [7].

While randomized controlled studies are considered as the gold standard of evidence for treatment effectiveness, other research models will be needed about the activity field of CAM when randomized controlled trials cannot be conducted. These models may include studies meeting randomized and nonrandomized conditions, observational cohort studies, case-control studies, studies evaluating the effectiveness of certain treatment packages as a whole, and studies measuring placebo or add-on effects [4].

Evidence-based medicine (EBM) can be defined as the systematic approach where physicians make their decisions by combining their experiences, preferences, and patient characteristics in light of the best available evidence. Today, opinion-based decision-making approach referring to individual clinical experience has been replaced by evidence-based decision-making practice because the former causes variations and inconsistencies among clinical practices, it leads to disconnections between clinical practice and medical research, and it has become difficult to follow all the sources in the rapidly growing medical literature as a result of the ever-increasing number of studies in the field of medicine [8]. EBM practices, which have replaced TM practices, necessitate that decisions should be supported with qualified and current researches with a systematic approach to make the right decisions in the diagnosis and treatment processes. It is very important to reach information about evidence by following a conscious approach in the process of EBM applications. Evidence is determined by carefully selecting from the findings of clinical trials carried out by using real patients as subjects and applying original, powerful research methods [9].

2. Ethical aspects of traditional medicine and complementary alternative medicine research

2.1 The value of a study: its social value

Value is defined as an evaluation done in a socially accepted manner, showing the positive or negative meanings of objects in the outside world for people and society (e.g., good and bad, beautiful and ugly, underlying in nature and society phenomena) [10]. Value, as a criterion, includes the distinction between what is and what should be, and always appears as something positive or negative. The value that something has in terms of producing the desired results or the value displayed by something that acts as a tool in reaching the desired result is called instrumental or pragmatic value [11]. It can be said that the results of studies, too, have instrumental value. It should be borne in mind that research has a social value through knowledge production.

When the value of a research project is analyzed for social and scientific reasons, it increases as much as it has the promise of improving health or increasing knowledge that is important for health.

Studies that have no social value can be seen as a waste of resources and may pose participants some risks. It is an ethical imperative that in clinical trials, volunteer participants are not at risk other than investigating socially and scientifically important results. For this reason, only therapies with the greatest scientific and social importance should be selected for research purposes. Ensuring that social and scientific value prevails over commercial value is an ongoing ethical responsibility. Preliminary studies on research are absolutely valuable. The information they provide can cumulatively improve healthcare. From time to time, additions to clinical applications can be made in this way [12].

Determining who will benefit from the research result is important in determining possible benefits. The potential value of each research should be outlined at the outset. The results of the research should not only be subject of scientific meetings and publications but should be disclosed to all stakeholders in an appropriate language and expression. The reason why a patient or a relative of a patient seeks remedy other than evidence-based medicine is that there is no treatment of the disease in question or probabilities are very low. We should also understand people who seek health alternatives. For TM and CAM, the social value of the research has a special importance in this sense. On the other hand, research should not be allowed to weaken the existing health system of the society. In TM and CAM research, researchers should carefully monitor the socio-cultural dimensions of disease experiences, practitioner-patient interactions, and mutual compliance with treatment instructions during the research process. Research in this field can be categorized as empirical, semi-empirical, descriptive, case presentation, and historical interviews. Regional differences, being patient-centered, using several approaches together, differences of research participant inclusion/exclusion criteria, and approaches of appropriate placebo selection can affect research outcomes. According to WHO, data supporting the global use of TM and CAM practices are insufficient. More researches are needed to integrate CAM and TM practices in healthcare systems and understand their efficiency, safety, and mechanisms [13].

2.2 Limitation of sources

To give a definition, the source is the sum of all inputs such as money, manpower, and equipment, which are essential for any production or utilized during production. The sources to be allocated for transplantation, dialysis, or cancer treatment and the share allocated to other services such as vaccination, infection treatment, and treatment of acute diseases should be balanced. The search for balance leads to discussions on how far scientific and technical possibilities should be used to support patients. However, the limited resources make this issue controversial, imposing the necessity to make a choice [14].

When the limitations of the sources are in question, it is of significance which CAM treatments to evaluate. Research into this area may aim at understanding the biological mechanism of disease or treatment, as well as primarily addressing common disease issues. Acupuncture is used in the case of back pain, headache, migraine, knee pain, facial pain, inducing labor, nausea, vomiting, post-operative pain, essential hypertension, depression, stroke, and so forth. Analgesic effects of acupuncture practices are outcomes of patients' expectations and interaction of neurochemical, physiological, and psychological factors. One of the weakest parts of acupuncture evaluation researches may be the lack of adequate control of placebo [15]. The potential benefit may be another important topic to focus on. It is also possible to center on existing evidence that interventions are effective. Besides the pre-research geographic records, existing records can be evaluated. Regardless of what subject to be addressed, evidence to justify funding for research should be disclosed. Having evidence at hand is crucial to reveal the reliability and effects of TM and CAM treatments. There are few studies comparing treatments specific to these areas [4].

2.3 Scientific validity

In the process of scientific research, the researcher is basically seeking an explanation and answer to a question. There are two important concepts in this regard. These are validity and reliability. Validity is a concept that determines whether the answer, which is the subject of the research, can be answered with the applied research method.

Predictive validity about how predictive the research tool is in real-life situations and the construct validity regarding how much the tool correlates with the theoretical psychosocial structure of the sample that is being measured should be carefully reviewed in studies to be validated. A valid sample for the research must be calculated correctly. Neutral data collection is necessary for unbiased measurement and evaluation of the result. Science means measurement. Doing wrong measurements will lead to unreliable evidence and treatments. For this reason, utmost attention should be paid to ensure that measurement tools give valid and reliable results [16].

Scientific principles and methods are needed to produce reliable results. Scientific validity should be in accordance with the research methodology [12]. The scientific validity range and what the conditions are for the current treatment must be clarified in CAM studies. This situation is important for the social value of the research, as well as epistemologically. According to the research results in relation to CAM treatments, labels of therapies such as approved/unapproved or overridden require an epistemological and ethical evaluation.

It is claimed that cupping has a therapeutic effect on many diseases that could not be conventionally treated. It is used in conditions such as enhancing immune system, fibromyalgia, chronic pain caused by rheumatic diseases, digestive system diseases, nausea, and vomiting. It was also mentioned that cupping could also reduce side effects of some medications [17]. Evidence-based research data are needed in treatment of diseases.

2.4 Fair topic selection

The right to health is about health. The term health in national and international sources of law does not only refer to not being ill but also expresses "a complete state of physical, mental and social well-being." The WHO Constitution and the 1978 Alma-Ata Declaration define health as "not only being not ill or disabled; instead, a complete state of well-being in physical, mental, and social terms." The committee, which is the authorized interpretation body of the United Nations Convention on Economic Social and Cultural Rights, can be seen to base its interpretations of the right to health on the definitions of WHO [18].

Health, which is the value protected by the right to health, is not only a value that requires service for improvement when it deteriorates but also a value that must be respected and maintained, just like body integrity. The example that people should be protected against activities harmful to their health can be considered in this context [19].

According to the United Nations Convention on Economic Social and Cultural Rights, the right to health should not be understood as just the right to be healthy. The freedom dimension of the right to health includes a person's control over their health and body and not being a subject of medical and experimental interventions that are not based on the person's consent or end up with torture, as well as gender and reproduction freedom of the person [20].

From this point of view, topic selection is important for all clinical trials. Equal sharing of benefits and burdens in research is an important requirement. Various headings such as the vulnerability of volunteers, risks, socioeconomic status, and gender inequality in high-risk research are important in determining the subject of the research. In the selection of the subject of the research, vulnerable groups should be carefully monitored. The use of CAM is believed to be closely related to sociodemographic variables such as gender, age, education, income, and health complaints. TM and CAM applications are often preferred by women [21]. This general information suggests that field-specific CAM and TM studies are expected to be carried out mostly on women, as opposed to conventional medicine. EBM information shows that there are important differences between men and women in terms of diagnosis, treatment, and disease prognosis of arthritis, heart diseases, and infectious diseases. These differences exist due to the biological (sex) and social (gender) differences between men and women. For example, according to research, the side effects of antihistamines, antibiotics, antiarrhythmics, and antipsychotics are more severe and/or more common in women than in men. The effectiveness of aspirin in the primary prevention of cardiovascular disease is another example of the existence of a significant gender difference. Therefore, both sexes must be included in studies.

2.5 Vulnerable: easily affected groups

Although there are many definitions for vulnerable and easily affected groups, they are known as disadvantaged groups of society. People with vulnerable and easily affected status had been one of the basic concepts studied in bioethics and health policies from the 1970s to 1990s. Over the years, so many groups, from old people who can be easily hurt to those with low levels of education, from individuals with insufficient resources to citizens without rights and a whole country open to exploitation, have been declared vulnerable that the scope of the term has expanded excessively [22]. There are three types of security vulnerabilities for this group. The first is vulnerability to physical harm, the second is vulnerability to damage to social stance or reputation, and the third is vulnerability to psychological and emotional distress. Vulnerable and easily affected groups include children suspected of the informed consent they deliver; individuals with inadequate mental capacity to give consent; individuals with learning difficulties, dementia, or similar ailments; individuals with cognitive impairments due to stroke; individuals with psychiatric or personality disorders; individuals who cannot socially give consent freely; students; individuals under care; members of the armed forces; juvenile criminals; prisoners and asylum-seekers; pregnant women; unconscious patients; and family members of researchers. The Belmont Report, the first document to identify vulnerable and easily affected groups, identifies individuals who cannot make decisions, are exposed to disproportionate incentives, and are involved in research due to administrative convenience as abusable [23].

The nineteenth article of the Declaration of Helsinki, titled Vulnerable Groups and Individuals, states that some groups and individuals are particularly vulnerable, and they may be more likely to be abused or damaged. Specially considered protective measures should be taken for all vulnerable groups and individuals. According to Article 20, medical research on vulnerable groups can be accepted provided that the research addresses the health needs and priorities of the group in question and it is not possible to conduct the research in a nonvulnerable group. Also, it involves the provision that the information, practices, and interventions obtained from the research should benefit the group in question. The eighth article of the UNESCO Universal Declaration on Bioethics and Human Rights (2005), which centers on the need to show respect to personal integrity and protect vulnerable and easily affected individuals and groups, emphasizes that human vulnerability should be taken into consideration in the application and development of scientific knowledge accumulation, medicine, and related technologies [24].

In research to be conducted on vulnerable and easily affected groups, risk/benefit measures, security, vulnerability, and assurances attached to privacy should be determined. It should be remembered that unethical use of personal data and privacy violations will affect the social texture of disadvantaged groups. Situations that involve sensitive issues related to stigmatization can also discredit individuals [25].

As for evidence of the effectiveness of TM and CAM approaches, the results of these studies have been found sufficient to construct a hypothesis to be tested by sound clinical trials. In many countries, TM and CAM applications are provided outside the national health system. Since the philosophy of CAM therapies is often seen as the mobilization of self-healing, there is a perception that they are safe. Due to this perception, volunteers can easily be found. These people do not think that they may be exposed to serious side effects during the research process. Side effects seen in CAM studies should be reported to the relevant authorities. As pointed out in many international documents (e.g., Helsinki Declaration, Article 12), medical research on volunteers should only be carried out by individuals with adequate ethical and scientific education, training, and qualifications. Research on patients or healthy volunteers should be conducted under the supervision of a competent and qualified physician or other healthcare professional.

2.6 Risk-benefit ratio

In any biomedical research on humans, the risks and burdens that participants may be exposed to and the possible benefits from the research must be balanced. Possible risks and burdens should always be minimal. This important rule stems from the principle of ethics that requires not to harm but to be useful. In studies involving interventions that may directly benefit the participant, higher risk and burden is acceptable. The potential risks that participants face can be not only physical but also psychological or social. Likewise, the possible benefits of the research to the participant may be of palliative as well as having a therapeutic nature [26].

According to the Nuremberg codes announced in 1947, the level of the risk to be taken in an experiment should never be more than the level to be determined according to the importance of solving the problem under the experiment for humanity. Appropriate measures should be taken to protect the subjects against low injury, disability, and possible death, and necessary environment and equipment should be provided. According to Articles 16, 17, and 18 under the Risk, Disadvantages, and Benefits heading of the Helsinki Declaration, before each medical research on humans, the estimated dangers and disadvantages that the research may cause compared to the benefits to be gained by the individuals or groups participating in the study of other individuals or groups who are affected by the disease that is the subject of the research should be carefully evaluated. Unless physicians make sure that risks are adequately assessed and they are satisfactorily addressed, they cannot participate in studies on human volunteers. According to the Convention for the Protection of Human Rights and Dignity of the Human Being with regard to the Application of Biology and Medicine, the dangers that the subjects of the research may be exposed to should not be disproportionate to the expected benefits of the research. Products such as plant extracts used in CAM research are likely to be pharmacologically complex. This may increase potential harm in research that tests to combine many compounds into a single product.

Use of *Hirudo medicinalis* has been a commonly used treatment method since sixth century BC. Some of the enzymes present in *Hirudo medicinalis*' saliva (e.g., calin, eglin, acetylcholine, and hyaluronidase) show anticoagulant, vasodilator, and anti-inflammatory effects. In many countries, *Hirudo medicinalis* is used in plastic surgery, traumatology, neurology, and ophthalmology fields. Point of application of *Hirudo medicinalis* can be chosen in different ways (for example, around the wound site, above the painful area). Researchers also try to make stimulating effects from on top of the acupuncture points [27]. Lack of a specific standard for TM and CAM research makes it difficult to address and assess the differences between practitioners and the risk-benefit ratio [28].

2.7 Cooperation

Without the involvement of researchers and host communities, a study is unlikely to have a lasting effect in developing countries. Without the investment of health policymakers, research results are unlikely to affect policymaking and scarce resource allocation. In almost all of the research, there is a collaborative communication between sponsors, policymakers, and research groups. In TM and CAM research, it is not always possible to predict the permanent effects of the research in question at the outset. It is difficult to be a stakeholder in the use of scarce resources without the investments of health policymakers. Emphasizing the importance of the health problem, evaluating the contribution of the research to the society, and planning, carrying out, and publishing the research all require cooperation.

Controlled clinical trials, one of the ways to access scientifically proven information, always need a competent and adequate research team with professional experience for the research. These studies are always carried out with a multidisciplinary approach. In this process, the follow-up of the volunteer participants can be maintained healthily with the work share plan within the research team. Records of the research can be taken and maintained following a protocol. Different situations arising during the research can be evaluated on time. In research based on compliance with patient rights, the safety of the patient and volunteer participants is ensured. The security profile of the applications can be followed. Patient and voluntary participant training can be maintained.

In addition to the abovementioned benefits of collaboration concerning clinical trials, there is also the opportunity that research results can be reviewed and they can be returned to the health sector as an investment. Cooperation is also inevitable for the population, in which the research has been conducted, to be able to obtain the expected benefit from the results, and to carefully follow the intellectual property rights and copyrights in the publication of the research report [29].

3. Informed consent

Human dignity is still an up to date topic as it has been in every period. As Baranzke [30] puts it, "biomedical ethics debates have raised a growing awareness that medicine may have a side that can violate human rights." Meeting people's health needs is a prerequisite for a dignified life. The existence of human dignity and what it means cannot be clearly understood anywhere other than extraordinary situations where it is seriously damaged [30]. Alan Gewirt bases human dignity on the person's opportunity to make a choice and take action based on these choices. Actions have two founding conditions and specific characteristics. Freedom is based on the control of the action performed by the person with all the information they need without any coercion. Happiness, on the other hand, can be described as the knowledge of having the skills and conditions necessary to fulfill various goals [31].

The right of the person, which is in parallel with the development of basic human rights, to make decisions about themselves, the right to determine their own destiny, has made up the rationale for a new ethical assignment to the physician in terms of giving the necessary information to the person to make an informed choice.

Informed consent can be considered as a concrete indicator of trust-based communication between physician/researcher and patient/volunteer participants. The informed consent is also an official document on the right of the patients/volunteer participants to determine their own future. It helps the communication between physician/researcher and patient/volunteer participants to conform to the law. Current literature focuses on research in which volunteering cannot be achieved in providing informed consent [32]. After the Second World War, Nuremberg Codes were developed to fill the legal and ethical gaps regarding the experiments, due to Nazi experiments carried out on humans. According to these codes, human subjects should definitely give their consent with their free will. In other words, the person participating in the experiment must have the quality to grant consent legally, must not be exposed to any sanction, deception, lie, threat or any other restriction or pressure, must be in a state to make decisions by making free choices, must be informed about the research and have enough information to make informed decision, and must understand everything.

According to the Helsinki Declaration (Article 25), the participation of individuals who can deliver informed consent in a study should be on a voluntary basis. Although it is deemed appropriate to consult family members or community leaders, no individual with the capacity to give informed consent can be included in any research study unless they agree with their free will. This statement has special importance in TM and CAM research in terms of not harming the participants. For instance, in countries like India, clinical research is based on the perception of the disease perceived through social values and power hierarchies in the village family based on regional values and practices and cultural values. Obtaining informed consent in these settings becomes difficult due to differences in local traditions in developing countries, including India [33].

According to the United Nations Convention on Civil and Political Rights, adopted by the United Nations on 16 December 1966, which aims to protect human rights and freedoms, no one can be subjected to torture or cruel, inhuman or degrading treatment or punishment. In particular, no one can undergo medical or scientific experiments without their free consent.

Article 16 of the Convention of Human Rights and Biomedicine on the Protection of Human Rights and Human Dignity in terms of the Application of Biology and Medicine, individuals who are the subjects of research must be informed about their rights and guarantees prescribed by law for their protection, the informed consent must be given clearly and in a specific manner, and this must be documented. According to Article 3 of the European Convention on Patient Rights (Rome 2002) titled "the right to information," each individual has the right to receive information about their health status, current health services and how they can benefit from them, and all scientific research and technological innovations.

3.1 Components of informed consent for research

Components of informed consent can be listed as voluntariness, autonomy, competence, clarity, and comprehensibility of the information supplied. Voluntariness is the situation where the person decides to do something with their free will for a purpose without expecting anything in return and realizes it. Respecting individual autonomy lies at the heart of volunteering. The concept of volunteering also includes the autonomy of a person and being free from compelling effects. For some people, volunteering can be defined as the presence of adequate information and the absence of psychological pressure and external restrictions.

The most important issue to consider regarding informed consent is whether the patient/volunteer has the capacity of giving consent. The orientation of capacity, informing, and consent to the authorized person is the first condition for a free and healthy will to form. In terms of medical intervention law, responsibilities such as respect for patient privacy, informing, and consent, which are accepted on the side of physicians, can be said to serve the patient to create an autonomous will.

The third article of the Categorical Imperative (act according to the principle of autonomy) refers to the self-governing of the will, which is ruled by the mind. It is not enough just to know what autonomy is. Active use of autonomy and the formation of personality will not be possible without moving to the autonomy stage. The basic condition of moral action is to be competent beyond being responsible. In any case, CAM studies are based on applications. Volunteer participants should always be at the center when investigating treatments. Additional efforts are necessary to inform these people if they have received multiple treatments and to ensure that the information provided is understood. In TM and CAM studies, therapeutic misunderstandings should be corrected and the components of belief and expectation should be taken into account in the informed consent process.

3.2 Information to be given about the research

The aim of giving information is to ensure that participants are in a position to decide on their own future and to increase the cooperation ability of the patient/volunteer participant.

The Belmont Report, which defines the ethical principles of human research, emphasizes that the form and context in which information is transferred is as important as the information itself. Information about the research should be clear, understandable, and accurate. It should be given to the person in a clear and comprehensible language. A person from the research team who is responsible for obtaining informed consent should provide the information. The information to be given about the research should first clearly reveal the purpose and duration of the research. Research procedures should be explained in detail. The risks and ailments related to the research, the possible duration of these conditions, and how they will be eliminated should be explained. The benefits expected from the research, alternative approaches, records regarding the research, who will reach the records and how, and the extent of confidentiality of the records should be explained. Research involving more than minimum risk should include available medical treatments for volunteers when needed; information about where to get more information, who to consult to in case of a research-related injury, and the amount of payment/compensation to be given to volunteers; explanation of who can be contacted for answers to questions about research and rights about the research (includes the patient representative and phone number of the research center); and information that participation is voluntary, the volunteer may quit the research at any time, and that there will be no loss of rights.

The information provided to volunteer participants regarding the research to obtain informed consent should be completely understood by the patients, but it has quite a complex content. This complexity poses a major barrier for nearly a quarter of adults with low literacy skills to grasping the information supplied. Research communities are responsible for developing innovative forms of consent, which increase patients' willingness to read and skills to understand. To speak of fully informed consent, it is important that the information provided is well understood and that the person can fully grasp what is to be done to their body and participate in the process. Readability is a language-specific concept developed in the United States (USA) in the early nineteenth century. Among the factors affecting the readability are average word length, word frequency, number of words with multiple syllables, average sentence length, number of words with multiple meanings, and average number of syllables. As the number of words in the sentence increases, the readability of a sentence decreases. Readability means full understanding of the text. Readability is highly impacted by issues such as the level of reading, the format of writing, the shortness of sentences and paragraphs, and the technical language. Therefore, to increase the readability, short sentences and paragraphs should be selected and text appropriate for the reading level should be created [34].

3.3 Person providing and provided with information

As a rule, the person to provide information is the person from the research team who undertakes the task of obtaining informed consent. The person to be provided with information is the person who is competent and adequate to give consent, who is invited to the research, and who meets the criteria for participation in the research. In the case of giving information and obtaining consent for the research, separate procedures are followed for children and vulnerable groups. As stated in the Helsinki Declaration, for a volunteer candidate who is deemed not having the capacity to grant an informed consent can give consent to participate in the research, an additional approval of a physician and the legal representative is necessary. If the person does not give consent, this should be respected. According to the World Medical Association, children, the family, and the legal representative have the right to actively participate in the medical practice decisions [35].

Pediatricians must take medical responsibility as a responsible researcher or assistant researcher in any research on children. One of the main points of Helsinki Criteria is that physicians who are specialists in the research topic take part in scientific research on people. Pediatricians should be included in the team in all studies on children. Responsibilities awaiting pediatricians who will take part in this matter are discussed in other approaches.

3.4 Time of the informing

The provision of information about the research should be done during the process of the invitation of the volunteers to the research without time pressure, and the volunteer should be given some time to decide.

4. TM and CAM research in Turkey

In Turkey, the frequency of use of TM and CAM applications in the general population and the distribution rate by methods are not known. Also, a few available studies include specific patient groups.

The first regulation governing traditional and complementary medicine practices in Turkey was the "Regulation of Acupuncture Treatment," which was published in 1991. The Department of Traditional, Complementary and Alternative Medicine Practices was established in 2012 under the Ministry of Health. It was renamed as the Department of Traditional and Complementary Medicine Practices in 2014, and the Regulation on Traditional and Complementary Medicine was published in the Official Gazette with the date October 27, 2014, and issue 29158. The regulation involves the definition of 15 traditional and complementary medicine applications (acupuncture, apitherapy, phytotherapy, hypnosis, leech applications, homeopathy, chiropractic, cup application, larva application, mesotherapy, osteopathy, prolotherapy, ozone application, reflexology, and music therapy). It also includes the personnel who can perform these applications, indication/contra-indication, and materials to be found at the application center [36].

Regulation on Traditional and Complementary Medicine Practices Clinical Research was published in the Official Gazette with the date March 9, 2019 and issue 30709. The purpose of this Regulation is to regulate the principles and procedures for conducting scientific research on people and protecting the rights of volunteers in the fields of traditional and complementary medicine within the framework of international agreements and good clinical practices. This regulation covers clinical studies on medicine, medicinal and biological products on humans, and clinical studies with cosmetic products and raw materials; observational drug studies; clinical studies of medical devices; stem cell clinical studies; noninterventional clinical studies; and traditional and complementary medicine applications. In cases where the research topic directly concerns children or is a clinical condition that can only be examined in children, or data obtained as a result of research on adult individuals have to be validated in children, the research is delivered to the assessment of the ethics committee after a child psychiatry specialist and/or child health and diseases specialist at the university or training and research hospital gives a positive opinion of the research on children if the research does not pose a predictable risk for the health of volunteers and there is a general medical opinion that the research will provide direct benefit to volunteers. Apart from children, another group of vulnerable—easily affected—subjects is pregnant, postpartum, or lactating women. In cases where the topic of the research is directly related to pregnant, postpartum, or lactating women or is a clinical condition that can only be examined in these women, and the research does not pose a predictable risk for the volunteer and the fetus or the infant, and there is a general medical opinion that the research will directly benefit the volunteers, then research on pregnant, postpartum, and lactating women may be allowed provided that the general research principles are followed. In cases where the subject of the research is directly related to the persons with limitations or is a condition that can only be examined in people with limitations, or where the current treatment options related to the disease of the person with limitations have completely been exhausted, the research does not pose a predictable risk for the health of the person with limitation, and there is a general medical opinion that research will provide a direct benefit to the people with limitations, then people with limitations can be the participants of TM and CAM research. The aforementioned regulation covers the provisions for starting, executing, stopping, and terminating the research, the responsibilities of the responsible researchers and supporters, and the details of the research product. Where and how the adverse events or serious adverse reactions will be reported detail the confidentiality of research records.

Good clinical practice, which is based on the principles of the current Helsinki Declaration, is an ethical and scientific quality standard for the design, execution, recording, and reporting of clinical trials on humans.

Good clinical practice assures the society that the rights, health, and privacy of the volunteers participating in the research are protected and that the data obtained from the research are reliable. The purpose of the Good Clinical Practices Guide is to provide a single standard to facilitate the international mutual acceptance of clinical data. Traditional and Complementary Medicine Practice Good Clinical Practice Manual (May 14, 2019) guides the collection of clinical data to be presented to the Health Services General Directorate and of their respective ethics committees and explains the details and principles of the traditional and complementary medicine practices clinical research that is carried out or scheduled to be conducted in Turkey [37].

Whether volunteer's participation will is accepted or not, these subjects are individuals whose free decision-making willpower can be affected due to the expectation of benefit of participation or the expectation of retaliation by individuals in a hierarchical structure if they refuse to participate. The examples of these individuals include people who are in a certain hierarchical structure such as medical, pharmacy, dentistry, and nursing students; hospital or laboratory staff working in the research setting; those working in the pharmaceutical industry; members of the armed forces; and soldiers and detainees. Also, the group involves patients with an incurable disease; those who live in nursing homes; unemployed or poor people; people who need urgent medical attention; and children or people who cannot give consent. Clinical research is defined as studies conducted on humans to reveal or confirm the clinical or pharmacological effects of one or more research products or traditional and complementary medicine methods, to identify adverse events or reactions, and to investigate their safety and efficacy. In scientific, medical, and ethical terms, TM and CAM research is evaluated in ethics committees assembled according to the Regulation on Traditional and Complementary Medicine Practices

Clinical Research. According to the current legislation in Turkey, ethics committee members are required to receive basic training on good clinical practices and clinical research before starting to work in these boards. If possible, an equal gender distribution among members should be provided in ethics committees.

5. Conclusion

Development of good empirical practices have been guiding ethical discussions and specifying a normative approach and framework for human research at the same time. CAM field necessitates evidence-based results on clinical research. Due to their approaches of preventing illnesses and providing treatment that have been preferred by many individuals, TM and CAM practices and research are expected to progress. Clinical research is essential in ensuring medical advances, and it establishes the bond between theory and practice in the field of medicine. In addition to determining effectiveness of a treatment, strength of the evidences on the medical application should be evaluated. The right path with CAM and TM clinical research and practices would include not to disacknowledge them, but to pave the way for the scientific research on the ones that have the potential to be beneficial and to bring the proven benefits to modern medicine. One challenge with these practices would be the fact that a considerable amount of CAM and TM data are not systematic and standardized, which have made them difficult to scientifically accept. In order to assist the international acceptance of CAM and TM data to pave the way for CAM and TM databases, meta-analyses and so on, good clinical practice guides being made specific to these areas is essential. For all human research, informed consent must be obtained from the voluntary participants. Throughout this process, it is imperative to show respect to participants' dignity, to omit forcing potential voluntary participants in any pecuniary and nonpecuniary way, to care on their confidentiality, and to pay utmost attention on privacy of their data. New treatment procedures and medications as results of clinical research are for public welfare, but the benefit of the voluntary participant must be prioritized.

When it comes to CAM and TM clinical research, a framework to their ethical evaluation is suggested herein. For clinical research on CAM and TM, legal arrangements are necessary regardless of geographical region, belief and cultural differences. Ethical committees for clinical practices and researches should be established specifically for CAM and TM research. In these committees, physicians from different fields of specialization who participated in clinical research conducted according to good clinical practices, physicians having expertise on CAM and TM, legal experts with specialization on medical law and patients' rights, pharmacology and pharmacognosy specialists, along with medical ethics and public health experts. Instructions should be prepared to cover how these committees to work and make decisions; number of members and their employment periods, duties and authorities in these committees; how the applications to these committees are to be done, and how the decisions are to be delivered to the researcher(s); within the legal framework of the countries that the ethical committees belong to.

Conflict of interest

The author declares that there is no conflict of interest.

Author details

Omur Sayligil
Eskisehir Osmangazi University, Faculty of Medicine, Department of History of Medicine and Ethics, Eskisehir, Turkey

*Address all correspondence to: sayligil@ogu.edu.tr

References

[1] World Health Organization. General Guidelines for Methodologies on Research and Evaluation of Traditional Medicine. Geneva: World Health Organization; 2000

[2] Ünal M, Dağdeviren HN. Traditional and complementary medicine methods. Eurasian Journal of Family Medicine. 2019;**8**(1):1-9. DOI: 10.33880/ejfm.2019080101

[3] Leach MJ, Canaway R, Hunter J. Evidence based practice in traditional & complementary medicine: An agenda for policy, practice, education and research. Complementary Therapies in Clinical Practice. 2018;**31**:38-46. DOI: 10.1016/j.ctcp.2018.01.011

[4] Institute of Medicine, Board on Health Promotion and Disease Prevention, Committee on the Use of Complementary and Alternative Medicine by the American Public. Complementary and Alternative Medicine in the United States. Washington, DC: The National Academies Press; 2005. DOI: 10.17226/11182

[5] Fras M, Strassl RP, Friehs H, Mllner M, Kundi M, Kaye AD. Use and acceptance of complementary and alternative medicine among the general population and medical personnel: A systematic review. The Ochsner Journal. 2012;**12**(1):45-56

[6] Kasapoğlu TM. Protection of Subject Rights in Clinical Research and Responsibility of Administration. Ankara: Seçkin Publishing; 2015. pp. 19-20

[7] Shah HM, Chung KC. Archie Cochrane and his vision for evidence-based medicine. Plastic and Reconstructive Surgery. 2009;**124**(3):982-988. DOI: 10.1097/PRS.0b013e3181b03928

[8] Kumar S. Evidence-based medicine: The new paradigm of clinical decision making. Indian Journal of Urology. 2000;**17**(1):73-78

[9] Uysal EK, Uçak NÖ. The evidence based information needs and information seeking behaviors of medical faculty. Bilgi Dunyasi (Information World). 2013;**14**(1):37-61

[10] Frolov I. Dictionary of Philosophy (Translated by Çalışlar A). İstanbul: Cem Publishing; 1991. p. 99

[11] Cevizci A. Dictionary of Philosophy. Istanbul: Paradigma Publishing; 1999. p. 201

[12] Emanuel ET, Wendler D, Grady C. What makes clinical research ethical. Journal of the American Medical Association. 2000;**283**:2701-2711. DOI: 10.1001/jama.283.20.2701

[13] Burton A, Smith M, Falkenberg T. Building WHO's global strategy for traditional medicine. European Journal of Integrative Medicine. 2015;**7**(1):13-15

[14] Akdur R. Health services and scattering of limited resources. Türkiye Klinikleri Journal of Medical Ethics-Law and History. 2000;**8**:38-45

[15] NIH Consensus Conference. Acupuncture. JAMA. 1998;**280**(17):1518-1524

[16] Aktürk Z, Acemoğlu A. Reliability and validity in medical research. Dicle Medical Journal. 2012;**39**(2):316-319. DOI: 10.5798/diclemedj.0921.2012.02.0150

[17] Qureshi NA, Ali GL, Abushanab TS, El-Olemy AT, Alqaed MS, El-Subai IS, et al. History of cupping (Hijama): A narrative review of literature. Journal of Integrative Medicine. 2017;**15**(3):172-181

[18] Declaration of Alma-Ata [Internet]. 1978. Available from: https://www.who.int/publications/almaata_declaration_en.pdf [Accessed: 02 January 2020]

[19] Temiz Ö. Right to health as a basic right in Turkish law. Ankara University Journal of SBF. 2014;**69**(1):166-188. DOI: 10.1501/SBFder_0000002307

[20] International Agreement [Internet]. 2003. Available from: https://humanrightscenter.bilgi.edu.tr/media/uploads/2015/08/03/EkonomikSosyalKulturelHaklar Sozlesmesi.pdf [Accessed: 02 January 2020]

[21] Kristoffersen AE, Stub T, Salamonsen A, Musial F, Hamberg K. Gender differences in prevalence and associations for use of CAM in a large population study. BMC Complementary and Alternative Medicine. 2014;**14**:463. DOI: 10.1186/1472-6882-14-463

[22] Beauchamp TL, Childress JF. Principles of Biomedical Ethics. 4th ed. New York, Oxford: Oxford University Press; 1994. p. 407

[23] The National Commission for the Protection of Human Subjects of Biomedical and Behavioral Research. The Belmont Report [Internet]. 1979. Available from: https://www.hhs.gov/ohrp/sites/default/files/the-belmont-report-508c_FINAL.pdf [Accessed: 02 February 2020]

[24] International Bioethics Committee. The Principle of Respect for Human Vulnerability and Personal Integrity Report of the International Bioethics Committee of UNESCO (IBC) [Internet]. 2013. Available from: https://unesdoc.unesco.org/ark:/48223/pf0000219494_eng [Accessed: 02 February 2020]

[25] White A, Boon H, Alraek T, Lewith G, Liu JP, Norheim AJ, et al. Reducing the risk of complementary and alternative medicine (CAM): Challenges and priorities. European Journal of Integrative Medicine. 2014;**6**(4):404-408. DOI: 10.1016/j.eujim.2013.09.006

[26] Steering Committee on Bioethics. Guide for Research Ethics Committee Members [Internet]. 2012. Available from: https://www.coe.int/t/dg3/healthbioethic/activities/02_biomedical_research_en/Guide/Guide_EN.pdf [Accessed: 01 December 2020]

[27] Yapıcı AK, Durmuş M, Tanyüksel M, Türkkan S, Tuzun HY, Arsenishvill A. Hirudo-medicinalis—Historical and biological background and their role in microsurgery: Review article. Hand and Microsurgery. 2017;**6**:34-38

[28] The Nuremberg Code [Internet]. 1949. Available from: https://history.nih.gov/research/downloads/nuremberg.pdf [Accessed: 02 January 2020]

[29] Emanuel EJ, Wendler D, Lillen J, Grady C. What makes clinical research in developing countries ethical? The benchmarks of ethical research. The Journal of Infectious Diseases. 2004;**189**(5):930-937. DOI: 10.1086/381709

[30] Baranzke H. Autonomy and Human Dignity in Western Biomedical Ethics Discussions (Translated by Temel MK). İstanbul: Betim Publishing; 2018

[31] Gürbüz N. Human Dignity in Biomedicine Law. İstanbul: XII Levha Publishing; 2014

[32] Roberts L, Kim JP. Does informed consent given by healthy individuals when enrolling in clinical research feel less voluntary than for ill individuals. Journal of Psychiatric Research. 2018;**103**:33-37. DOI: 10.1016/j.jpsychires.2018.04.018

[33] De Costa A, D'souza N, Krishnan S, Chhabra MS, Shihaam I, Goswami K.

Community based trials and informed consent in rural North India. Journal of Medical Ethics. 2004;**30**:318-323. DOI: 10.1136/jme.2002.001065

[34] Güvenbaş M, Şaylıgil Ö. Is it necessary to develop a series of rational standards for informed consent? Türkiye Klinikleri Journal of Medical Ethics-Law and History. 2019;**27**(3):277-284. DOI: 10.5336/mdethic.2018-62938

[35] World Medical Association. Declaration of Ottawa on the Right of the child to Healthcare. Canada: 50th World Medical Assembly; 1998

[36] Öztürk H, Şaylıgil Ö. Ethical viewpoint of the regulation on traditional and complementary medicine. Türkiye Klinikleri Journal of Medical Ethics-Law and History. 2016;**24**(1):1-10. DOI: 10.5336/mdethic.2016-50492

[37] Essentials and Details on Traditional and Complementary Medicine [Internet]. 2019. Available from: https://dosyamerkez.saglik.gov.tr/Eklenti/30689getat-klinik-arastirma-kilavuzpdf.pdf?0 [Accessed: 02 February 2020]

Chapter 8

Evaluation of the Research Protocol by Ethical Committee

Paulo Santos, Pedro Teixeira, Helena Beça and Alberto Hespanhol

Abstract

Nowadays, the submission of a research project to an ethical committee and its approval is mandatory. However, researchers often overlook this obligation, because they are too engaged in the design and the process of construction of the study, because of the common tight deadlines, and many times because some devaluation of the role of the committee. Based on our experience of 10 years working in an ethical committee, we propose a way to get close researchers and evaluators, respecting their own aims but bringing them together as partners in the investigation process, protecting patients' values, at the same time that makes it possible to implement strategies to answer to the research question and to create useful knowledge. Our aim is to smoothen the way researchers look to the ethical committee and, at the same time, to make them understand what really is at stake. Ethics should be a commitment for all and not an obligation.

Keywords: ethics, ethics committees, research, ethical analysis, beneficence, personal autonomy

1. Introduction

Medicine is born from the human need to survive to diseases. At first, someone within the tribe began to realize the constellation of symptoms and signs that defined the diagnosis, for which a proper treatment could make the difference in the course of the disease. The so-defined medical act gave him the mastery over life and death. The awareness of this power has given rise to a set of self-regulating norms, early transmitted to the disciples who applied to learn the noble art of healing. This ethical code became known as the Hippocratic oath of the School of Kos, in honor to Hippocrates, the father of scientific medicine, known for having received himself the knowledge from the hands of Asclepius the Greek god of medicine [1].

Based on a solid knowledge derived from the scientific method, medicine commits itself to the patient and society in simple but basic principles of beneficence, non-maleficence, justice, truth, confidentiality, and respect for the human being. This commitment should be enough to ensure unblemished medical practice, both in the care and, no less important, in the research and experimentation.

But the twentieth century and the horrors exposed by the Holocaust of the Second Great War came to demonstrate that it was not enough.

On August 8, 1945, the countries that formed the Allied Forces in World War II signed the constitution agreement of the International Military Tribunal, to prosecute the Nazi officers on charges of committing peace crimes, war crimes, and crimes against humanity (Control Council Law No. 10: Punishment of persons guilty of war crimes, crimes against peace and humanity). The initial trial in the city of Nuremberg, Germany, was followed by 12 other trials until 1949. The first of these trials became known as the medical case and resulted in the conviction of 16 of the 23 defendants for their involvement, among others, in research projects (high-altitude experiments; freezing experiments; malaria experiments; mustard gas experiments; sulfanilamide experiments; bone, muscle, and nerve regeneration experiments; bone transplantation experiments; seawater experiments; epidemic hepatitis experiments; sterilization experiments; vaccination experiments for yellow fever, smallpox, typhus and other rickettsiosis, paratyphoid A and B, cholera, and diphtheria; poison experiments; and incendiary bomb experiments) [2].

The question wasn't the possibility of conducting research in human beings but the way and circumstances under which it was done. This court had to define what were the permissible medical experiments, in accordance with ethics, morals, and law. The 10 principles emanating from this court formed the first code of ethical appreciation for research involving humans [2], later developed and extended in its application by the World Medical Association's Declaration of Helsinki (1964) [3]. Nevertheless, ethical evaluation was still a commitment of the investigators. The case of Tuskegee, USA, syphilis experience (1932–1972) has warned of the need for direct follow-up and the establishment of independent committees able to ensure the appraisal, evaluation, and guidance of research protocols, as proposed by the following Belmont Report [4]. The increasing complexity of ethical problems with the advance of knowledge has dictated the structuring of responses at the national level. In 1983, in France, President François Mitterrand established the "Consultatif National d'Ethique Committee," the first ethics committee for health and life sciences, with the mission to opine on ethical and social problems arising from progress in the fields of biology, medicine, and health. The increased specificity of experimental research in the new treatments has led to the establishment of research ethics committees with the highest technical capacity to fulfill their mission (US Institutional Review Boards, and the Ethics Committees for Clinical Research in Europe).

2. The ethical principle

The ethics committees are now multidisciplinary boards, including medicine, nursing, social work, law, pastoral care, healthcare administration, and various specialty areas. Their role includes the ethics education, policy formation and review, ethics consultation, and research ethics [5]. In clinical investigation, every protocol must be submitted previously to the beginning of the study for consideration, comment, guidance, and approval. The ethics committees must be independent of the researcher, the promoter, and the sponsor and transparent in their function [6].

The ethics committees for health, especially in the context of clinical research, are thus born not from an internalization of the need for self-regulation but from an external, regulatory, and legal imposition. Since the beginning, they assume a problem-solving police nature that they rarely escape, as their mission includes the laws and regulations as well as applicable international norms and standards. It is common that the ethics committees focus much attention on legal and procedural aspects, answering to this feature, more than applying an individualistic appraisal of the factual project. Consequently, the researchers look at them more as obstacles to the execution of projects than as partners in their implementation.

Ethic must be above the law, respecting it, but discussing it and framing it towards the specific case [7], keeping in mind the protection of the human being. The primacy of ethics, which compromises us all, is certainly not the primacy of the ethics committees, as if they were the exclusive holders of the absolute truth, or its juridical version. The ethical appraisals must return to the Hippocratic matrix of the basic ethical principles to find their guiding path, combining the need for innovation and development in health, through research strategies, with respect for the human being in his or her dignity and vulnerability, in health or in illness, in the daily care they need, or in participating in a research project.

The ethical principles were defined in 1971 by Thomas Beauchamp and James Childress [8]. Beneficence, non-maleficence, justice, and autonomy had become the basis of bioethical decision-making in the last 50 years. Despite the heated discussion about their interpretation, importance, and role of each one, they remain the most internationally accepted. Nevertheless, there's some controversy about their inability to answer to all the challenges posed by the complexity of current biomedical decisions.

The same principles were adopted by clinical research as a key for ethical evaluation, ensuring the protection of the rights of participants. But there are several differences in their application.

Beneficence refers to the promotion of the best practices to improve the patient. However, many times, the beneficiary of the results in a research project is not the participant but the all other population that will receive the medicine under tests [9]. The principle of non-maleficence derives from the Hippocratic sentence of "primum non nocere," which means that doctor abstains from practice that put patients under danger. Although we may argue that the risk of harm is under control, we cannot guarantee for certain even in observational research. The principle of justice and equity reminds us to provide the best treatments to those who really need them. The randomization of the sample removes any selective allocation criteria. The autonomy respects the patient's freedom of choice, based on a sufficient given information. The informed consent freely expressed gives legitimacy to the inclusion of participants. Although the guaranties of the possibility of self-exclusion, no one really wants that to happen, which may compromise the very principle. The definition of autonomy has evolved over the last years, as the boundaries for information and self-determination, leading to structured forms almost widely accepted, but depersonalized and eventually far from the participant.

Over time, other ethical principles have been defined and applied in scientific research. As ethical principles, by definition, require action, new procedures have been adopted to ensure respect for the participants in investigation protocols.

In 1978, the Belmont Report of the National Commission for the Protection of Human Subjects of Biomedical and Behavioral Research [4] described the ethical principles for research in human beings: beneficence, justice, and the respect for the person. The respect for the participant implies that researcher must recognize the subjects' autonomy to their own will and assume the duty to protect the most vulnerable.

In the 1998 Barcelona Declaration, a panel of experts defined the principles of autonomy, dignity, integrity, and vulnerability [10]. Autonomy comes up as an ideal to reach. Dignity is an intrinsic value of the individual, meeting himself with others. Integrity is the right to inviolability, implying the respect for privacy, personal ideas and expectations, and for patient's understanding of his own life and illness. Vulnerability expresses the susceptibility to be hurt. It is commonly understood as the condition of a patient before the threat of disease. In investigation, vulnerability refers to the fragility of participants before the methods. It implies the duty of not to harm the integrity of the participants and, at the same time, to protect their integrity. Classically, we consider children, pregnant women, and elders, but there are many other ways of turning vulnerable, such as the invitation to participate in a research of the doctor to his patient.

More than the statement in the ethical issues of a paper, assuring that the authors were committed with Declaration of Helsinki and Oviedo Convention for the protection of human rights and dignity of the human being with regard to the application of biology and medicine, researchers should incorporate the ethical principles since the formulation of the research question and in the entire definition of study design and results analysis, in the assumptions they make, and in the decisions they take.

3. Checklist of the ethical appraisal

The ethical appraisal of a research protocol often starts with a descriptive characterization of the study. Checking the presence or absence of certain elements in the protocol may assist this task. **Table 1** presents the checklist of ethical appraisal. It does not pretend to be a definitive tool of decision, since the ethical appraisal of a research protocol is not an assignment that may be reduced to a checklist. This table should be perceived as a summary tool to help structure and guide critical thinking regarding ethical research assessment.

Overall the research protocol proposal is assessed for its merit and integrity alongside with the description of appropriate and rigorous methods and procedures committed to non-maleficence. The use of sound scientific methods is warranted. Although this is not the main focus of the consideration it is important to assure that the use of resources and enrolment of participants is based on solid scientific grounds. Additionally, appropriate academic conduct in terms of references

Ethical appraisal			
Research team	Yes	No	**Merit and integrity**
Individual researchers curriculum vitae			• Research team CVs are expected to demonstrate capacity to develop the study
Problem	Yes	No	
Coherent rationale supported by literature			• Relevance and feasibility of the study is expected
Questions	Yes	No	
Clear and answerable research questions			
Design	Yes	No	**Appropriate and rigorous**
Experimental			• Coherence between research questions and research design is expected
Pre-post intervention			
Control group			• Efficient resource allocation is expected
Observational			
Cohort			• Appropriate bias identification and mitigation strategies are expected
Case–control			
Cross-sectional			
Qualitative			
Validation (e.g., scale development)			
Mixed methodology			
Other			

Ethical appraisal			
Participants enrolment	Yes	No	**Non-maleficence**
Clear description of the procedure to enroll participants			• Consideration of benefits and costs (risks) weighting for individual and for community interests is expected • Commitment to respect the integrity, dignity and autonomy of participants is expected • Appropriate informed consent presentation is expected • Appropriate procedure of data protection is expected • Appropriate procedure for adverse event reporting is expected
Data collection	Yes	No	
Gathering data from people			
Questionnaire			
Individual interview			
Group interview (e.g., focus group)			
Biological sampling (blood, urine, saliva, sputum, feces, semen, tissue or other bodily fluids)			
Gathering data from clinical records			
Anonymized or pseudonymized data			
Written informed consent from participants			
Data storage	Yes	No	
Safety measures for database management			
Participant safety	Yes	No	
Written informed consent from participants			
Procedure for reporting of adverse events			
Identification of possible risks or harm to participants (e.g., stress or anxiety)			
Insurance (intervention studies, if applicable)			
Data analysis	Yes	No	**Merit and integrity**
Exploratory			• Detailed explanation of statistical analytical plan is expected (viz., identifying all variables and statistical tests use) • Commitment to report and make results public is expected • Declaration of any conflict of interest is expected
Prespecified			
Reporting	Yes	No	
Publication or registry of the protocol			
Compromise to send a report of the research			
Intention to publish the study			
Funding	Yes	No	
Funding protocol (if applicable)			

Table 1.
Checklist of the ethical appraisal.

and authorship authorizations in the use of tools and instruments for research is required. The main focus of attention is usually placed in all the interactions with participants. From first contact where the study is presented to enrollment in the study followed by all the activities required to the end of the study and eventual subsequent follow-up.

Author details

Paulo Santos[1,2,6]*, Pedro Teixeira[3,4,6], Helena Beça[5,6] and Alberto Hespanhol[1,2,6]

1 Department of Medicine of Community, Information and Health Decision Sciences, Faculty of Medicine, University of Porto, Portugal

2 CINTESIS, Center for Health Technology and Services Research, University of Porto, Portugal

3 ICVS/3B's, PT Government Associate Laboratory, Life and Health Sciences Research Institute (ICVS), Portugal

4 School of Medicine, University of Minho, Portugal

5 Family Health Unit of Espinho, ACeS Espinho-Gaia, Portugal

6 Ethics Committee of the Administração Regional de Saúde do Norte, ARS Norte, Portugal

*Address all correspondence to: psantosdr@med.up.pt

References

[1] Hipocratis opera vera at adscripta. Vol. Quartus. Lausanne; 1771. pp. 197-199

[2] Trials of War Criminals before the Nuernberg Military Tribunals under Control Council Law n. 10. Volume II: The medical case; 1949

[3] Declaration of Helsinki: Recommendations guiding doctors in clinical research; 1964

[4] Sims JM. A brief review of the Belmont report. Dimensions of Critical Care Nursing. 2010;**29**(4):173-174

[5] McGee G, Caplan AL, Spanogle JP, Asch DA. A national study of ethics committees. The American Journal of Bioethics. 2001;**1**(4):60-64

[6] WMA Declaration of Helsinki—Ethical principles for medical research involving human subjects; 2018

[7] Sikes P, Piper H. Ethical research, academic freedom and the role of ethics committees and review procedures in educational research. International Journal of Research & Method in Education. 2010;**33**(3):205-213

[8] Beauchamp TL, Childress JF. Principles of Biomedical Ethics. 8th ed. New York: Oxford University Press; 2019

[9] Grady C. Chapter 2—Ethical principles in clinical research. In: Gallin JI, Ognibene FP, Johnson LL, editors. Principles and Practice of Clinical Research. 4th ed. Boston: Academic Press; 2018. pp. 19-31

[10] Kemp P, Dahl RJ. The Barcelona declaration. Towards an integrated approach to basic ethical principles. Synthesis Philosophica. 2008;**23**(2):239-251

[11] Santos P, Hespanhol A. Informed consent for case reports—An ethical perspective. Acta Bioethica. 2017;**23**(2):271-278

[12] Teixeira PM. p-Hacking—A call for ethics. Pulmonology. 2018;**24**(3):207-208

[13] Cain DM, Detsky AS. Everyone's a little bit biased (even physicians). JAMA: The Journal of the American Medical Association. 2008;**299**(24):2893-2895

[14] Dekking SA, van der Graaf R, van Delden JJ. Strengths and weaknesses of guideline approaches to safeguard voluntary informed consent of patients within a dependent relationship. BMC Medicine. 2014;**12**:52

[15] Jamrozik K. Research ethics paperwork: What is the plot we seem to have lost? BMJ. 2004;**329**(7460):286-287

[16] Estabelece os princípios e regras aplicáveis às comissões de ética que funcionam nas instituições de saúde, nas instituições de ensino superior e em centros de investigação biomédica que desenvolvam investigação clínica. Decreto-Lei n.° 80/2018. Sect. 198/2018. Série I de 2018-10-15; 2018

Chapter 9

Ethics in Laboratory Medicine: An Overview of Considerations for Ethical Issues

Neerja Aggarwal, Pawan Kumar Kare and Sudip Kumar Datta

Abstract

Several ethical issues exist within the diagnostic medical laboratory. The major ethical challenges such as; consent, confidentiality, codes of conduct, conflict of interest, lab utilisation, proficiency, and direct access testing are some times more prevalent in resource-limited settings. Presently, decisions regarding diagnosis and patient's treatment are commonly taken on the basis of outcomes and interpretations of laboratory test results. Therefore, ethics plays a significant role in laboratory medicine. Apart from the lab results, laboratory staff is another important aspect of the laboratory. Hence, it is highly recommended that knowledge of ethics helps to protect confidence; operational integrity, capability, impartiality, and safety of the staff. Many countries and their professional societies have developed policies and guidance material with regard to ethical issues in the area of laboratory medicine. The organizations specially; International of federation of clinical chemistry (IFCC), American Association of Clinical Chemistry (AACC) and International Organization for Standardization (ISO) have defined ethical recommendations for clinical laboratories. They are, in general, outlined the responsibilities of laboratory professionals towards their profession, the patient, and the society. However, implication of ethical standards and guidelines are vary between different cultures, geographies, and according to available resources. In this chapter, we have mentioned the ethical consideration of IFCC, AACC and ISO 15189:2012 with regard to laboratory medicine and also addressed the various ethical issues that arises day to day in laboratory medicine in the current scenario.

Keywords: laboratory, ethics, issues, principles

1. Introduction

In the current time, ethical concerns exist everywhere whether it is a medical field or life science. Lab Medicine and biomedical research, both fields are interconnected by laboratory testing where new results, remaining patient's blood sample, and genetic testing, etc. are some of the major ethical issues that commonly exist. Ethical issues plays very crucial role in laboratory medicine. Therefore, it is required for laboratories to strictly follow ethical principles.

The field of ethics involves 'a set of principles of right conduct' [1] and bio-ethics is well-defined as a branch of applied ethics that studies the philosophical, social and legal issues arising in medicine and life sciences. IFCC-task force has suggested that all the area of medicine to fulfill with ethical standards and guidelines and the field of lab medicine is no exemption. According to the IFCC verdict, prognosis as well as medications associated with certain medical conditions is usually determined by outcome, results and analysis of laboratory tests [2]. When we talk about the laboratory system, staff comes at first as they are directly linked in interaction with patients and their care. Apart from laboratory staff, everyone who is involved on the way is equally responsible for maintaining laboratory ethical values. Henceforth, it is highly obligatory to evade any such activity that would downgrade the expertise, neutrality, outcomes, operational truthfulness or patient's confidence in laboratory. Laboratory staff's behavior and etiquettes also comes in this category, thus, their actions should be in a professional way for example, wearing laboratory coat/apron, proper dressed-up, phones should be turned silent/OFF during the time of testing and not discussing any report with clients and others. Hence, various international and national guidelines and declarations have been evolved with time to time and thus critically upgraded the practice of bio-ethics in the field of biomedical research. Compliance with these guidelines confirms the autonomy, dignity and well-being of participants as well as the integrity and credibility of research results [3].

2. Evolution of ethics

Evolution of biologically-centered ethical guidelines in medical or biomedical research has upgraded the understanding of ethics over the years. Various guidelines and declarations evolved over the period, including international and national, are mentioned as below:

2.1 International guidelines

a. **Nuremberg Code, 1947:** This code was the initiation of modern ethical morals. It introduced the discussion on rationale and explanation of research risks or benefits analysis. Initially there was no ethical conduct for the research involving biological subjects, the time before World War II. Nuremberg, during 1947, was the first to establish ethical principles for such researches which delineates the necessity of competent and trained staff, participant's consent, and circumstances under which research should be discontinued [4].

b. **Declaration of Geneva, 1948:** It comes into existence soon after Nuremberg code which emphasized guidelines for ethical issues related to clinical medicine. Soon after Nuremberg code, Declaration of Geneva was conscripted and accepted by World Medical Association (WMA) in 1948. It was actually a physician's oath that was proposed as an amendment of Hippocratic Oath which was the assertion of physician's commitment towards his duty for humanity in medicine. Its concept is applicable to clinical medicine unlike Nuremberg code [5].

c. **Declaration of Helsinki, 1964, Revised 1983, 1989, 1996, 2000, 2008, 2013**: It is obliged to make individual rights as informed decisions, duties of investigators and research participant's welfare as well. It was developed by World Medical Association (WMA) which includes the set of ethical moralities for conducting research involving humans for medical community. Ten principles of Nuremberg code and Declaration of Geneva were tied in a single document named as 'Declaration of Helsinki' [6].

d. **Belmont report, 1978–1979:** The Belmont Report was generated by the United States of America (USA) National Commission for the Protection of Human Subjects of Biomedical and Behavioral Research in 1978. It is one of the key work concerning ethics and healthcare research and explains the ethical guidelines for experiments involving human participants. It three basic principles includes the respect for participants, justice and beneficence. In research these basic principles obliged to consider. It also saves the rights of participants in clinical and experimental researches. It also describes the approval of study by the Institutional ethical committee and ensures that participant should at least get nominal care for their medical condition [7].

e. **CIOMS guidelines [Council for International organizations of Medical Sciences], 1992–1993, Revised 2002**: International ethical guidelines were announced by CIOMS for epidemiological studies in 1991 and for researches involving human participants were announced in 1993. It focusses on the pharmacovigilance, reporting of adverse drug effects along with protection of research participants [8].

2.2 National guidelines

a. National ethical guidelines for Health Research and biomedical sciences involving Human Contributors; ICMR, 2017.

In 2017, Indian Council of Medical Research introduced ethical guidelines for research on Human Participants. In India, it is mandatory for all research organizations to strictly follow these guidelines in letter for all types of biomedical research involving human beings, along with complete documentation to protect safety and wellbeing of all participants [9].

3. Principles of ethics

The important three core ethical principles are discussed in all documents. These are as below (**Figure 1**):

a. **Respect for persons:** We must respect patient and their self-respect. There is freedom of decisions to the each participant of the study. It is an obligation to respect the decisions made by people concerning their own lives. This is respecting human dignity. We must not interfere with the decisions of competent adults, and also actively empower others for whom we are responsible.

b. **Beneficence**: It refers to our duties in the best interests of the patients or research participants. The goal is to maximize benefits and minimize harms, the latter sometimes Latinate as *non-maleficence.* Everyone must be fair and correct in all their actions and must take positive steps to prevent harm.

Figure 1.
Core principles of ethics.

c. **Justice**: It is an obligation to provide all participants with whatever they are deserve. Basically, we have an obligation to treat all people equally, fairly, and impartially. All individuals should have an opportunity to participate in research unless contraindicated and we must not impose unfair burdens. All doubts of research participants should be cleared by concerned staff. We should make available all the safety concerns as demanded by research participants.

4. Various international ethical considerations

Similar to other fields of medicine, laboratory medicine is obliged to adhere to high ethical standards. With the advancement of medical science in the area of laboratory medicine, special ethical considerations should be taken in addition to the general ethical framework followed in biomedical research. Various policies and guidelines related to ethical issues are being developed time to time by several countries or related societies.

4.1 Ethical consideration in ISO 15189

The International Organization for Standardization (ISO) that created ISO 15189:2012 "Medical laboratories-Requirements for quality and competence" in 2012 [10]. Its section 4.1.1.3 elaborated the ethical conduct required in laboratories. ISO 15189 is technically applicable for laboratory equipment, personnel, environmental conditions, consumables, pre- and post-examination processes, reporting and release of laboratory results, and lab information management. As per ISO 15189 standards, the core principles that stated in documents are: (i) there should not be participation in any activities that would diminish confidence in the laboratory's competence, impartiality, judgment or operational integrity; (ii) management and personnel are free from any undue commercial, financial, or others pressure and influences that may adversely affect the quality of work; (iii) where potential conflicts in competing interests exist, they shall be openly and appropriately declared; (iv) there are appropriate procedures to ensure that staff treat human samples, tissues or remains according to relevant legal requirements; (v) confidentiality of information is maintained.

4.2 Ethical consideration in AACC

The American Association for Clinical Chemistry (AACC) has also recommended fifteen principles of ethical conduct for laboratories. The major highlights are that [11]: (i) to be honest in all professional accomplishments, and retain the high level of personal veracity; (ii) need to avoid any scientific or professional delinquency; (iii) should report any professional that is degrading the standards of laboratory and professionalism that would affect patients care; (iv) to maintain high quality reagents, equipments and consumables. Also, they must confirm the reliability of test reports and quality of confidentiality of reports; (v) respect the privacy and confidentiality of protected health information; (vi) continuously endeavor to augment the professional qualifications, knowledge, and skills, and present them accurately; (vii) encourages the safety of patients, staff and the environment; (viii) must disclose the actual conflicts of interests; (ix) encourage open and honest discussion among physicians, other healthcare providers and/or facility managers; (x) fulfill the appropriate laws and pursue to change whenever they seem contrary to patient's interests.

4.3 Ethical consideration in IFCC

Despite the importance of bio-ethics in lab medicine, still there are lacunae in education training focused on ethics in laboratory. To address this issue, IFCC has recently constituted a task force on ethics (TF-E) to rationalize the documents and spread the education and training on ethics [12]. This task force (TF-E) has created a toolkit which serves as a repository of documents developed worldwide in the kingdom of laboratory ethics [13]. Although the members of the IFCC Task Force on Ethics also contribute to achieve the goal of ethics education in the field of laboratory medicine through the publications on the topic of ethics in collaboration with the electronic journal of International Federation of Clinical Chemistry (eJIFCC).

5. Codes of ethics

Professional personnel of a medical laboratory are bound by the ethical codes of their respective profession. A code of ethics may be described as an expression of basic values –the principles and standards by which we should conduct ourselves. Several laboratory professional societies and organizations have developed codes of ethics, with common principles of conduct which act as guidelines to professional members of those organizations [14]. The International Federation of Biomedical Laboratory Science (IFBLS) suggests to maintain strict confidentiality of patient information and test results; safeguard the dignity and privacy of patients and above all be accountable for the quality and integrity of clinical laboratory services being provided [15]. In same line, the American Society of Clinical Pathologists (ASCP) has also advised laboratory staff to treat patients and colleagues with respect, care and thoughtfulness; perform duties in an accurate, precise, timely and responsible manner; and safeguard patient information as confidential, within the limits of the law.

6. Ethical issues in laboratory

There are several ethical issues in laboratory (**Figure 2**). These issues have divided into three phases according to the laboratory work distribution.

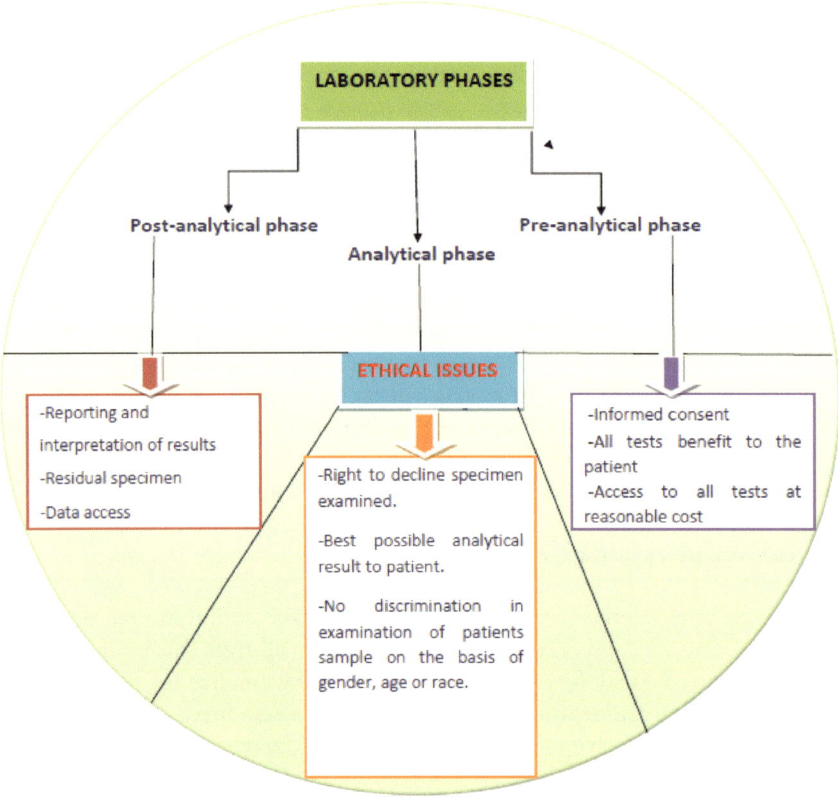

Figure 2.
Ethical issues in laboratory phases.

Pre-analytical phase issues are related to patient's interaction, specimen collection, sample receiving and its transport. Analytical phase issues are usually related to quality control, whereas, post-analytical phase issues are related to reporting of results, keeping and maintaining records [16].

6.1 Pre-analytical phase

Clinicians ordering laboratory tests is also comes under the most important ethical obligations. The laboratory personnel are required to act every time to confirm whether the tests, which are referred by a clinician, are being met with the diseased person requesting the tests or not. However, it is commonly assumed that clinicians are referring laboratory tests so as to benefit the patient without any financial interests. In this phase, there is collective responsibility of many people including nurse, healthcare providers, researcher, or the technical staff collecting the samples. Their role includes:

1. Identification of a patient with respect to the tests ordered.

2. Proper collection, labelling, and handling of samples till the tests are performed.

Three basic ethical principles in pre-analytical phase are:

a. **Respect for persons:** Consent must be understood by the patient. It may be expressed if a participant is asked for written agreement. It may be implied when patient is comfortably sits and allows his sample to be taken. Informed consent may lead to an ethical problem if participant is incompetent owing to age, mental status, or critical illness. The patient's right to refuse to get tests done should be appreciated. In special cases, healthcare professionals should be obliged to consult the institutional guidelines. Any information regarding patient's demographics, their visit to the testing laboratory, the tests that were ordered, and the requirement for these tests, should be provided only to suitable personnel. At every step of sample handling, from sample collection to data entry, confidentiality should be maintained.

b. **Beneficence:** All tests performed/referred must benefit to patient. Any adverse event during or after sample collection have to be managed by trained workers, with the help of standard operating procedures (SOPs). Collection of samples should be done as per universally recommended precautions so as to protect the patient and the healthcare worker. The additional samples should not be drawn from the patient without informing and getting the permission from Institutional ethics committee. The specimens should be well-labeled with minimum two unique identifiers. Transportation of samples should be done to protect the integrity of the sample.

c. **Justice:** It provides access to several laboratory tests at reasonable cost. The laboratory should evaluate the need to introduce new tests and the opportunities to discontinue older tests when better tests are available. No preference should be given to some patients in order to facilitate or accelerate the collection procedure at the cost of others.

6.2 Analytical phase

In laboratories settings confidentiality, quality and competency are essential. During this, confidentiality is almost a by-product of laboratory automation which uses automated code readers, automated analysis, as well as auto-verification and also names of patients are normally given a unique sequential number for processing. Maintaining confidentiality is more challenging during the analytical phase in small laboratories as compared to larger ones, as smaller laboratories perform manual testing. However, it is most important to maintain ethical standards by each laboratory in conducting patient's testing. The three principles in this phase are as follows:

a. **Respect for persons**: After collection and processing of samples, patients have the right to refuse to have their specimens examined. In such a case, confidentially should be appreciated and preserved. Special care must be taken to preserve confidentiality in point-of-care testing as much as possible.

b. **Beneficence**: The aim of the laboratory is to make available the best possible result to patient. This is accomplished via good laboratory practices (GLP) and maintaining high professional standards. Good laboratory practice should involve the establishment of demanding quality assurance program including quality control analysis, proficiency testing and accreditation of laboratory. In this regard "a wrong result is worse than no result" is a critical guiding principle. Good laboratory practice (GLP) refuses to evaluate or account a result in the presence of poor sample integrity, improper or poor labeling or any other insufficiency that may lead to compromise with the test result. In this regard, Acceptability of "difficult to obtain" (such as cerebrospinal fluid) samples may be taken as special case, and departmental facilities should develop some suitable policy on examination and records of such specimens where specimen integrity or identification is being compromised. All laboratories should maintain proper authorization as required by their country or region. Only qualified, properly skilled and regularly re-accredited employees should be authorized to execute point-of-care testing (POCT).

c. **Justice**: There should be no discrimination in the examination of patient's samples on the basis of gender, age or race; otherwise it would be an injustice. All samples should be treated likewise. It is recognized, that laboratories must develop some appropriate provision for STAT or priority testing. Laboratories should also state which tests are included and their expected turnaround times. It is anticipated that all specimens are being analyzed correctly in a timely routine.

6.3 Post analytical phase

This phase includes reporting and interpretation of tests results, storage of residual sample, and access to the data. All laboratories should have a procedure for storage of a specimen that is analyte dependent. An essential part of good laboratory practice is to archive the results either in electronic and/or hard copy format. Documents that can be archived include request forms, raw analytical as well as quality control data, results, and reports. Guidelines on retention or destruction of medical records along with remaining sample retention and its dispose of should be kept in place. Policy manual should also mention the strategies on the identification of authorized personnel such as doctors, patients, and laboratory staff; that would be allowed to access medical records. Besides this, the patient should have the right to give consent to access by others (such as family members), if required. Applying the basic ethical principles in post analytical phase as follows:

a. **Respect for persons:** Patients reasonably expect that their specimens will not be used beyond the testing prescribed by a clinician and solely used for only prescribed testing for medical purposes. However, in the world substantial differences are there concerning the confidentiality of results. In many areas, the patient and referring clinician are the only authentic recipients of laboratory data. Exceptions are there in case if patients are not able to receive or understand the tests reports. In other areas, the patient's family is also considered to be genuine recipients of a patient's test reports. Laboratories must develop a strategy for results dissemination so as to respect local customs. Reliable communication methods are to be used, and security should be protected in conveying the results regardless of the channel of communications including, hand deliveries by messengers. The local ethics committee or board should also have provision on any further testing in residual specimen (except for the samples used in validation processes), and patient consent may be obligatory.

b. **Beneficence:** Results misinterpretation may harm the patients and it could be reduced only when a skilled staff would interpret the reports; to minimize this harm. The reporting should be in proper time with correct and all necessary information so that clinician gets the true interpretation. Furthermore, a complete report usually covers the name of the test executed, a suitable reference interval (which might be age or gender specific), units of measurement if needed, and a remarking that the result is within or outside the reference interval. As per laboratory conditions, turnaround time (TAT) should be minimum for any test but it should not compromise the legitimacy of the results. Timely access to test results is very essential for the welfare of patients, particularly in emergency conditions and delaying the access to results in case of non-payment may harm the patient and ethically also not correct. Hence, delaying in reporting should be avoided. Errors should be notified to clinicians immediately after they came into notice and test results should be rectified or repeat tests should be done, whichever required. Finally, incorrect results should be still accessible but marked as erroneous and corrected result should be mentioned on the report.

c. **Justice:** Although reporting of test results must be consistent for all patients, speedy reporting may possibly be demanded for some results, including "critical" and "significant-risk" results. Instructions for quick reporting must apply irrespective of the source of sample as well as the patient's financial capability and not disclosing the results just due to payment should also be avoided. Remaining patient's samples should not be used further without patient's knowledge which is very common these days. There is much discussion in the literature about who owns patient specimens and whether patients should share in profits if financial gains are derived from leftover samples. However, rules and practices differ among different regions and institutions.

7. Conclusion

Finally, it has been observed that it is necessary to incorporate the core principles and guidelines of bioethics in the areas of laboratory medicine. Any laboratory involving human participants should follow international standards and practices of ethics. Laboratories shall not engage in practices restricted by law and should uphold the reputation of their profession. It is required to develop an ethics policy and add it to the laboratory's quality assurance manual. Development and implementation of an ethics training program for laboratory staff should be done in such a way that it would promote the development of the professional life of laboratory staff, highlighting human values and responsibility, honesty in their work. This will surely initiates and encourages the change of paradigm with the aim of increasing knowledge keeping in mind ethical principles in daily procedures.

Conflict of interest

The authors declare no conflict of interest.

Author details

Neerja Aggarwal[1], Pawan Kumar Kare[2]* and Sudip Kumar Datta[3]

1 Department of Medicine, University College of Medical Sciences, Delhi, India

2 Department of Medical Biochemistry, Gandhi Medical College, Bhopal, India

3 Department of Laboratory Medicine, AIIMS, New Delhi, India

*Address all correspondence to: pawankare4@gmail.com

References

[1] Mathur R, Swaminathan S. National ethical guidelines for biomedical & health research involving human participants, 2017: A commentary. The Indian journal of medical research. 2018; 148(3):279.

[2] Bruns DE, Burtis CA, Gronowski AM, McQueen MJ, Newman A, Jonson JJ, IFCC Task Force on Ethics. Variability of ethics education in laboratory medicine training programs: results of an international survey. Clin Chim Acta. 2015; 442:115-8.

[3] Beshir L. Research ethics committees in laboratory medicine. eJIFCC. 2020; 31 (4): 282-291.

[4] The Nuremberg Code [Internet]. [cited 2020 Jun 27]. Available from: http://www.cirp.org/library/ethics/nurem- berg/

[5] WMA - The World Medical Association-Declaration of Geneva [Internet]. [cited 2020 Jun 27]. Available from: https://www.wma.net/what-we-do/medical-ethics/declaration-of-geneva/

[6] WMA - The World Medical Association-Declaration of Helsinki [Internet]. [cited 2020 Jun 27]. Available from: https://www.wma.net/what-we-do/medical-ethics/declaration-of-helsinki/

[7] Read the Belmont Report [Internet]. HHS.gov. 2018 [cited 2020 Jun 27]. Available from: https://www.hhs.gov/ohrp/regulations-and-policy/belmont-report/read-the belmont-report/index.html

[8] International Ethical Guidelines for Health-related Research Involving Humans, Fourth Edition. Geneva. Council for International Organizations of Medical Sciences (CIOMS); 2016.

[9] National ethical guidelines for Health Research and biomedical sciences involving Human Contributors; ICMR, 2017. Available from https://www.icmr.nic.in/sites/def ault/files/guidelines/ICMR_Ethical_Guidelines_2017.pdf:

[10] ISO 15189:2012 (en), Medical laboratories-Requirements for quality and competence [Internet]. [Cited 2020 Jun 27]. Available from: https://www.iso.org/obp/ui/# iso: std: iso: 15189: ed-3:v2: en.

[11] AACC Ethics guidelines. American Association of Clinical Chemistry. https://www.aacc.org/about-aacc/governance/ethic-guidelines. Accessed 9 January 2016.

[12] Task Force on Ethics (TF-E) - IFCC [Internet]. [cited 2020 Oct 7]. Available from: https://www.ifcc.org/taskforce - ethics/

[13] TF-E Toolkit-IFCC [Internet]. [cited 2020 Oct 7]. Available from: https://www.ifcc.org/taskforceethics/toolkit/

[14] Datta SK. Ethics in laboratory medicine: perspectives and challenges in resource limited settings. eJIFCC. 2020; 31 (4): 274-281.

[15] Gallicchio VS, Gallicchio LM. Ethics and Quality Assurance in Biomedical Laboratory Science. 2013; (2):4.

[16] Sekadde-Kigondu CB, Higgins T, Bruns DE. Ethical Considerations in Clinical Chemistry and Laboratory Medicine. IFCC. 2016; (1):1-15.

Chapter 10

Biomedical Ethics and Communicative Maxims: Case Studies in Outpatient Health

Jonathan Comyn de Rothewelle

Abstract

Effectual and ethical healthcare communication is essential in medicine. Health communication not only includes taking medical histories and communicating diagnoses with patients, but its scope is also far broader. In light of recent research that suggests the importance of communication in health, this case study argues more ethical and communicative oversight that may be merited. This case study examines orthonyms associated with dental clinics. The orthonym, or proper noun, as a form of healthcare communication is a communicative practice influencing outpatient health. This communicative entity was selected as it is previously unstudied and adequately narrow so as to be analyzed without tangent. This chapter endorses the amalgamation of communicative maxims and bioethical principles as a backbone for effective and ethical healthcare communication. A framework uniting these maxims and principles is provided.

Keywords: bioethics, health communication, biomedical ethics, communicative maxims, medicine

1. Introduction

In healthcare, communication can be the difference between life and death. Communication happens in a variety of contexts from explaining infectious disease to the masses to messaging providers through online health portals—no matter its function, communication a modality for effective treatment.

Aside from communicating clearly and concisely when taking medical histories and explaining treatment options, studies have shown the importance of effective language [1]. Patient safety and quality of care may be enhanced or endangered, owing fault to whether or not communication is effectual [2]. The language used during clinical interactions may be correlated with health outcomes, for example, patients whose *L1*[1] is not the language of operation at the hospital in which they are receiving treatment have a higher chance of readmission [3].

A patient centered experience built through effective communication may build stronger patient-practitioner relationships, enhance decision making, and reduce patient uncertainty [4]. Clear communication is not only key to the effective exchange of health information, but it is also important when practicing clinical empathy [5]. In addition, adequate communication between patients and

[1] A patient's L1 is their first language or mother tongue.

practitioners promotes higher patient comprehension and therefore may yield higher treatment adherence [6].

Higher treatment adherence promotes patient satisfaction and increases the probability of better clinical outcomes [6]. As communication styles differ between cultural groups, it is important for practitioners to appreciate nuanced differences in communicative styles. As language influences cognition, including that of the patient, it is self-evident that healthcare communication should be interpreted as an influencing factor in patient care and thus be governed by bioethical praxis [7–12].

2. The *lingua franca* of bioethics

Bioethics is founded on the ethical care of patients. With particular insights and influences from various philosophies, all theories of biomedical ethics try to arrive at the best way to honorably and justly treat patients [13]. Recent research suggests the study of bioethics would benefit from methodological study from all perspectives, whether bioscience or the humanities, as the patient in which it seeks to serve may be influenced by more worldviews than one [14].

Philosophers T. Beauchamp and J. Childress laid important framework for bioethical research and analysis [15]. This chapter uses, in brief, their work as a scaffold for bioethical standards that is further combined with communicative maxims. Beauchamp and Childress describe four principles for ethical medical practice including respect for autonomy, beneficence, nonmaleficence, and justice [15].

The principle of *beneficence* states that the clinician has a moral obligation to do all they can to benefit the patient. The principle of *nonmaleficence* declares that clinicians should ensure latent harm from treatments does not outweigh potential benefits, and that patients are not unnecessarily exposed to hurt. The principle of respect for *autonomy* articulates that the patient should always have a choice and play a role in their treatment. The principle of *justice* expresses that clinicians should always aim to do the most good for the most people and distribute resources fairly [15].

As language is the *de facto lingua franca* of bioethics—that is there would be no ability to express nor debate the virtuous versus the corrupt without language itself—it is axiomatic that communicative mores should be included in any sort of debate concerning itself with the benevolence of the patient. Notwithstanding how small or impartial the orthonym may appear in the context of the outpatient clinic, it is a communicative form within healthcare and therefore necessitates bioethical oversight.

Bioethics, to be properly applicable to that which is human, must concern itself with all things humanistic. That is, bioethics must not be so overtly preoccupied with stem-cell research, that it overlooks the outwardly mundane yet strikingly influential lexemes that affect the cognition and treatment of the everyday patient. The study of proper names in healthcare lends to the larger study of healthcare communication which has been proven unremittingly important to the patient. From increased patient satisfaction to higher treatment adherence, proper linguistic form in healthcare contexts is worth deeper evaluation.

3. Communication in healthcare

As language, ethics, and cognition are inherently intertwined [16], there is a need to study all forms of communicative value [including that of the orthonym] from various perspectives including bioethics [14]. Communication between clinics and communities must respect and abide by ethical institutions to do no harm [17]. Health communication, in order to ethically serve the patient, must be clear and concise, honest, and not sensationalize information; health communication should "adhere to the principles of beneficence, nonmaleficence, respect for personal autonomy, and justice" [17, 18].

When it comes to language, British philosopher H. P. Grice, created a rubric for effectual discourse that this chapter argues should be used in conjunction with the principles of bioethics set forth by T. Beauchamp and J. Childress [15] in the context of healthcare communication. According to H.P. Grice, the language used in communication should include what is relevant and necessary given the environment of the discourse. More specifically effective communication should be appropriate in manner, relevance, quality, and quantity according to its context [5, 19].

Along with communication in general, proper names structure social spaces [20]. The ability to structure social spaces could be interpreted through association, the cognitive process of linking the abstract with the physical. This creates cultural narrative reinforced by language ideologies which are held in the collective consciousness of community members [20]. Concerning bioethics, the institutions within the collective consciousness of a culture are what determines what is morally right or aversive and may already be heavily influential in medicine [21].

4. Onomastic analysis

Relative to their cultures, orthonyms denote people, places, and things [20]. There is some evidence to suggest that descriptions alone cannot uniquely distinguish an object, place, or person, therefore requiring a proper noun to represent the physical object [20, 22–24]. However, time changes the meaning of language and the ethics of a community can change over time; proper nouns are context-dependent [22].

Within the analysis of proper nouns, social semiotics provides a framework for the context and the constituent parts [25]. Through the analysis of an orthonym, ideologies surrounding healthcare can be interpreted [26, 27]. The orthonym can also be interpreted as assigning roles to the creator verses the viewers [26, 28]. Throughout this chapter, time is spent on how the creators (most likely dentists) perceive themselves and viewers (patients) and how the discourse may be interpreted as potentially influencing the viewer's (patient's) cognition.

This creator-viewer relationship demonstrates the reflective influence communication and communicators have on each other [5, 29]. It is widely reported that communication plays a role in cognition, which in any event within healthcare ought to be upheld to bioethical principles [30].

5. A trip to the dentist

Before proceeding to the case study portion of this discussion on the intersection of communicative maxims and bioethical principles, this section will exemplify a case where the violation of communicative maxims ultimately leads to the undermining of biomedical ethics. The following anecdote illustrates a hypothetical patient experience at "Luxe Cosmetic Dentistry," and aims to begin this chapter's discussion on bioethical principles within healthcare communication at the micro-level.

A patient is browsing listings of local dentists on the internet. Without a referral, they rely on things such as start ratings and the names of various clinics.

The patient stumbles upon "Luxe Cosmetic Dentistry." What comes to mind when reading the name "Luxe Cosmetic Dentistry?" The patient's mind may bring up images of beautiful teeth, linen clothed, perfectly tanned people on a sand beach, living the life of luxury—an image straight out of The Condé Nast Traveler. And that would be the purpose! Names are chosen specifically to draw on the human mind's ability to draw from memories, feelings, and associations. This orthonym, or proper noun, relies on the patient's idea of luxuriousness to paint an image of the dental clinic the patient will find. This patient, upon reading the name of this dental clinic, and imagining how pleasant of an experience a visit might be, decides to book an appointment. But when the patient arrives, this is what they find (**Figure 1**).

When the patient enters the waiting room, they are shocked, even angry. This is nothing like what they were expecting. Based on the patient's framework about what a dental clinic titled "Luxe Cosmetic Dentistry" should look like, the patient experiences a state of dissonance. The patient decides to continue with the appointment, but the patient's care may be at risk. When patients feel uncomfortable, they may be less likely to communicate properly with practitioners about their health. In this case, the patient just wants to get the appointment over with.

If the language used properly reflected its referent, the dental clinic may have more accurately looked like the following (**Figure 2**). If this were the case, the language would have clearly communicated, set patient expectations, and set the tone for the appointment—all before the patient started speaking with the doctor. Thus, in this analogy, the language of the orthonym has directly affected the patient's experience, and may go on to have a negative impact on patient care. And in such a case, this form of healthcare communication has violated bioethical principles.

By violating the aforementioned communicative maxims the principles of bioethics were also violated. As insignificant as the name of an outpatient clinic may seem, it is clear that this type of communication within a healthcare context could influence patient care and ought to be guided by bioethical principles. In this anecdote one example of a violation was presented, this chapter will examine the multiple way in which orthonyms can uphold or violate bioethical principles.

Figure 1.
Probably not the dental clinic the patient was expecting [31].

Figure 2.
Perhaps something more along the lines of what a patient might expect from a dental clinic titled "Luxe Cosmetic Dentistry" [32].

6. Methodology

Healthcare communication is most considerably studied on a larger scale. To add more breadth to the field of health communication research it was desired that a previously unstudied form of communication was selected. To aid in the ease of application of various principles, communicative values with succinctness were also preferred. The proper noun was selected as it met both criteria.

Using Google's map apparatus, clinics in Chicago were searched. The names of 50 dental clinics were recorded with attempt to select equal portions of dental offices on Chicago's North, West, and South sides. Each of the clinics on the list was assigned a number [1-50]. Using the Random Number Service's random integer generator, 10 numbers were selected from integers 1–50 [33]. These 10 numbers and the dental clinics they represent were designated for the analysis.

The medical dental clinics selected are not intended to represent a complete sampling but to offer localized analysis. The City of Chicago is composed of distinct neighborhoods, each with specific socio-economic characteristics; various neighborhoods were selected to offer diversity to the study. Socioeconomic data of the various neighborhoods were gathered from The Statistical Atlas which sources from the United States Census [34].

The data was then analyzed in the context of the aforementioned four bioethical principles and the four communicative maxims to assess ethical compliance and communicative value. The above rubrics (**Tables 1** and **2**), which offer concise applications of the theories of Grice [19], Beauchamp and Childress [15], Parli [20], and Webber [22], and others were created to aid this case study.

Along with broad-reaching analysis of orthonyms specifically and healthcare communication in general, this analysis will offer focused insight. In Illinois, where this research has been conducted, there are specific guidelines that dental practices must follow when representing themselves through the orthonym. These laws are a preventative measure should outpatient dental clinics attempt to misrepresent their practice with potential consequences on patient health.

The Illinois Dental Practice Act [37] states that dental clinics may not include titles or specialties in their orthonym that they are not certified to perform, use words that misrepresent or cause the patient to misinterpret services provided in order to gain more patients, practice under a false name, allow another uncertified individual or clinic to practice under their name, and must include disclaimers when appropriate.[2]

In addition to regional regulations limiting the naming of dental clinics, the American Dental Association sets national regulations for dentists [38]. These standards (**Table 3**), as influenced by bioethical principles, state that an orthonym that is misleading in any way is considered to be unethical.[3]

The Lexeme	What lexemes compose the orthonym? What are the possible and perceived definitions of these lexemes? [20, 35]
The Environment	What is the socioeconomic composition of the community? How is cultural context reflected within the orthonym? [23] What association habits contextually define this discourse? [22–24, 36]
The Significance	What significance is assigned through association habits? [22, 36] How does this discourse position itself within the community? How might this discourse influence the patient and the practitioner? [26, 27]

Table 1.
A brief guide for onomastic analysis and discussion.

Does the discourse follow the Four Communicative Maxims?	Does the discourse comply with the Four Bioethical Principles?
1. Manner: Is this communication in a manner that is appropriate within a healthcare setting?	1. Beneficence: Does this communication work toward benefitting the patient?
2. Relevance: Is this form of communication relevant to the task at hand?	2. Nonmaleficence: Does this communication disallow harm to the patient?
3: Quality: Is this communication of quality, that is does it provide false information?	3. Respect for Autonomy: Does this communication promote the patient's informed involvement in their health?
4. Quantity: Is this communication appropriate in amount for which the medium requires?	4. Justice: Does this communication represent fair distribution of health services in the community?

**If the answer to any of the above is* "no," *how might the discourse be altered so as to better fit communicative and ethical standards within healthcare communication?*

Table 2.
A rubric for analysis of communicative value and bioethical compliance within healthcare communication.

ADA Principles of Ethics & Code of Professional Conduct [Revised 2018]
5.G. Name of Practice Since the name under which a dentist conducts his or her practice may be a factor in the selection process of the patient, the use of a trade name or an assumed name that is false or misleading in any material respect is unethical. Use of the name of a dentist no longer actively associated with the practice may be continued for a period not to exceed 1 year.

Table 3.
Excerpt from the American Dental Association's document on ethical standards.

[2] The full version of the *Illinois Dental Practice Act* can be found at www.ilga.gov.

[3] The full version of the American Dental Association's *Principles of Ethics & Code of Professional Conduct (2018)* can be found at www.ada.org.

7. Data and analysis

In this case study, the analysis of ten medical dental clinic names will be categorized in units for ease of reading. The units include explanations of orthonyms that: (a) list medical credential, (b) describe the ideal patient, (c) focus on the patient experience, (d) refer to health outcomes and (e) appeal to patient identity (**Table 4**). Each section will anecdotally analyze clinic names used in this case study.

Communicative purpose	Description
Orthonyms that list medical credential	Orthonyms in this category contain lexical entries that position the clinic or dentist as qualified to operate [ex: D.M.D. or D.D.S.]
Orthonyms that describe the ideal patient	Orthonyms in this category contain lexical entries that describe the type of patient they seek to treat [ex: pediatric]
Orthonyms that focus on the patient experience	Orthonyms in this category contain lexical entries that describe services offered [ex: salon]
Orthonyms that refer to health outcomes	Orthonyms in this category contain lexical entries that refer to patient health outcomes [ex: perfect smile]
Orthonyms that appeal to patient identity	Orthonyms in this category contain lexical entries that build patient rapport through identity [ex: ownership]

This table shows the major categories of orthonyms in the region of data collection.

Table 4.
Regional trends in dental clinic names.

7.1 Orthonyms that list medical credential

Medical credential is important—it is what determines whether or not someone should be operating on a patient. Whether the proper name uses the word 'doctor' or lists dental degrees such as D.D.S. or D.M.D., or specialties such as periodontics or maxillofacial surgery, these orthonyms are communicating with the patient that the dentist they are seeing is indeed qualified to be practicing dentistry. Due to their nature, the misuse of lexemes in this category would be highly unethical.

Dr. Joseph Watson, DDS. This orthonym is simple and direct. It is made up of the doctor's name followed by their medical credential. The words used in the orthonym are obvious in their meanings and do not require the patient's analysis. This language conforms to discursive maxims in that it uses only what is necessary to convey only what information is essential. This dental clinic is located in the South Shore Neighborhood of Chicago on Chicago's Southside, and is 2.8% White, 1.8% Hispanic, 93.3% Black, 0.4% Asian, 1.3% Mixed, and 0.4% other. The median household income in this neighborhood is $27,900.

This orthonym states medical credential and education level twice. This may be a reinforcement to the patient that the dentist is an expert and qualified to do their job by using "Dr." and "DDS" in the clinic's name. This may also be interpreted as the dentist positioning themselves as of greater education level perhaps serving to increase the gap between the patient and the practitioner. This orthonym adheres to both communicative maxims and bioethical principles: It communicates well and does not appear to cause harm.

Dr. Louis C Rutland III. This orthonym takes a similar form as the previous. It is short and direct and does not require interpretation. The words used in this orthonym are orthonyms on their own combined with a marker of education level. This orthonym is short and gives only the necessary information in accordance with discursive maxims [19]. Dr. Rutland's office is located in the East Chatham neighborhood of Chicago on Chicago's Southside. This neighborhood is 2.7% White, 1.2% Hispanic, 94.4% Black, 0.8% Asian, 0.5% Mixed, and 0.3% Other, and has a median annual income of $23,800. Furthermore, this orthonym is similar to that of Dr. Joseph Watson, DDS and the socioeconomic composition of the surrounding community is also very similar. As community reflects cultural values, and cultural values are the collection of simultaneously performed identities, this orthonym might be interpreted as influenced by and influencing the neighborhood.

The use of "Dr." could be interpreted as a means of positioning the dentist as doctorally educated as well as informing the patients that they have the proper medical credential. As previously mentioned, communication has the power to influence within medical contexts. This orthonym could be interpreted as building trust with the patient before they even walk into the door. For this reason, it could be interpreted ethically and communicatively satisfactory.

7.2 Orthonyms that describe the ideal patient

Describing the ideal patient attracts only the patients that a dentist can properly serve. For example, a dental clinic that uses the word "pediatric" is probably not somewhere that maxillofacial surgery is performed. This could be viewed as cutting to the chase, as it were, and prepping patients for what to expect when entering a specific dental clinic. Ethical problems may occur when services are misrepresented in this way causing the patient some level of hurt.

1st Family Dental of Andersonville. This orthonym gives a *prima facie* description of the ideal patient through the use of the lexical entry "family." It is made up of patient descriptors as well as an area marker. There is no need for interpretation here. For example, one would not interpret this to be a luxury dental office, but, rather one where parents and children can get the treatment they need. This dental office is located in the Andersonville neighborhood on Chicago's Northside and is 68.4% White, 10.5% Hispanic, 7.4% Black, 10% Asian, 3.6% Mixed, and 0.2% Other. Habitants of Andersonville have a median household income around $76,500.

The words that make up this orthonym that are of particular note include "1st", "Family", and "Andersonville." By including first in the orthonym, this dental office could be perceived as claiming their territory. The word "family" is also important and could be understood as a marker for the clinic's ideal patient. By using the word family in the orthonym, the creators are designating general services for the whole family and effectively warding off patients who need more specialized procedures such a periodontal surgery. This orthonym could also be interpreted as positioning itself as a family friendly place within the community, one that brings to mind toys in the waiting room and the occasional crying child. While the communicative values are ethical, they could be clearer [19]. Family is culturally defined and a dentist may find ethical dilemmas within. Ethical principles may also be overlooked should this clinic not actually be a family dental practice, whether implicitly or explicitly, which would then flout both communicative and ethical principles.

Sonrisa Family Dental. This dental clinic's name, half in Spanish and half in English, yields important information about this clinic. The words in this orthonym include "sonrisa" and family dental. Family dental warrants no explanation, though to some readers "sonrisa" may be an unfamiliar term. Sonrisa means smile in Spanish. This orthonym code switches between Spanish and English to say what may be translated as "Smile Family Dental." For the person in the community in which this clinic is set, it may be viewed as following the norms of effective discourse. It is short and explains its patient base: families. As a perfect example of discourse being shaped by and shaping community, if this dental office was located in a different neighborhood the definition of the word "sonrisa" may go amiss.

The Sonrisa Family Dental selected for this study is located in McKinley Park. Sonrisa Family Dental appears to have multiple locations across Chicago yielding slight complication and potential inaccuracy to this research. McKinley Park is located in Chicago's Southwest side and is 16.9% White, 62.7% Hispanic, .9% Black, 19% Asian, 0.7% Mixed, and 0.8% Other. This clinic has a large population of potential Spanish-speaking patients surrounding it, where other neighborhoods where this clinic exists, such as Archer Heights on Chicago's Southwest side has a majority Black population [67%] with Hispanic being the second largest at 23.7%.

By using a Spanish word in the orthonym, the potential patient may note that this clinic may be a place they can use the Spanish language. This may also be observed as serving to frame this medical dental clinic as a safe place for community members who do not know English to seek medical help. As discourse may influence worldview, this orthonym may develop ideas or thoughts in the mind of the patient about the type of treatment they are to receive [bilingual] before they actually walk into the clinic. This orthonym falls in line with bioethical principles that seek to serve the benefit of the patient so long as the orthonym accurately describes the practice, and only in the context of a patient who understands Spanish.

Montrose Tooth Fairy. This orthonym is describing the ideal patient in a more abstract way. The words in this orthonym do not explicitly state that it is a dental office at all, nor do they attempt to describe any sort of medical credential. In fact, a child may even interpret this establishment as a place to bring a fallen tooth for a monetary reimbursement. Arguably this discourse, though it requires some extrapolation, may be somewhat effective, once the patient realizes that this dental clinic probably offers pediatric services.

Montrose Tooth Fairy is located in the Uptown neighborhood on Chicago's Northside. The population of Uptown is 44.9% White, 12.9% Hispanic, 28% Black, 11.5% Asian, 2.1% Mixed, and 0.6% Other. The median household income in this community is $37,600 a year. As this orthonym is more metaphorical than literal, the patient's mind may be perceived as subconsciously left to explore the intended meaning.

Along with describing the ideal patient, this orthonym could be understood as relaxing the potential patient. For example, a child going to this dental office may have warm feelings about the tooth fairy and may not experience as high of levels of nervousness when visiting for cleanings or other procedures. As with the previously mentioned examples of healthcare communication, this communication hesitantly falls in line with bioethical principles given it is an accurate representation. If, for example, this clinic turned out to be a place where children were not welcomed and pediatrics was not a specialty, this could be viewed as maleficent. A more explicit orthonym that requires less analysis could be more bioethically appropriate.

7.3 Orthonyms that focus on the patient experience

Orthonyms that are focused on the patient experience are found as taking creative liberties within healthcare communication. For example, a dental clinic using the word "salon" in their orthonym would strike a different image in the perspective patient's imagination than a clinic simply named "dental office." This form of language verges on art and may be beyond the scope of healthcare communication. Ethical concerns may arise for those clinics that use superfluous adjectives that may lead patients to expect something beyond the practitioner's scope.

Art of Modern Dentistry. The name "Art of Modern Dentistry" sets this clinic apart from the previous. By alluding itself to art, the words used in this proper

American Dental. This orthonym may be appealing to a sense of identity as well as patriotism. This dental clinic is located in located in Avondale on Chicago's Westside. American Dental appears to have multiple locations with the Avondale clinic selected through the random selection process. Avondale is 35.7% White, 54.5% Hispanic, 2.1% Black, 5.5% Asian, 2.0% Mixed, and 0.2% Other, with a median household income of $54,4000. Through inserting identity markers such as "American" in its orthonym, this dental clinic could be positing itself as a place for patients who identify with American values or have a strong sense of American identity. By including unnecessary values in its name, this clinic may be ignoring the maxim of relevance; the nationality may not be an essential lexeme in discourse surrounding dental care. Where cultural competence may be necessary during the patient interaction, nationality outside of the interaction may be entirely auxiliary. In fact, this form of identity marker may serve as a potential deterrent to patients who may feel sequestered from "American" culture.

My Dentist Chicago. The name of this clinic alludes to ownership and uses language to frame itself within the community. Arguably, My Dentist Chicago could be considered as using the "my" in its orthonym to promote a sense of ownership with its customers. This dentist is located in Beverly View on Chicago's Southside and is 14.3% White, 1.1% Hispanic, 82.4% Black, 1.1%, Asian, 1.1% Mixed, and 0% Other. Beverly View has a median household income of $70,300. Where the word Chicago may be viewed as a relevant location marker, the creation of a sense of ownership is a known communicative technique to create a stronger bond with patients. This could be taken as a revenue-generating practice and may therefore not be in the best ineptest of the patient. Communicatively the orthonym does not disregard communicative maxims, in fact the phrase "my dentist" may be an utterance of natural occurrence in language. However, bioethical principles must always question whether something is being used to increase patient-base.

8. Discussion

One of the larger trends observed is the difference of the language used in orthonyms of clinics in upper middle class and wealthy communities versus the language used in lower socioeconomic communities. Lexical values used in proper names in those communities that are in the middle class and wealthy category include words that appeal to a luxurious experience.

From words like modern, art, design, and salon, these clinics are creating specific images in the community's mind that influence perceptions of dentistry. This relationship is reflexive, the community could be interpreted as maintaining these ideologies surrounding their health, which may influence the public health of the community. Something with the power to influence public health, even on a micro-scale, must be under bioethical guidance. In this example, healthcare communication could be interpreted as widening the health equity gap, therefore breaking the bioethical principle of justice. The use of such sensationalized health communication may also show disregard for the principle of nonmaleficence: It may decrease harm to the individual and the community to have this form of healthcare communication more strongly regulated.

In lieu of health communication in wealthy communities, there are many clinics in communities of lesser socioeconomic standing that seem to bypass glamorous lexical entries. The language surrounding dental clinics in these communities may be communicating something different. In these communities clinics may be named after the dentist themselves, or have other lexical entries that appeal to things like patriotism, ownership, and heritage language.

One major trend includes naming a clinic after the dentist who owns or founded the practice along with their medical credential. Through bypassing appeals to lavishness, clinics with names that are more straightforward and only contain the name of the provider. Their qualifications could be viewed as positioning in authoritative positions within their communities. This case study argues that this may be the form of orthonym that is in least defiance of communicative maxims and bioethical principles.

Ethical questions start to arise when noting the difference between communicative practices in wealthy communities versus this communicative practice which does exist in both ends of the stratum, though more concentrated in lower- and working-class communities. This form of communication is more straightforward and to the point. Because of this it could be taken as protecting the patient's autonomy and practicing beneficence. By not using persuasive or otherwise auxiliary language to attract patients, the patient may have more agency to pick a dentist in an unbiased way. It could also be upholding the principle of beneficence in this same way; by not using persuasive language that may be deceitful, the patient is protected.

However, it could also be argued that using linguistic markers like these to send cognitive cues to patients is a form of asserting authority over patients. Orthonyms with lexical entries such as "doctor," "DDS," and "DMD," can be found across socioeconomic boundaries. In this case the patient is no longer getting distracted or persuaded by lexemes, but the practitioner and practice may be using this form of health communication to build and maintain barriers between the patient and the practitioner. This practice as used across a community or region may flout the principle of nonmaleficence. By increasing barriers patients may struggle to effectively communicate with their practitioners.

Another observation within this case study includes names that describe health outcomes and the ideal patient. Both forms of orthonym were dispersed commonly throughout the sample, not only existing in an area of certain socioeconomic status. "Big Smile Dental" and "Perfect Smile Dental Spa" both appealed to the desired health outcomes of patients; through the use of perfect smile and big smile these clinics are advocating their ability to give patients what they want, a perfect or big smile. In another example, "Montrose Tooth Fairy," "Sonrisa Family Dental," and "1st Family Dental of Andersonville," are quite unconcealed as to whom they are attracting as patients: children and families.

By using such markers in their proper nouns, they are effectively filtering out any patients that are looking for a spa experience, and welcoming in families and children. Alluding to health outcomes and the ideal patient for the practice do fall within the guidelines for effective communication—these modalities help patients pick a clinic to seek treatment at. However, in the same light, these practices may also be going against bioethical principles. For example, alluding to cosmetic outcomes as health outcomes may press patients to pursue procedures that are not necessary with potentially harmful side effects such as chemical burns that result from whitening procedures. This form of communication may promote the violation of the principle of nonmaleficence.

Through this case study, this research provides a foundation for the future study of clinic names as to how they might influence the patient experience, and the combination of communicative maxims with biomedical ethical principles. For example, by seeing "Perfect Smile Dental Spa" on the front door of a clinic, it could be argued that the patient has already been predisposed to feel a certain way about their experience at that clinic. The ways a patient feels about their treatment are argued to be an extension of the treatment itself. The authority of language in this application is difficult to measure due to its penchants on cognition, but is interpreted as influencing the cognition of patients nonetheless. One may ask, *If the pre-dispositions on the patient's experience are good, why does it matter?* Not only is it relevant as to whether or not these communicative values are ethical or unethical, it is worth thinking of this form of communication on a spectrum. One communicative form may be more ethical than another on the spectrum, and by allowing the spectrum to exist potential divergences may occur.

If a patient is predisposed to perceive care in a certain way many violations of bioethical principles may occur. If a patient is predisposed to perceive their healthcare as a positive experience, when they were actually receiving below-standard care, it poses an ethical problem. On the other hand, if a patient is predisposed to be unsatisfied with what is actually a highly-effective treatment, their adherence to that treatment and relationship with the clinician may suffer. Along with this influence language has on healthcare, language may also assign value, or perhaps assign who is worthy to receive certain experiences as a patient. For example, the general volume of words that create a sense of luxury or relaxation in dental clinic orthonyms, including spa and salon, are indicative of the care the patient may receive at that location. As previously mentioned, the language used within a community may influence perception and cognition in a cyclical manner.

This research posits that there could be a *juste milieu* for naming in a bioethically sensible manner. Clinic names that include a combination of *the practitioner's name, medical credential,* and *specialty* could be considered as *ad hoc* guidance for clinic naming procedures. This case study argues that orthonyms such as *Dr. de Rothewelle, Periodontist, Dr. de Rothewelle, DMD, Orthodontic Associates,* and similar most clearly observe communicative and biomedical ethical standards.

Guidelines for Selecting a Clinic Name

In order to communicate efficiently and effectively and maintain the principles of bioethics namely to protect the patient, research recommends including only one or more of the following units.

I. **The Practitioner's Name.** Limited to combination of given and surname, or surname exclusively. May include the honorific "Dr."	II. **Medical Credential.** Academic degrees required to practice such as D.D.S., D.M.D., or foreign equivalents.	III. **Specialization.** Specialties as approved by the governing body such as periodontal, orthodontic, and maxillofacial.

By including only the above communicative values, the patient will receive information that is appropriate in manner, relevance, quantity, and quality. These communicative values promote on a microscale the bioethical principles of beneficence, nonmaleficence, justice, and autonomy.

**A note on auxiliary words: lexemes of this type, such as office, practice, associates* et cetera *may be used if doing so enhances communication.*

***A note on the use of honorifics: as in many contexts (i.e. education, research, theology, et cetera) "Dr." may be used as an honorific not denoting one's profession as a doctor. For this reason, the use of "Dr." should be followed by granted medical credential.*

Due to the role of communication's influence on the patient experience, health outcomes, and treatment satisfaction, further analyses should be extended through all communicative modalities with the potential to influence cognition and patient care. This includes units of micro-communicative value such as the proper noun. This research further suggests that the communicative maxims and bioethical principles be further analyzed in symbiosis to enhance healthcare communication.

9. Conclusion

Due to language's ability to influence cognition, language's place within healthcare is of vital importance. Whether it be the communication of treatment instructions, or maintenance of doctor-patient rapport, not only is effective communication essential, but communication that aligns with communicative maxims and bioethical principles within healthcare settings is necessary. After a micro-analysis of one facet of healthcare communication, this case study found that the orthonym may be an influencing factor in patient care. Ascribing the potential correlation between this type of communication and health outcomes, this research suggests that a bioethical approach needs to be adopted and invites sustained study of bioethics as is applicable to healthcare communication through the symbiosis of communicative maxims and bioethical principles.

Acknowledgements

Thanks to my colleagues at The Reading Center, and within The Department of Communication. Gratitude to those who provided feedback on this research at the 2020 International Conference on Medical Humanities at University of Oxford. Special recognition to Destiny Mitchell and Mary Hall for language editing, and Shelley Sandow for a little inspiration.

Author details

Jonathan Comyn de Rothewelle
The Department of Communication, The Reading Center, Harry S Truman College, Chicago, Illinois, United States

*Address all correspondence to: jcomynderothewelle@gmail.com

References

[1] Partida Y. Language and health care. Diabetes Spectrum. 2012;**25**(1):19-22. DOI: 10.2337/diaspect.25.1.1

[2] Schyve PM. Language differences as a barrier to quality and safety in health care: The joint commission perspective. Journal of General Internal Medicine. 2007;**22**(Suppl 2):360-361

[3] Karliner LS, Kim SE, Meltzer DO, Auerbach AD. Language barriers and hospital care. Journal of Hospital Medicine. 2010;**5**:276-282. DOI: 10.1002/jhm.658

[4] Epstein RM, Franks P, Shields CG, et al. Patient-centered communication and diagnostic testing. Annals of Family Medicine. 2005;**3**(5):415-421. DOI: 10.1370/afm.348

[5] de Rothewelle JC. Comics and medical narrative: A visual semiotic dissection of graphic medicine. Journal of Graphic Novels and Comics. 2018:1-27. DOI: 10.1080/21504857.2018.1530271

[6] Henry SG, Matthias MS. Patient-clinician communication about pain: A conceptual model and narrative review. Pain Medicine. 2018;**19**(11):2154-2165. DOI: 10.1093/pm/pny003

[7] Boroditsky L. Linguistic Relativity, Encyclopedia of Cognitive Science. London: Boroditsky; 2003

[8] Chomsky N. Language and Mind. Cambridge, United Kingdom: Cambridge University Press; 2006

[9] Pinker S. The Language Instinct: How the Mind Creates Language. Perennial; 1994

[10] Sapir E. The status of linguistics as a science. Language. 1929;**5**(4):207-214

[11] De Gruyter W. History of Linguistic Thought and Contemporary Linguistics; 1975

[12] Trabant J. How relativistic are Humboldt's "Weltansichten"? In: Martin P, Verspoor MH, editors. Explorations in Linguistic Relativity. 2000

[13] Donaldson TM. In: Clark PA, editor. Ethical Resources for the Clinician: Principles, Values and Other Theories, Contemporary Issues in Bioethics. IntechOpen; 2012. DOI: 10.5772/34750

[14] Morales-González Á, Tirado-Lule JM, González-Cisneros A, López-De-León EO, Sanchez-Morales A, Manzanilla-Granados HM. In: Morales-González JA, Nájera MEA, editors. Bioethics in Education, Reflections on Bioethics. London, United Kingdom: IntechOpen; 2018. DOI: 10.5772/intechopen.74519. Available from: https://www.intechopen.com/books/reflections-on-bioethics/bioethics-in-education

[15] Beauchamp T, Childress J. Principles of Biomedical Ethics. 6th ed. Oxford University Press; 2009

[16] Alba-Juez L. Perspectives on Discourse Analysis: Theory and Practice. Newcastle upon Tyne, United Kingdom: Cambridge Scholars Publishing; 2009

[17] Popa F. The value of ethics in health communication. Journal of Medicine and Life. 2008;**1**(4):363-364

[18] Strasser T, Gallagher J. The ethics of health communication. World Health Forum. 1994;**15**(2):175-177

[19] Grice HP. Logic and conversation. In: Cole P, Morgan JL, editors. Syntax and

Semantics, Speech Acts. Vol. 3. New York: Academic Press; 1975. pp. 41-58

[20] Parli U. Proper name as an object of semiotic research. Sign Systems Studies. 2011;**39**(2)

[21] Boas F. Race, Language and Culture. New York: Free Press; 1966

[22] Webber ET. Proper names and persons: Peirce's semiotic consideration of proper names. Transactions of the Charles S Pierce Society. 2008;**44**(2)

[23] Pierce CS. In: Hartshorne C, Weiss P, editors. 1931-1936. The Collected Papers. Vol. 1-6. Cambridge, MA: Harvard University Press; 2007. DOI: 10.1007/s11606-007-0365-3

[24] Saussure F. Course in General Linguistics. New York: Philosophical Library; 1959

[25] Thibault PJ. Editorial: Social Semiotics. In: The Semiotic Review of Books. Vol. 4. Peterborough, Ontario, Canada: Trent University; 1993. p. 1

[26] Davies B, Harre R. Positioning: The discursive production of selves. Journal for the Theory of Social Behaviour. 1990;**20**(1):43-63. DOI: 10.1111/j.1468-5914.1990.tb00174.x

[27] Schiffrin D. Approaches to discourse. In: Blackwell Textbooks in Linguistics. Oxford: Blackwell; 1994

[28] Tirado F, Galvez A. Positioning theory and discourse analysis: Some tools for social interaction analysis. Forum: Qualitative Social Research. Sozialforschung. 2007;**8**(2):1-28

[29] de Rothewelle JC. Visual pathologies: The semiotics of the patient and the practitioner in comics. In: Primary Care. IntechOpen; 2019

[30] Borah P. Media effects theory. In: The International Encyclopedia of Political Communication. Hoboken, New Jersey: John Wiley & Sons, Inc.; 2016. pp. 1-12. DOI: 10.1002/9781118541555.wbiepc156

[31] Toth A. Remodeling your practice. American Dental Association New Student Blog. 2015. Available from: https://newdentistblog.ada.org/remodeling-your-practice/

[32] Tung E, Lui Q. White Dental Spa Homepage. Available from: https://www.whitedentalsmile.com [Accessed: 16 July 2020]

[33] Haahr M, Haahr S. Random Integer Generator. Randomness and Integrity Services Ltd. Available from: random.org [Accessed: 16 July 2020]

[34] Statistical Atlas. Cedar Lake Ventures, Inc. Demographic data from U.S. Census Bureau. Available from: statisticalatlas.com [Accessed: 16 July 2020]

[35] Austin JL. In: Urmson JO, Sbisá M, editors. How To Do Things With Words. 2nd ed. Cambridge, MA: Harvard University Press; 1962

[36] Pape H. Pierce and Russel on proper names. Transactions of the Charles S. Peirce Society. 1982;**18**(Fall 4):339-348

[37] Professions, occupations, and business operations (225 ILCS 25/). Illinois Dental Practice Act. [Accessed: 04 June 2020]. Available from: http://www.ilga.gov/legislation/ilcs/ilcs3.asp?ActID=1296&ChapterID=24

[38] Principles of Ethics & Code of Professional Conduct. The American Dental Association Council on Ethics, Bylaws and Judicial Affairs. 2018. [Accessed: 16 July 2020] at ada.org

Chapter 11

Parkinsonism and Potential of Mucuna Beans

Suresh S. Suryawanshi, Prajakta P. Kamble, Vishwas A. Bapat and Jyoti P. Jadhav

Abstract

Parkinson's is a neurodegenerative disease, which is common all over the world. Various aspect like damages of reactive oxygen species, excitotoxicity, mitochondrial dysfunction, and inflammation-facilitated cell damages are included in the etiology of disease. Good-balanced nutrition is an important part involved in the body health maintenance and reduction in the risk of chronic diseases. Genus *Mucuna* falls under family Fabaceae, containing high contents of L-DOPA (commonly used as an anti-Parkinson drug). Plant-based medicines are the superfluous source of polyphenols, flavonoids, carotenoids, antioxidants (ROS and RNS), terpenoids, isoflavonoids, and other biologically active phytochemicals. All these molecules have health beneficial effects with superlative pharmaceutical values. The existing chapter summaries to determine the influence of different nutritional, anti-nutritional, and medicinal potential of the *Mucuna* species present in India and its significance in the management of Parkinson's disease (Shaking Palsy) as well as other medicinal values. It also covers various treatment models used in studying the Parkinson's disease like *Drosophila melanogaster*, zebrafish, mice, rat, and humans. This chapter also focuses light on the neurosurgical treatments used in the treatment of Parkinson's disease. This study concluded that the use of *Mucuna* seeds for the treatment of Parkinson's disease is the best choice besides chemical drugs and other therapies.

Keywords: Kampavata, L-DOPA, *Mucuna*, neuroprotective, experimental models, neurosurgical remedy

1. Introduction

Research in plant-based medicine is gaining much more attention due to its lack of side effects, its large availability, and its multiple medicinal applications. In the past few years, this has also increased people's awareness toward functional food, thus enhancing the consumption rate of legumes enormously [1]. The attention toward *Mucuna* improved colossally after 1937 when it was initially revealed that a huge quantity of levodopa is present in it. Genus *Mucuna,* which is familiar by different names (like: velvet beans, sea beans, cowitch, buffalo beans, cowhage, atmagupta, kapikachu, and dopa bean), is one of the conventional medicine performing crucial roles in both health and disease management. *Mucuna* is an excellent source of proteins, starch, micronutrients, dietary fiber, and bioactive compounds (L-DOPA), which play a great role in the traditional as well as modern medicine all

IntechOpen

over the world against Parkinson's disease. The genus *Mucuna* belongs to the family *Leguminosae* distributed throughout the tropical and subtropical regions of various countries of the world comprising hundreds of species [2]. Recently, Pulikkalpura et al. [3] and Patil et al. [4, 5] studied the occurrence and biochemical activities of various *Mucuna* species in India.

Plant-based remedies are most beneficial remedies, having a cumulative effect of phytochemical and other bioactive components obtained from plants. *Mucuna* has various actions like antioxidant activity, anti-inflammatory activity, wound-healing activity, and activity against snake bite. Along with all this, *Mucuna* is also rich in nutritional and anti-nutritional compounds (rich in minerals) [6]. *Mucuna* is also well known for nematicidic effects and also possesses notable allelopathic activity, which was reported by Gliessman et al. [7]. In countries like Guatemala and Mexico, it is used as a coffee substitute. Along with that, in some countries (lower hills of the eastern country and Himalayas), it is commonly consumed as vegetable beans after frying and boiling. *Mucuna* seed powder has been used in active management of Parkinson's disease in several countries due to its L-DOPA content.

In early health care system of India, Parkinson's disease (PD) is known as "Kampavata", which is a common neurological disorder related to neuromelanin containing nigrostriatal dopaminergic neuronal loss. Kampa means tremor and vata means lack of muscular movement. It was found via door-to-door survey in Bangalore (Karnataka, India) that 76 per 100, 000 (age adjusted) and 33 per 100,000 (crude prevalence) people suffered from Parkinsonism [8]. Out of that, 5–10% of individuals having PD belong to families with history of genetic disorders [9, 10]. The disease commonly occurs in the age group of 60–80. A survey on Parsi community in Mumbai shows that they were diminutively stable to PD [8]. Diagnosis of disease is sometimes difficult by clinical method, which involves analysis by considering large number of motor and non-motor symptoms in PD patients. PD is caused due to decrease in dopaminergic neurons in the substantia nigra pars compacta (SNpc) part of the brain, which leads to motor symptoms including rigidity, bradykinesia, tremor, and development of some non-motor symptoms in later stages of the disease. Although there is significant number of improvements in the medicinal and surgical actions against PD, scientists have not yet identified definite targets for disease management. According to research by Ramanan and Saykin, neurodegeneration is divided in to four main levels out of which aggregation and accumulation of abnormal or misfolded mutated proteins is crucial [11]. Some reports examine that oxidative stress (OS) and neuroinflammation (NIF) are the main reasons of neurodegenerative disease. Thus, to tackle these neurodegenerative disorders, researchers have diverted their attention toward finding oxidative stress enzymes and pathogenesis of PD. Abraham et al. studied 115 cases of PD and concluded that catalase (CAT), superoxide dismutase (SOD), glutathione peroxidase (G-Px), and glucose-6-phosphate dehydrogenase (G6PD) levels are present in greater quantities in PD patients as compared to the control patients [11]. Various oxidative stress enzymes, pro-inflammatory cytokines, proteinase, reactive oxygen intermediates, and complement proteins are secreted by the immune cells against the neuronal cell damage response, which leads to inflammation. Along with mitochondrial dysfunction, altered proteolysis, oxidative damage of cells and Lewy bodies' formation are some of the symptoms of Parkinson's disease.

There are various strategies for the management for PD but there is no complete cure for this disease. The only management of the lowered dopamine levels is to control various metabolic inhibitions and enhancement pathways, preventing degeneration of neurons and other non-dopaminergic (surgelogical) treatment. L-DOPA (L-3,4 dihydroxy phenylalanine) is a non-protein amino acid used in the treatment of Parkinson's disease. It acts as precursor of dopamine, norepinephrine (noradrenaline), and epinephrine (adrenaline). It is synthesized by plants, animals, and microbes (bacterial and fungus) and has a different role in each source. Chemical synthesis of L-DOPA is easy but the purification of L and D forms is slightly complicated and expensive. Synthetic L-DOPA is considered as anti-nutritional factor and its consumption causes abdominal distention, vomiting, dyskinesia, nausea, etc. This happens when L-DOPA is converted to dopamine in the peripheral nervous system by enzyme dopa decarboxylase. Whereas, L-dopa is also used as the treatment against infertility recovering the spermatogenic harm to some extent. Numerous other secondary metabolites produced by plants like flavonoids and phenolics illustrate strong anti-inflammatory and antioxidant potentials, which have a significant role in the management of health and disease of human beings.

The current book chapter is focused on the use of plant source of L-DOPA and neuroprotective potential of magical dopa beans and their special prominence in Parkinson's disease treatment. Along with this we have also focused on the models and other treatments used in treating Parkinson's disease.

2. Treatment strategies of Parkinson's disease

2.1 Neuroprotective potential of *Mucuna*

Parkinson's disease (PD) was initially discovered by Dr. James Parkinson in 1817; it is a chronic neurological disorder triggered by a progressive loss of dopaminergic neurons present in the nigrostriatal part of the brain and found to be common in U.S [12]. In 1970, only few effective drugs were available for treatment of the PD but there is no such therapy yet that completely treats PD. Only thing we can do is to stop the progression of Parkinson's disease or delay the of PD by replacement of dopaminergic neuron or by mimicking the neuron by using substituent. Management of PD is mainly divided into two categories: first involves improving symptomatic treatment of motor and non-motor types of symptoms and second will be addressing potential causes of PD. Firstly in 1978 Vaidya et al., published the report that PD can be treated by *Mucuna* extract, a natural source of levodopa having better activity than the synthetic version of levodopa drug [13]. Similar studies were reported in 1990 and 1994 by Kempster et al. and Rabey et al. [14, 15]. L-DOPA is a precursor of dopamine (**Figure 1**), norepinephrine (noradrenaline),

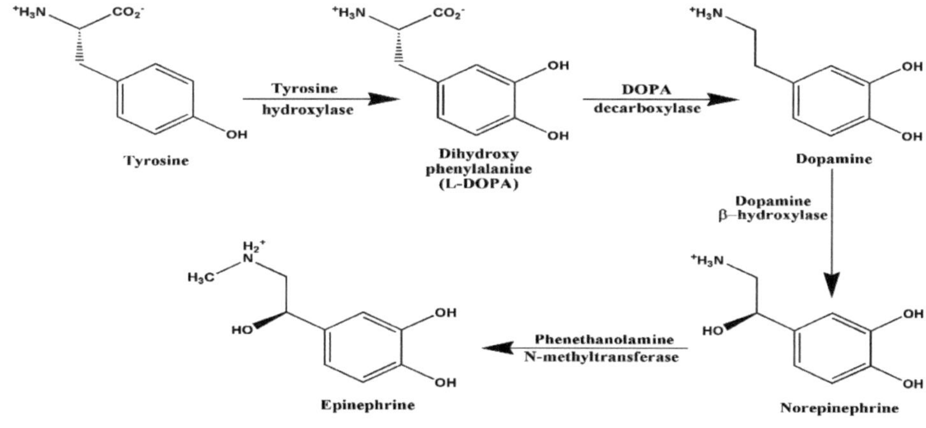

Figure 1.
Synthesis pathway of L-DOPA, dopamine and further metabolites.

and epinephrine (adrenaline), together known as catecholamines. The dopamine produced cannot cross the blood–brain barrier but L-DOPA can. Outside the brain, L-DOPA can directly be converted to 3-O-methyldopa (3-OMD) by catechol-O-methyl transferase (COM T) and then further to vanillactic acid (VLA), which leads to primary same side effect. To avoid this conversion, standard clinical practices use DOPA decarboxylase inhibitor such as carbidopa or benserazide and often a catechol-O-methyl transferase (COMT) inhibitor [16–18]. L-DOPA present in *Mucuna* plant (anti-Parkinson's drug) [19–21] helps to produce dopamine. Along with L-DOPA, the reactive oxygen species (ROS) and reactive nitrogen species (RNS) produced by *Mucuna* are stress-producing free radicals playing a great role in the physiological functioning of the body [21–30]. The content of antioxidant compounds using different solvents in different species of *Mucuna,* the concentration of antioxidants and other phytochemicals are extremely different. Ethanolic extract of *Mucuna* seed shows good antioxidant activity due to high phenolic content as compared to methanol, water, and acetone [31]. Some reports also conclude that water is as universal solvent, which shows the significant quantity of phenolic, flavonoids, and strong antioxidants which have the ability to scavenge free radicals using different assays. LCMS (liquid chromatography mass spectrophotometry) report of four different species **Table 1** shows that there are various components like phenolic flavonoids and bioactive compounds present in the *Mucuna* that are responsible for the production of reactive species [32]. Along with L-DOPA and antioxidants, other secondary metabolites like phenolics, flavonoids, vitamins, enzymes, and protein also have a cumulative effect in the management of PD. Few reports on *Mucuna* show correlation between L-DOPA, protein, and carbohydrates. The use of plants for the treatment of PD is more beneficial than chemically manufactured L-dopa due to its high potential required in the levo and dextro form purification. It is also studied that various compounds present in *Mucuna* are responsible for the antimicrobial action, which can be utilized in dealing with various infectious diseases and ulcers [31, 32, 44]. Experiments on various plant pathogens suggest that methanolic extract of *Mucuna pruriens* seeds showed the highest antimicrobial activity [45]. A similar type of study done by Pujari et al. also determined that methanol extracts of *Mucuna pruriens* seeds were found to have the best inhibiting activity among all scrutinized pathogens as compared to ethanol and acetone solvents. But alcoholic extract of *Mucuna pruriens* (L.) leaves has significant antioxidant and antibacterial

Sr.no	*Mucuna* species	Reference
1	*Mucuna imbricata*	[32]
2	*Mucuna macrocarpa*	[33]
3	*Mucuna monosperma*	[34]
4	*Mucuna Bactetia*	[35]
5	*Mucuna sanjepee*	[36]
6	*Mucuna Autripuria*	[37]
7	*Mucuna Latiparica*	[5]
8	*Mucuna prurience*	[3, 6, 20, 21, 24, 38–43]
9	*M. nigricans*	[5]
10	*M. gigentea*	[5]
11	*Dhanwantari*	[5]

Table 1.
Different species of Mucuna *studied till date.*

activity [45]. Dopaminergic agonists or dopamine replacement therapy is a common and most effective way to cure PD. It decreases the signs of disease by sustaining the level of dopamine; however, it cannot regenerate or halt the degeneration. It only replaces or mimics dopamine by inhibiting its breakdown. Apomorphine, bromocriptine, pergolide, piribedil, pramipexole, and ropinirole are some dopaminergic agonists mainly used to heal the PD. All these bioactive compounds present in the *Mucuna* species have cumulative effect in the treatment of PD. *Mucuna pruriens* is a species from the Fabaceae family and Faboideae subfamily. *M. pruriens* is an annual twinning plant in bushes, hedges, and one of the popular medicinal plants indigenous to tropical countries like India [42]. It is useful in relieving inflammation, delirium, neuropathy, cephalalgia, and general debility, nephropathy, dysmenorrhea, amenorrhea, ulcers, constipation, elephantiasis, consumption, helminthiasis, fever, and dropsy. The trichomes of pods contain serotonin and mucunain. The trichomes are used as anthelmintic. Seeds contain glutathione, gallic acid, levodopa (4-3, 4-dihydroxy phenylalanine), lecithin, prurenine, prurenidine, glycosides, nicotine, minerals, and dark brown viscous oil [42].

2.2 Experimental models studied for PD

Natural products are valuable sources of bioactive compounds that can be exploited for novel therapeutic potential in PD pathogenesis. There are number of publications reported till now dealing with experiments on hundreds of compounds from various plant species for their different activity in curing the Parkinson's disease [46]. However, rapid screening of plant-derived natural products and characterization of bioactive compounds is a challenging job. This problem was combated by using *Drosophila melanogaster* and zebrafish as experimental models at initial stages then followed by studies on various experimental models like, *C. elegans,* mice/rat, and also cell lines (e.g., murine BV-2 microglia and human SH-SY5Y neuroblastoma cells). Few verdicts using different models are listed underneath.

2.2.1 Drosophila melanogaster

Drosophila melanogaster, universally familiar as the fruit fly, have turned up as an outstanding model for human neurodegenerative diseases, comprising PD. Due to their high degree of conserved molecular pathways with mammalian models, Drosophila PD models serve to be an inexpensive solution to pilot stages of target validation in the drug discovery pipeline. Fruit fly acts as a screening platform to evaluate the therapeutic potential of phytochemicals from natural extracts against PD [47]. *Drosophila melanogaster* is a persuasive tool to explore molecular facets and physiopathology of Parkinson's disease (PD) [48]. There are studies that compare the effects of L-DOPA vs. MP extract using a Drosophila model of autosomal recessive PD in which flies carried a mutation in the PTEN-induced putative kinase 1 (PINK-1) gene [49]. Their observations illustrates that *Drosophila* fed on MP had a significantly extended lifespan, showed a restored olfactory response and improved climbing behavior compared to flies that consumed L-DOPA.

*2.2.2 Zebra fish (*Danio rerio*)*

Owing to its large number of favorable properties, Zebra fish has been used widely as experimental animal for various diseases. Zebra fish are inexpensive, easy to conserve, develop rapidly, and breed in large quantities. Larval zebra fish are also extensively used in toxicity screens since they have a permeable skin through which substances added in the rearing medium are effortlessly taken up. This permits for greater control over dosage and ease of administering substances to large numbers of animals. Furthermore, larval behaviors can be exploited in assays to test the effects of the treatment. Being a vertebrate, the central nervous system of larval zebra fish expressions an extremely homologous to humans. Therefore, toxicology studies performed on larval zebra fish can be very helpful in deciding the putative targets in humans [50]. Thus *Danio rerio* commonly known as zebrafish is a charming popular animal model for treatments in neuropharmacology and pharmacogenetics. Both the adult and larval zebrafish are presently studied to increase the understanding of central nervous system's function and dysfunction [51]. There are various studied reports by Gerlai et al. on the latent learning, behavior, and memory alteration of adult zebrafish [51–55]. Apart from fish, there are several other experimental models.

2.2.3 Mice/rat/rodents

Many clinical trials have been done on herbal extract and their isolated compound, which has opened a new scenario in the area of PD. The models of PD are induced by different chemical compounds like 6-OHDA, rotenone, and MPTP based on the aim of the experiment. Singh et al. investigated the effect of ethanolic extract of *Mucuna pruriens* (Mp) on the reduction of oxidative stress level, nitric oxide (NO), and subsequent influence on lipid peroxidation in paraquat (PQ)-induced Parkinsonian mouse model. MTTP-induced Parkinson mouse models were also studied by them for the reduction of estrogen by *Mucuna pruriens* [26, 55, 56]. Yadav et al. demonstrated that Mp seed extract reduces oxidative stress in nigrostriatal tissue and improves neurobehavioral activity and the expression of tyrosine hydroxylase (TH) in SN and striatum of the brain in PQ-intoxicated mice [56]. Apoptotic pathways of dopaminergic neurons in the PQ mouse model were also found to be inhibited by the neuroprotective activity of *Mucuna pruriens* [57]. Experiments also reveal that the *Mucuna pruriens* has a resilient anti-inflammatory property that diminishes the neuroinflammation by plummeting inducible nitric oxide synthase expression in Parkinsonian mice model. Symptomatic and neuroprotective efficacy in rodent model of Parkinson's disease using *Mucuna pruriens* seed extract were studied by Kasture and Pontis [58]. Significance of *M. pruriens* in sperm parameters and sexual behavior of streptozotocin-induced diabetic male rat were studied by Sekar Suresh et al. [59]. Earlier efforts exhibited the capability of *Mucuna pruriens* seeds extract to induce contralateral turning behavior in the 6-hydroxydopamine (6-OHDA)-lesion persuaded rat model of PD [60]. Likewise, the potential of *Mucuna* seed powder extract significantly improved activity of brain mitochondrial complex-I without disturbing total monoamine oxidase activity *in vitro* [61]. *Mucuna pruriens* also has the potential to amend immune components like tumor necrosis factor-α (TNF-α), interleukin-6 (IL-6), interferon-λ (IFN-λ), interleukin-1β (IL-1β), inducible nitric oxide synthase (iNOS), and interleukin-2 (IL-2) in the central nervous system, thereby, averting the progression of dopaminergic neurons degeneration, in PD [62]. Moreover *M. pruriens* was also assessed for levodopa pharmacokinetics by Sarrafchi et al., in a double-blinded clinical and pharmacological study. They observed that administration of *M. pruriens* at doses of 2.5, 5.0, or 10.0 g/kg/day for 52 weeks significantly increased the dopamine content of the cortex in animal model of PD. Therefore, they concluded that *M. pruriens* seed powder with natural source of levodopa probably has benefits over conventional levodopa preparations in treatment of PD due to its longer action and speedy onset without increase in adverse effect [63]. Thus, *M. pruriens* would seem to be a remarkable commercially viable alternative to standard L-dopa [64]. Likewise, an improvement in motor skills and dyskinesia analogous to those induced by equivalent doses of L-DOPA was also found to be induced by *M. pruriens* preparations in rat and macaque monkey models of PD [61]. Therefore, all these studies strongly advocate the use of *Mucuna pruriens*: a treasurable herbal plant for treatment of PD.

2.2.4 Human trials

Copious studies have scrutinized the effect of MP in PD patients. The bioavailability of L-DOPA in the central nervous system is about one-fifth when pooled with carbidopa or benserazide due to nonexistence of DOPA decarboxylase inhibitor (DDCI) in *Mucuna* [61–65]. Even then, the first report in 1978 revealed that 23 PD patients were treated with MP with similar effect and better tolerance than L-DOPA/benserazide. This study was followed by an open study where 60 PD patients in Hoehn and Yahr stage I–IV were treated with MP preparations for over 12 weeks, which led to momentous headways in both UPDRS score and the Hoehn and Yahr stage. This captivated the attention on the registration of *Mucuna* preparation (Zandopa™) as a treatment for PD in India [61]. Cilia et al. executed a crossover study (N = 14) on PD patients in an advanced stage with motor fluctuations and peak-of-dose dyskinesia. The observations indicated better motor improvement on *Mucuna* powder consumption in comparison to the effect imparted by intake of an equivalent dose of L-DOPA/benserazide [66]. Innumerable experiments are still under investigation by several assemblages of researchers from all over the world to understand the PD progression and come up with innovative strategies to treat Parkinson's disease [61, 62].

2.3 Other drugs and their effects

Gargantuan improvement has been done to treat the PD for over half-century. Diverse individuals from different places have come up with different treatments. Miscellaneous categories of drugs are being used apart from levodopa. Yet to date there are no complete cure strategies for PD. But some symptomatic palliative treatments are being implemented to slow down the disease progression. There are various enzymes used in the management of PD. Enzymes monoamine oxidase-B (MAO-B) and catechol-o-methyl-transferase (COMT) are normally involved in metabolism of dopamine. Therefore, few inhibitors of the two enzymes (MAO-B and COMT) have been extensively studied. Inhibitors like selegiline, rasagiline, tolcapone, and entacapone are being used for the disease modification in PD from previous few years. N-propargyl-methamphetamine is known as selegiline, which is an irreversible inhibitor of MAO-B. Action of selegiline was studied in 1-methyl-4-phenyl-1,2,3,6-tetrahydropyridine (MPTP)-induced Parkinsonism in monkeys [67]. It was used at concentration of 10 mg/day, but they found that it loses its selectivity at higher concentrations of dosage [67]. Various other reports show the neuroprotective potential of selegiline but there is no such report proving that selegiline has "disease-modification" effects [68–70]. The *in vitro* and *in vivo* experimental Parkinsonian models indicate that rasagiline (N-propargyl-1-(R)-aminoindan) acted as irreversible MAOB inhibitor exhibiting anti-apoptotic effect [68–70].

There are various studies proving that the dopamine agonist or levodopa has stronger symptomatic benefits as the MAOB inhibitor; however, there was no evidence of direct comparison between them [71–74]. Older people provide more rapid onset improvement than younger patients. Dopamine agonist is more prominent in younger people with dyskinesia and older people with orthostatic hypotension and CNS effects (hallucinations). Pharmacokinetic studies revel that COMT inhibitors prevent degradation of peripheral levodopa by extending half-life and also permit it to cross the blood–brain barrier in higher concentrations. There are large number of compounds used to treat motor symptoms and motor complications occurring due to dopaminergic mode of action as reported in review by Oertel et al. Whereas **carbidopa** (modified form of L-dopa soluble), **opicapone** (COMT-inhibitor), **safinamide** (MAO-B-inhibitor), **droxidopa** (NMDA-receptor antagonist), **istradefylline** (noradrenaline precursor), **tozadenant** (Adenosine 2A receptor antagonist), **pimavanserin** (5HT2A inverse agonist), **and donepezil** (Acetyl choline esterase inhibitor) are some of the most common compounds used in the treatment internationally [16]. Along with benzhexol and orphenadrine, anticholinergic drugs are also recommended as they reduce the effect of acetylcholine in the brain by antagonizing cholinergic receptors and restore the acetylcholine/dopamine balance within the brain. They also prevent hyperkinesia.

3. Neurosurgical treatments

Apart from chemical drugs, there are some physical therapies involved in the treatment of PD that have given special attenuation toward movement (motor) symptoms of patient. In this treatment, parts of the brain involved in the progression of disease are either removed, bombard with electric impulse, or subjected to neuroimaging [75]. In **pallidotomy and thalamotomy,** the globus pallidus part of the brain, which is overactive in the PD patients, resulting in slackening down the body movements, is surgically destroyed permanently. This destruction of globus pallidus significantly reduces tremor, bradykinesia, balance problems, and writing problems, and eliminates rigidity, while thalamus part is involved in the involuntary movement (like tremor) [76, 77]. **Deep brain stimulation** (DBS) is also one of the unconventional treatments used, in which brain pace markers (microelectrode) are applied where an electrical impulse passes over the electrodes to the specific part of the brain. DBS decreases the secondary difficulties elevated due to dopaminergic replacement therapy. There are survival disadvantages of pallidotomy and thalamotomy cases due to dysphagia, hypophonia, and dysarthria [78]. DBS is having significant results over thalamotomy because it does not need hardware and have very fewer side effects and relatively lowers the risk of complications. Besides all these pharmacological and surgical treatments, few other strategies like speech therapy, mediations are being currently explored along with it for the treatment of PD. Apart from singular therapy, doctors are now recommending a combination of treatments to impart cumulative effect in the treatment of PD.

4. Conclusion

Mucuna is a natural, rich source of precursor of dopamine that acts as gold standard for the treatment of Parkinson's disease to control body movements, hormonal balance, emotion, and memory. The *Mucuna* is a pharmaceutically and biochemically valuable plant used from the early days, having high market value due to its large amount of bioactive compounds. It also contains a maximum amount of phenolics, flavonoids, and antioxidants, which have a cumulative role in the release of oxidative stress produced by body systems. *Drosophila melanogaster*, zebrafish, mice/rat, and human models were used to check the potential of *Mucuna* seed powder on rotenone, MPTP-induced Parkinson's model, and it was concluded that *Mucuna* seeds have great values in the treatment of PD. The book chapter also covers various other drugs and neurosurgical treatments used in Parkinson's disease. It also gives an introduction to various other species (apart from *Mucuna FSTA*) used in the PD treatment and lowers the burden on commonly used(Mucuna pruriens) species.

Author details

Suresh S. Suryawanshi[1], Prajakta P. Kamble[2], Vishwas A. Bapat[3] and Jyoti P. Jadhav[1,3]*

1 Department of Biochemistry, Shivaji University, Vidyanagar, Kolhapur, Maharashtra, India

2 Department of Microbiology, Shivaji University, Vidyanagar, Kolhapur, Maharashtra, India

3 Department of Biotechnology, Shivaji University, Vidyanagar, Kolhapur, Maharashtra, India

*Address all correspondence to: biochemjpj@gmail.com

References

[1] Luthria DL, Pastor-Corrales MA. Phenolic acids content of fifteen dry edible bean (*Phaseolus vulgaris* L.) varieties. Journal of Food Composition and Analysis. 2006;**19**:205-211. DOI: 10.1016/j.jfca.2005.09.003

[2] Jeff J, Toby R, Michael J, Doyle JJ, Herendeen P, Hughes C, et al. Legume phylogeny and classification in the 21st century: Progress, prospects and lessons for other species-rich clades. Taxon. 2013;**62**:217-248. DOI: 10.12705/622.8

[3] Pulikkalpura H, Kurup R, Mathew PJ, Baby S. Levodopa in *Mucuna pruriens* and its degradation. Scientific Reports. 2015;**5**:2-10. DOI: 10.1038/srep11078

[4] Gaikwad SV, Gurav RV, Yadav SR. Karyotype studies in *Mucuna macrocarpa* wall and *Mucuna sanjappae* Aitawade et Yadav (Fabaceae) from India. Chromosome Botany. 2017;**12**:52-55. DOI: 10.3199/iscb.12.52

[5] Patil R, Aware C, Gaikwad S, Rajebhosale M, Bapat V, Yadav S, et al. RP-HPLC analysis of anti-Parkinson's drug l-DOPA content in *Mucuna* species from Indian subcontinent. Proceedings of the National Academy of Sciences, India, Section B. 2019;**89**:1413-1420. DOI: 10.1007/s40011-018-01071-9

[6] Suryawanshi SS, Kamble PP, Bapat VA, Jadhav JP. Bioactive components of magical velvet beans. Legume;**2**(2020). DOI: 10.5772/intechopen.92124

[7] Gliessman SR, Garcia RE, Amador MA. The ecological basis for the application of traditional agricultural technology in the management of tropical agro-ecosystems. Agro-Ecosystems. 1981;**7**:173-185. DOI: 10.1016/0304-3746(81)90001-9

[8] Gourie-Devi M, Gururaj G, Satishchandra P, Subbakrishna DK. Prevalence of neurological disorders in Bangalore, India: A community-based study with a comparison between urban and rural areas. Neuroepidemiology. 2004;**23**:261-268. DOI: 10.1159/000080090

[9] Cookson MR, Xiromerisiou G, Singleton A. How genetics research in Parkinson's disease is enhancing understanding of the common idiopathic forms of the disease. Current Opinion in Neurology. 2005;**18**:706-711. DOI: 10.1097/01.wco.0000186841.43505.e6

[10] Warner TT, Schapira AHV. Genetic and environmental factors in the cause of Parkinson's disease. Annals of Neurology. 2003;**53**:16-25. DOI: 10.1002/ana.10487

[11] Ramanan VK, Saykin AJ. Pathways to neurodegeneration: Mechanistic insights from GWAS in Alzheimer's disease, Parkinson's disease, and related disorders. American Journal of Neurodegenerative Disease. 2013;**2**:145-175

[12] Hughes RC. Parkinson's disease and its management. BMJ. 1994;**308**:281. DOI: 10.1136/bmj.308.6923.281

[13] Li H, Yu C, Chen R, Li J, Li J. Novel ionic liquid-type Gemini surfactants: Synthesis, surface property and antimicrobial activity. Colloids and Surfaces A: Physicochemical and Engineering Aspects. 2012;**395**:116-124. DOI: 10.1016/j.colsurfa.2011.12.014

[14] Kempster PA, Bogetic Z, Secombei JW, Martin HD, Balazs ND, Wahlqvist ML. Motor effects of broad beans (*Vicia faba*) in Parkinson's disease: Single dose studies. Asia Pacific Journal of Clinical Nutrition. 1993;**2**:85-89. Available from: http://www.ncbi.nlm.nih.gov/pubmed/24352104

[15] Manyam BV. Paralysis agitans and levodopa in Ayurveda: Ancient Indian medical treatise. Movement Disorders. 1990;**5**:47-48. DOI: 10.1002/mds.870050112

[16] Oertel WH. Recent advances in treating Parkinson's disease. F1000Research. 2017;**6**:1-14. DOI: 10.12688/f1000research.10100.1

[17] Ellis JM, Fell MJ. Current approaches to the treatment of Parkinson's disease. Bioorganic & Medicinal Chemistry Letters. 2017;**27**:4247-4255. DOI: 10.1016/j.bmcl.2017.07.075

[18] Barclay CL, Suchowersky O, Rivest J. COMT inhibitors in Parkinson's disease. The Canadian Journal of Neurological Sciences. 2008;**26**:34-38

[19] Lieu CA, Venkiteswaran K, Gilmour TP, Rao AN, Petticoffer AC, Gilbert EV, et al. The antiparkinsonian and antidyskinetic mechanisms of *Mucuna pruriens* in the MPTP-treated nonhuman primate. Evidence-based Complementary and Alternative Medicine. 2012;**2012**. DOI: 10.1155/2012/840247

[20] Lampariello L, Cortelazzo A, Guerranti R, Sticozzi C, Valacchi G. The magic velvet bean of *Mucuna pruriens*. Journal of Traditional and Complementary Medicine. 2012;**2**:331-339. DOI: 10.1016/S2225-4110(16)30119-5

[21] Yadav MK, Upadhyay P, Purohit S, Pandey BL, Shah H. Phytochemistry and pharmacological activity of *Mucuna pruriens*: A review. International Journal of Green Pharmacy. 2017;**11**:69-73

[22] Wong SP, Leong LP, William Koh JH. Antioxidant activities of aqueous extracts of selected plants. Food Chemistry. 2006;**99**:775-783. DOI: 10.1016/j.foodchem.2005.07.058

[23] Murthy SN, Malgaonkar MM, Shirolkar AR, Pawar SD, Sangvikar S. A comparative assessment of pharmacologically active principles and antioxidant activity of commonly occurring *Mucuna* sp. in India. International Journal of Ayurveda and Pharma Research. 2015;**3**:8-13

[24] Longhi JG, Perez E, de Lima JJ, Cndido LMB. In vitro evaluation of *Mucuna pruriens* (L.) DC. antioxidant activity. Brazilian Journal of Pharmaceutical Sciences. 2011;**47**:535-544. DOI: 10.1590/S1984-82502011000300011

[25] Aitawade MM, Yadav SR. *Mucuna sanjappae*, a new species from the North-Western Ghats, India. Kew Bulletin. 2012;**67**:539-543. DOI: 10.1007/s12225-012-9369-1

[26] Yadav SK, Rai SN, Singh SP. *Mucuna pruriens* reduces inducible nitric oxide synthase expression in Parkinsonian mice model. Journal of Chemical Neuroanatomy. 2017;**80**:1-10. DOI: 10.1016/j.jchemneu.2016.11.009

[27] Oseni OM, Pande V, Nailwal TK. A review on plant tissue culture, a technique for propagation and conservation of endangered plant species. International Journal of Current Microbiology and Applied Sciences. 2018;7:3778-3786. DOI: 10.20546/ijcmas.2018.707.438

[28] Misra L, Wagner H. Extraction of bioactive principles from *Mucuna pruriens* seeds. Indian Journal of Biochemistry & Biophysics. 2007;**44**:56-60

[29] Vadivel V, Biesalski HK. Bioactive compounds in velvet bean seeds: Effect of certain indigenous processing methods. International Journal of Food Properties. 2012;**15**:1069-1085. DOI: 10.1080/10942912.2010.513466

[30] Kavitha C, Thangamani C. Amazing bean *Mucuna pruriens*: A comprehensive review. Journal of Medicinal Plants Research. 2014;**8**:138-143. DOI: 10.5897/JMPR2013.5036

[31] Mastan SA, Ramayya PJ, Naidu LM, Mallikarjuna K. Antimicrobial activity of various extracts of *Mucuna pruriens* leaves. Biomedical and Pharmacology Journal. 2009;**2**:55-60

[32] Suryawanshi SS, Rane MR, Kshirsagar PR, Kamble PP, Jadhav JP. Antioxidant, antimicrobial activity with mineral composition and LCMS based phytochemical evaluation of some *Mucuna* species from India. International Journal of Pharma and Bio Sciences. 2019;**9**:312-324

[33] Aware C, Patil R, Gaikwad S, Yadav S, Bapat V, Jadhav J. Evaluation of L-dopa, proximate composition with in vitro anti-inflammatory and antioxidant activity of *Mucuna macrocarpa* beans: A future drug for Parkinson treatment. Asian Pacific Journal of Tropical Biomedicine. 2017;**7**:1097-1106. DOI: 10.1016/j.apjtb.2017.10.012

[34] Inamdar S, Joshi S, Jadhav J, Bapat V. Innovative use of intact seeds of *Mucuna monosperma* Wight for improved yield of L-DOPA. Natural Products and Bioprospecting. 2012;**2**:16-20. DOI: 10.1007/s13659-011-0051-3

[35] Rane M, Suryawanshi S, Patil R, Aware C, Jadhav R, Gaikwad S, et al. Exploring the proximate composition, antioxidant, anti-Parkinson's and anti-inflammatory potential of two neglected and underutilized *Mucuna* species from India. South African Journal of Botany. 2019;**124**:304-310. DOI: 10.1016/j.sajb.2019.04.030

[36] Patil RR, Rane MR, Bapat VA, Jadhav JP. Phytochemical analysis and antioxidant activity of *Mucuna sanjappae*: A possible implementation in the Parkinson's disease treatment. Journal of Pharmaceutical and Medical Research. 2016;**2**:48-51

[37] Patil RR, Aware C, Gaikwad S, Yadav S. RP-HPLC analysis of anti-Parkinson's drug l-DOPA content in *Mucuna* species from Indian subcontinent. Proceedings of the National Academy of Sciences, India, Section B: Biological Sciences. 2019. DOI: 10.1007/s40011-018-01071-9

[38] Rai SN, Birla H, Singh SS, Zahra W, Patil RR, Jadhav JP, et al. *Mucuna pruriens* protects against MPTP intoxicated neuroinflammation in Parkinson's disease through NF-$κ$B/pAKT signaling pathways. Frontiers in Aging Neuroscience. 2017;**9**:1-14. DOI: 10.3389/fnagi.2017.00421

[39] Contin M, Lopane G, Passini A, Poli F, Iannello C, Guarino M. *Mucuna pruriens* in Parkinson disease. Clinical Neuropharmacology. 2015;**38**:201-203. DOI: 10.1097/WNF.0000000000000098

[40] Fathima KR, Soris PT, Mohan VR. Nutritional and antinutritional assessment of *Mucuna pruriens* (L.) DC var. pruriens an underutilized tribal pulse. 2010;**1**:79-89

[41] Duangnin N, Phitak T, Pothacharoen P, Kongtawelert P. In vitro and in vivo investigation of natural compounds from seed extract of *Mucuna pruriens* lacking L-DOPA for the treatment of erectile dysfunction. Asian Pacific Journal of Tropical Medicine. 2017;**10**:238-252. DOI: 10.1016/j.apjtm.2017.03.001

[42] Suresh S, Prakash S, et al. The Journal of Sexual Medicine. 2011;**8**:1943-1956. DOI: 10.1111/j.1743-6109.2011.02221.x

[43] Sathiyanarayanan L, Arulmozhi S. *Mucuna pruriens* Linn.—A

comprehensive review. Pharmacognosy Reviews. 2007;**1**:157-162

[44] Mohan VR. Antibacterial activity of *Mucuna pruriens* (L.) Dc. var. pruriens—An ethnomedicinal plant. Journal of Scientific Research and Reports. 2011;**1**:69-72

[45] Rayavarapu KA, Kaladhar DSVGK. Evaluation of antimicrobial activity of *Mucuna pruriens* on plant pathogens. Asian Journal of Biochemical and Pharmaceutical Research. 2014;**1**:593-600

[46] Dey A, De JN. Possible Anti-Parkinson's Disease Therapeutics from Nature: A Review. Elsevier B.V.; 2015. DOI: 10.1016/B978-0-444-63460-3.00009-2

[47] Pokrzywa M, Pawełek K, El W, Sarbak S, Chorell E, Almqvist F, et al. Effects of small-molecule amyloid modulators on a *Drosophila* model of Parkinson's disease. 2017:1-21

[48] Maitra U. Using *Drosophila* as a platform for drug discovery from natural products in Parkinson's disease. 2019:867-879. DOI: 10.1039/c9md00099b

[49] Poddighe S, De Rose F, Marotta R, Ruffilli R, Fanti M, Secci PP, et al. *Mucuna pruriens* (Velvet bean) rescues motor, olfactory, mitochondrial and synaptic impairment in PINK1B9 *Drosophila melanogaster* genetic model of Parkinson's disease. PLoS One. 2014;**9**. DOI: 10.1371/journal.pone.0110802

[50] Bakthavatsalam S, Das Sharma S, Sonawane M, Thirumalai V. A zebrafish model of manganism reveals reversible and treatable symptoms that are independent of neurotoxicity. Disease Models & Mechanisms. 2014:1239-1251. DOI: 10.1242/dmm.016683

[51] Kalueff AV, Stewart AM, Gerlai R. Zebrafish as an emerging model for studying complex brain disorders. Trends in Pharmacological Sciences. 2014;**35**:63-75. DOI: 10.1016/j.tips.2013.12.002

[52] Gerlai R. Associative Learning in Zebrafish (*Danio rerio*). 3rd ed. Elsevier Ltd; 2011. DOI: 10.1016/B978-0-12-387036-0.00012-8

[53] Fernandes Y, Gerlai R. Long-term behavioral changes in response to early developmental exposure to ethanol in zebrafish. Alcoholism, Clinical and Experimental Research. 2009;**33**:601-609. DOI: 10.1111/j.1530-0277.2008.00874.x.Long-Term

[54] Fernandesa Y, Trana S, Abrahamb E, Gerlaia R. Embryonic alcohol exposure impairs associative learning performance in adult zebrafish. Behavioural Brain Research. 2015;**15**:181-187. DOI: 10.1016/j.bbr.2014.02.035.Embryonic

[55] Gómez-laplaza LM, Gerlai R. Latent learning in Zebrafish (*Danio rerio*). Behavioural Brain Research. 2011;**208**:509-515. DOI: 10.1016/j.bbr.2009.12.031.Latent

[56] Yadav SK, Prakash J, Chouhan S, Singh SP. *Mucuna pruriens* seed extract reduces oxidative stress in nigrostriatal tissue and improves neurobehavioral activity in paraquat-induced Parkinsonian mouse model. Neurochemistry International. 2013;**62**:1039-1047. DOI: 10.1016/j.neuint.2013.03.015

[57] Rai SN, Birla H, Zahra W, Sen Singh S, Singh SP. The role of *Mucuna pruriens* and *Withania somnifera* in the neuroprotection and treatment of Parkinson's disease. SOJ Neurology. 2018;**5**:1-7

[58] Kasture S, Pontis S, Pinna A, Schintu N, Spina L, Longoni R, et al.

Assessment of symptomatic and neuroprotective efficacy of *Mucuna pruriens* seed extract in rodent model of Parkinson's disease. Neurotoxicity Research. 2009;**15**:111-122. DOI: 10.1007/s12640-009-9011-7

[59] Suresh S, Prakash S. Effect of *Mucuna pruriens* (Linn.) on sexual behavior and sperm parameters in streptozotocin-induced diabetic male rat. The Journal of Sexual Medicine. 2012;**9**:3066-3078. DOI: 10.1111/j.1743-6109.2010.01831.x

[60] Hussain G, Manyam BV. *Mucuna pruriens* proves more effective than l-DOPA in Parkinson's disease animal model. 1997;**423**:419-423

[61] Liebert MA. An alternative medicine treatment for Parkinson's disease: Results of a multicenter clinical trial HP-200 in Parkinson's Disease Study Group. Journal of Alternative and Complementary Medicine. 1995;**1**:249-255

[62] Rijntjes M. Knowing your beans in Parkinson's disease: A critical assessment of current knowledge about different beans and their compounds in the treatment of Parkinson's disease. Parkinson's Disease. 2019;**2019**:1-9

[63] Marsden CD, Parkes JD, Rees JE. A year's comparison of treatment of patients with Parkinson's disease with levodopa combined with carbidopa versus treatment with levodopa alone. Lancet. 1973;**2**:1-4

[64] Katzenshlager R, Evans A, Manson A, Palsalos PN, Ratnaraj N, Watt H, et al. *Mucuna pruriens* in Parkinson's disease: A double blind clinical and pharmacological study. Journal of Neurology, Neurosurgery, and Psychiatry. 2004;**75**:1672-1677. DOI: 10.1136/jnnp.2003.028761

[65] Mars H. Modification of levodopa effect by systemic decarboxylase inhibition. Archives of Neurology. 1973;**28**:91-95. DOI: 10.1001/archneur.1973.00490200039004

[66] Cilia R, Laguna J, Cassani E, Cereda E, Raspini B, Barichella M, et al. Daily intake of *Mucuna pruriens* in advanced Parkinson's disease: A 16-week, noninferiority, randomized, crossover, pilot study. Parkinsonism & Related Disorders. 2018;**49**:60-66. DOI: 10.1016/j.parkreldis.2018.01.014

[67] del S. López-Cortez M, Rosales-Martínez P, Arellano-Cárdenas S, Cornejo-Mazón M. Antioxidants properties and effect of processing methods on bioactive compounds of legumes. Grain Legumes. 2016. DOI: 10.5772/63757

[68] Maruyama W, Takahashi T, Naoi M. (−)-Deprenyl protects human dopaminergic neuroblastoma SH-SY5Y cells from apoptosis induced by peroxynitrite and nitric oxide. Journal of Neurochemistry. 2002;**70**:2510-2515. DOI: 10.1046/j.1471-4159.1998.70062510.x

[69] Munirathinam S, Lakshmana MK, Raju TR. (−) Deprenyl attenuates aluminium induced neurotoxicity in primary cortical cultures. Neurodegeneration. 1996;**5**:161-167. DOI: 10.1006/neur.1996.0023

[70] Kamada T, Morimoto K, Ohde H, Chow T, Hiroi T, Imaoka S, et al. Metabolism of selegiline hydrochloride, a selective monoamine B-type inhibitor, in human liver microsomes. Drug Metabolism and Pharmacokinetics. 2002;**17**:199-206. DOI: 10.2133/dmpk.17.199

[71] Holloway RG. Pramipexole vs levodopa as initial treatment for Parkinson disease: A 4-year randomized controlled trial. Archives of Neurology.

2004;**61**:1044-1053. DOI: 10.1001/archneur.61.7.1044

[72] Teo KC, Ho SL. Monoamine oxidase-B (MAO-B) inhibitors: Implications for disease-modification in Parkinson's disease. Translational Neurodegeneration. 2013;**2**:1. DOI: 10.1186/2047-9158-2-19

[73] Adler CH, Sethi KD, Hauser RA, Davis TL, Hammerstad JP, Bertoni J, et al. Ropinirole for the treatment of early Parkinson's disease. Neurology. 1997;**49**:393-399. DOI: 10.1212/WNL.49.2.393

[74] Parkinson Study Group. A controlled trial of rasagiline in early Parkinson disease. The tempo study. Archives of Neurology. 2012;**59**:1937-1943

[75] Munhoz RP, Picillo M, Fox SH, Bruno V, Panisset M, Honey CR, et al. Eligibility criteria for deep brain stimulation in Parkinson's disease, tremor, and dystonia. The Canadian Journal of Neurological Sciences. 2016;**43**:462-471. DOI: 10.1017/cjn.2016.35

[76] Chao Y, Gang L, Na ZL, Ming WY, Zhong WS, Mian WS. Surgical management of Parkinson's disease: Update and review. Interventional Neuroradiology. 2007;**13**:359-368. DOI: 10.1177/159101990701300407

[77] Lee D, Dallapiazza R, De Vloo P, Lozano A. Current surgical treatments for Parkinson's disease and potential therapeutic targets. Neural Regeneration Research. 2018;**13**:1342-1345. DOI: 10.4103/1673-5374.235220

[78] Mehanna R, Fernandez HH, Wagle Shukla A, Bajwa JA. Deep brain stimulation in Parkinson's disease. Parkinson's Disease. 2018;**2018**:3-5. DOI: 10.1155/2018/9625291

Section 2

Public Health

Chapter 12

Ethical Values in Radiation Protection

Hiromichi Fumoto

Abstract

The subject of bioethics probably first began appearing in radiation protection terminology when the reference was being made to the survivors of the atomic bombs in Hiroshima and Nagasaki. This chapter, therefore, referring to the history of radiation protection since X-ray and radium radiation sources, addresses the nightmare of atomic bombs based on a review of original data and endeavors to determine what the role of ethics is in the radiation protection system as applied to our daily lives constituent to these horrific events. Somatic effects, as differentiated from genetic effects, or late somatic effects are discussed, and an introduction to stochastic effects is also made. It should be noted that a linear no-threshold (LNT) model has been widely applied to radiation protection systems in its pragmatism to be applied to regulatory authorities. However, the radiation detriment below 50 mSv/y is not clearly explained so far. Even though it is only a model, some countries couple LNT with stochastic effects, believing that "lesser is better" as far as radiation exposure is concerned, with criteria reaching as low as tens of micro Sieverts/year, which is equivalent to one two-hundredth of the average exposure received from nature in our living environment.

Keywords: atomic bomb survivors, somatic effects, late somatic effects, stochastic effects, linear no-threshold, environment

1. Introduction

In this chapter, the ethical values in radiation protection are reviewed.

Soon after the discovery of X-rays and radium, radiation protection was applied to those who engaged with radiation sources [1, 2]. In the Manhattan Project, which developed the atomic bomb, the idea of free release was introduced as a precursor of exemption or clearance concepts [3]. The next tragedy witnessed by all the people in the world generated the idea to protect the general public from unexpected radiation exposures.

Hiroshima and Nagasaki survivors' data became the basis of peaceful use of nuclear energy in later days. Utilitarian ethics led the radiation protection society first, up to the Chernobyl Accident. The accident had changed the utilitarian ethics into individual right-oriented ethics in radiation protection [4].

Nevertheless, Japan is one of the states which insisted on utilitarian ethics up to now. However, world nuclear society is shifting its base on radiation protection to lean on ethical values, as high as ever have been discussed before [5].

2. Dawn of radiation protection

Radiation protection has a long history since the first recommendation of the International Committee on Radiation Protection (hereinafter referred to as ICRP) for X-ray and radium protection in 1934 [1, 2].

The permissible dose is chosen to be 0.2 r/day (600 mSv/y), referring to 1/100 of the exposure to induce erythema [2]. Although the ICRP recommendation for radium is limited only to the radiation shield from the radium sources, 1 year ahead of the recommendation adopted in 1933, Robley D. Evans, US scientist, evaluated the maximum load of radium in the body at 10 Ci (3700 Bq). He made an excellent effort to extract the samples of radium painters or radium injected patients to evaluate the permissible radium level in our body with a load of activities [6].

Radiation effects on our body consist of external exposures and internal exposures. External exposures like X-rays, in which ionizing rays pass through our body and giving their energy to our tissues, which may cause harm occasionally. Internal exposures were coming from radioactive elements inhaled or digested into our body, stayed in a certain period, and gave exposures to our bodies.

It must be noted that as early as the 1930s, the basis of controlling external radiation and internal radiation was already established. Therefore, we are generally free of radiation detriment to our health due to great efforts by our preceding generations.

However, all of the controlled personnel were professionals or patients who gave their consent to radiotherapy.

In the Manhattan Project, a secret code of atomic bomb development project, they used a considerable amount of uranium resources. They produced uranium, irradiated them in their nuclear reactor, and extracted plutonium for military purposes. During the development of plutonium separation, they needed to use many stages of chemical units and fission products, radioactive substances produced by the fission of uranium are distributed throughout the plant.

This was the first experience of a human encountering such a significant amount of man-made radioactive substances. The controlled area was set-up to eliminate the infinite migration of radioactive substances to the outer area or environment. Radiation measurement instruments must protect workers in the controlled area. Although radiation protection for workers by radiation exposure was established for X-ray technicians in 1934, they needed to establish other criteria for goods, articles, components to be removed from the controlled area to create a normal environment. This was the origin of clearance or exemption in the later establishment of radiation protection, although the ethical values sometimes seem misinterpreted. It will be shown later in this document.

3. Age of Manhattan project and its entrails

In the Manhattan Project, the concerns were mainly bulky office equipment such as desks and tools for fixing processing units. Even if the project had the highest priority, the number of radiation measurement instruments was limited. The radiation protection experts were requested to give the most effective means of free release criteria, which they call and are still calling the process of removing, for example, office equipment from controlled areas to the USA's outer areas.

The contamination was expected to be on the surface of the equipment. Therefore, they decided to determine the surface-specific radioactivity. **Table 1** shows the criteria given by AEC in 1974 [3].

Nuclide[a]	Average[b,c]	Maximum[b,d]	Removable[b,e]
U-nat, U-235, U-238, and associated decay products	5000 dpm α/100 cm^2	15,000 dpm α/100 cm^2	1000 dpm α/100 cm^2
Transuranics, Ra-226, Ra-228, Th-230, Th-228, Pa-231, Ac-227, I-125, I-129	100 dpm /100 cm^2	300 dpm /100 cm^2	20 dpm/100 cm^2
Th-nat, Th-232, Sr-90, Ra-223, Ra-224, U-232, I-126, I-131, I-133	1000 dpm /100 cm^2	3000 dpm / 100 cm^2	200 dpm / 100 cm^2
Beta-gamma emitters (nuclides with decay modes other than alpha emission or spontaneous fission) except Sr-90 and others noted above.	5000 dpm β-γ/100 cm^2	15,000 dpm β-γ/100 cm^2	1000 dpm β-γ/100 cm^2

[a]*Where surface contamination by both alpha- and beta-gamma-emitting nuclides exists, the limits established for alpha- and beta-gamma-emitting nuclides should be applied independently.*
[b]*As used in this table, dpm (disintegrations per minute) means the rate of emission by radioactive material as determined by correcting the counts per minute observed by an appropriate detector by background, efficiency, and geometric factors associated with the instrumentation.*
[c]*Measurements of average contaminant should not be averaged over more than 1 square meter. For objects of less surface area, the average should be derived for each such objects.*
[d]*The maximum contamination level applies to an area of not more than 100 cm^2.*
[e]*The amount of removable radioactive material per 100 cm^2 of surface area should be determined by wiping that area with dry filter or soft absorbent paper, applying moderate pressure, and assessing the amount of radioactive material on the wipe with an appropriate instrument of know efficiency. When removable contamination on objects of less surface area is determined, the pertinent levels should be reduced proportionally, and the entire surface should be wiped.*

Table 1.
Acceptable surface contamination levels.

Removable radioactive material per 100 cm^2 was measured by smear methods assuming a certain efficiency, since most of the contamination was expected to be on the surfaces in the Manhattan Project.

Since the measurement was made manually, highly ethical values were requested to radiation protection experts, but they managed to have no disputes in their free release practices.

In 1986, NRC launched the Below Regulatory Concern (BRC) policy to enhance the recycling of waste materials. However, it gave rise to metal industry and cement industry's concern with radiation safety on its implementation, and it was in the moratorium. However, the free release was still implemented [6].

This means our highly ethical valued conduct can overcome our worry about unknown radiation detriment. On the other hand, the proposal based on utilitarian ethical value had been gone in the USA. Before the Chernobyl accident, the world scene of the nuclear industry was led by utilitarian ethics. In 1982, the International Atomic Energy Agency (IAEA) announced that General Principles for Exemptions of Radioactive Substances would be established [7].

Up to that moment, the exemption, as defined by the specific concentration of radioactivity, specified only two levels: one for natural origin radioactive element and the other for the artificial radioactive element [6]. Soon after, the exemption was provided by exposure dose per year, and corresponding specific concentration was named as the exemption levels given to individual radioactive elements.

The radiation exposure dose of several tens of micro Sievert per year was adopted for exemption in 1988 [8]. This is the reflection of utilitarian ethics in radiation protection society. However, the timing was the worst. The world witnessed that the general public could be exposed to significant radiation levels, aside from atomic bomb victims.

Thus, the International Commission on Radiological Protection (ICRP) admitted that their ethical values were based on utilitarian ethical values and changed their thinking from a utilitarian ethical value to an individual-oriented philosophy [4].

The public opinion in the USA objected to the unlimited circulation of very low slightly contaminated waste from the nuclear industry. However, they trusted the site workers who verified waste with an instrument to let the waste freely released.

The public acceptance comes from establishing a long safe record since the Manhattan Project that no one has been receiving any health detriment by the conduct of free releases.

This is one of the examples that the general public cannot accept the utilitarian interpretation of the lowest risk at the happiness of the majority. However, the trust in the workers full of ethical dignity allowed people to accept the release of verified waste to the conventional environment from nuclear facilities.

4. Hiroshima/Nagasaki and its entrails

We witnessed that the general public suffered from radiation diseases. Atomic Bomb Casualty Commission (ABCC) was established to undertake the long-term study of the survivors of the atomic bombings of Hiroshima and Nagasaki [9]. First, the purpose was for military support to use the data for infantry at a stage of atomic bomb deployment. Thereafter, with the statement of atoms for peace, the data was gradually declassified for the public with the purpose of radiation protection.

The life span study of atomic bomb survivors had been continued, and the carcinogenic effects of low-LET radiation exposure in humans had been piled up during the term between 1958 and 1998. One of the examples is an excess relative risk of solid cancer with colon exposures [10].

Figure 1 shows the linear-averaged excess relative risk (ERR) [10].

On this base, the linear no-threshold (LNT) hypothesis for radiation protection was proposed and adopted as a base of regulation for radiation protection in a pragmatic way [11]. After the Hiroshima and Nagasaki tragedy, the general public must be protected, and for the Atomic Bomb victims, LNT sounded more comfortable since it is a way of conservatism. If we choose the LNT hypothesis, it cannot show threshold values. The radiation protection experts will face difficulties in persuading the general public to feel comfortable even though their additional exposure to the natural background is small.

At the beginning of the LNT campaign, the Rockefeller Foundation gave financial support to Biological Effects of Atomic Radiation (BEAR) of the US Academy of Science (NAS). Most of the Genetics Panel members, one of the committees in BEAR, believed that any doses of radiation were harmful, irreversible, cumulative, and linearly acting [12]. This suggested that those who believed atoms for peace had chosen utilitarian ethics in radiation with LNT assumption in their global application to nuclear industries.

Radiation effects are categorized as somatic, also known as deterministic, or late somatic, also known as stochastic effects [13]. Somatic or deterministic effects

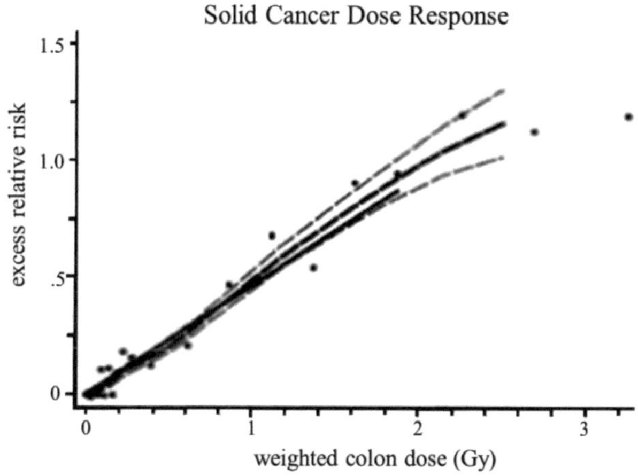

Figure 1.
Solid cancer dose-response function. The thick solid line is the fitted linear gender-averaged excess relative risk (ERR) dose-response at age 70 after exposure at age 30 based on data in the 0- to 2-Gy dose range. The points are nonparametric estimates of the ERR in dose categories. The thick dashed line is a nonparametric smooth of the category-specific estimates, and the thin dashed lines are one standard error above and below this smooth [10] ©2020 radiation research society.

mean that the effects appear soon after the explosion. In contrast, late somatic or stochastic effects mean that the effects appear after a certain period and follow a random process. In some countries, the stochastic effect is translated into a probability effect, which sounds more scientific than ethical.

Mechanism of detriment inducement in such levels of doses believed to have a stochastic effect below 50 mSv/y cannot be explained [11].

Hereditary risk estimates, observed over years, decreases the confidence to quantify the hereditary risk [13].

The radiation protection standard in 1934 had been revised, according to the data obtained in Hiroshima and Nagasaki. In 1990, ICRP recommended exposure levels for an occupationally exposed person. The total effective dose received in a full working life should not exceed about 1 Sv received and a limit on the effective dose of 20 mSv per year. Averaged over five years, with the further provision that the effective dose should not exceed 50 mSv in any single year [14].

The number of 1 Sv from the estimation with solid linear extrapolation of the estimates at 1 Sv could be used to estimate solid cancer risks at lower doses [15].

As for the threshold of 50 mSv in any single year, the author would like to show one of the samples given to the Fukushima Accident. United Nations Scientific Committee on the Effects of Atomic Radiation (UNSCEAR) 2017 white paper suggested that "among the 173 workers with doses greater than 100 mSv (mainly from external exposure), the Committee had considered it unlikely that such increased incidence of cancer due to irradiation would be discernible" [16].

Those sentences or expressions are very common in the society of radiation protection; namely, they are the best to express themselves concerning the threshold with their dignity of ethical dimensions. The development of nuclear energy had been completed in such a short period from 1942 to 1945 with such a large scale of development as President Harry S. Truman stated, "We have spent two billion dollars on the greatest scientific gamble in history-and won" [17].

Since the general public was exposed to the levels over fatal doses in Hiroshima and Nagasaki, the situation of radiation protection surrounded by society changed themselves dramatically from protecting occupational radiation personnel to protecting the general public. For the best ethically dignified experts, the sample statement shown above would be agreed on by the general public the most in terms of low dose carcinogenesis effects.

Since the general public is not radiation controlled, the dose limit for the general public is down from 20 mSv/year to 1 mSv/year. However, ICRP alerted that the dose limit shall not be regarded as the criteria to judge the situation is safe or not [18].

5. Fukushima and its entrails

Japan was the only country to keep utilitarian ethical values, even after the Chernobyl Accident. The general public in Japan had been listening to voices saying that nuclear was safe, different from Three Mile Ireland (TMI), which had an accident in 1974 and Chernobyl, in 1986. Due to the unexpected Tsunami, the radiation levels in the adjacent area easily exceeded the levels to achieve the happiness of majorities. Since utilitarian ethics promised the residents that nuclear power plants would have no significant increase in radiation levels. However, neighboring societies could not accept the detectable increase in radiation levels.

Since Japan's government sets the levels of doses for safety as low as 1 mSv for the public, waiving the pressure of media to announce evacuation to all the residents, whether they are young or old at a nursing home, 1600 accidental deaths were piled other than radiation effects. This is the saga of the LNT application in Fukushima [12].

As for the fallout, ICRP states that although fallout has never been explicitly excluded in legislation, not requiring additional controls are defined on a case-by-case basis instead of defining a category. It further states that for the codified systems, the concepts of exclusion and exemption are very useful [18].

Japan adopted exemption and exclusion for naturally radioactive substances, although they cannot be applied to man-made radioactive substances. Thus, once the soil was found to be contaminated by the fallout of Fukushima nuclear plants, the residents asked for it to be removed. This led to the tremendous amount of very slightly contaminated soil, piled up to 15 million tons, or another estimated 28 million tons of contaminated soil, collected and waiting for reuse [19].

Exclusion and exemption are vital concepts in the implementation of ethical values. A cordial system is governing radiation dose levels at a rigid value. Such swift development of nuclear energy hinders us from understanding radiation control for the general public in a cordial manner. However, it must be done by our unconscious societal rationales provided by virtue and wisdom acquired through our experience and daily ethical conduct.

As is indicated by ICRP, exclusion and exemption are essential elements in radiation protection. Japan has to move from such an obstinate cordial protection system to a flexible system following our ethical values.

6. Discussions

As does the other ordinary industries using hazardous substances did, radiation protection was started by systemizing protection means for occupational personnel. However, due to the atomic bomb explosion, the protection of the general public is requested. Entering the atoms for peace phase, radiation protection for the public becomes a critical issue in their use of nuclear energy for peaceful purposes.

In this initial stage in the civil development of nuclear energy, LNT's adoption was chosen to persuade the general public to have their consent on using the power from nuclear power plants. Whether the founders of LNT were conscious or not, it attributes the ethical values to utilitarianism, which was applicable before witnessing the Chernobyl Accident.

Nine to 22 years exposures of apartment residents given by accidental contamination of 60Co to the recycled steel in Taiwan were analyzed, and no mortality rate increase was observed, contrary to the LNT hypothesis [20].

The author believes that LNT is an assumption of two-edged swords. Our sincere endeavors with prudence and dignity shall allow our community to understand the safe, sustainable, and comfortable environment coexisting with radiation.

The Fukushima Accident gave attention to the severe accident beyond design-based accidents. As is stated in the German Ethics Commission for a Safe Energy Supply, a nuclear disaster in the worst-case scenario is unknown or can no longer be assessed [21]. The Ethics Commission concluded that less risky energy sources could replace nuclear capacity following ecological, economic, and social compatibility; however, the author believes that nuclear energy still has the potentials to contribute to our society.

Severe accidents beyond design-based accidents cannot be assessed, as is the case with design-based accidents that provide scenarios in the anticipated consequences of an accident. Thus, criteria for levels of radiation exposure could not be supplied to severe accidents in nature. However, the magnitude of an accident can be assumed in the hypothetical worst scenario. The author believes that the justification for nuclear energy options can be given even to Japan with its ethical dignity, nuclear energy can survive.

7. Conclusion

Radiation protection had been soundly developed since gradually, the researchers, medical doctors, and X-ray operators realized the health detriment caused by radiation exposure. Due to the exposure of the people in the Hiroshima and Nagasaki atomic bombing, the general public is very keen and sometimes reluctant to accept the permissible level of radiation exposure.

Our society has chosen the protection system with ethical values instead of systematic coding since the threshold cannot be provided or mutually agreed upon by the LNT hypothesis. However, LNT itself dominates the position of regulatory principles and never tries to give the post to a more rational and scientific one with either threshold or linear-quadratic models [22]. Radiation protection experts have shown much scientific evidence, regulatory authorities insist on keeping LNT at the center of regulation. Only experts with ethical dignity can explain both sides of LNT, namely, pragmatic in radiation protection in normal practices and generation of unnecessary worry, dispute, chaos, and social expenditures, as we have experienced in the TMI, Chernobyl, and Fukushima Accidents. Inexhaustible efforts of dissemination by radiation protection experts will advance our society to understand the defects of LNT and regain our peaceful daily life surrounded by radioactive substances.

International Radiation Protection Association (IRPA) will soon deliver a statement on 'Reasonableness' in Optimization of Protection. Its draft referred to the balancing of fundamental ethical values and the importance of optimized processes based on realistic assessments of doses [5].

In July 6, 2020, the European Commission decided to appoint its Joint Research Centre (JRC) as the group of experts to assess nuclear under the sustainable finance taxonomy. FORATOM, the European nuclear trade body, has welcomed it [23].

Starting July 1, 2020, Germany, Portugal, and Slovenia are in cooperation, referred to as a Trio, by the EU's presidency. JRC participation in sustainable finance taxonomy suggests the possibility of a reconsideration of nuclear energy, which might further lead to be involved in the framework of Sustainable Development Goals (SDGs) or Environment, Society, and Governances (ESGs).

The author believes that nuclear energy can be back as an essential source of energy, and our ethical conduct of dignity gives way to receive the understanding of the general public not in long time intervals as ordinary people are expecting now.

Acknowledgements

The author is very grateful to Mr. Bernie Bridger for his valuable comments and reviews during the completion of this paper. His substantial support and encouragement are much appreciated.

The author also thanks Ms. MItos Galutan for her patience in reading through the manuscript for corrections. Without her contribution, this article would not be completed.

Author details

Hiromichi Fumoto
Japan Inspection Co., Ltd., Chuo-ku, Tokyo, Japan

*Address all correspondence to: h.fumoto@nihonkensa.co.jp

References

[1] International Recommendations for X-ray and Radium Protection, Revised by the International X-ray and Radium Protection Commission at the Fourth International Congress of Radiology, Zurich, July 1934. p. 1 [From The British Journal of Radiology, Vol. VII, New Series, No. 83,NOVEMBER, 1934]

[2] Kosako T, editor. Nuclear Energy Textbook, Radiological Safety. Vol. 1. Tokyo: Ohmsha; 2013. ISBN 978-4-274-21323-6; p. 10 (In Japanese)

[3] US Atomic Energy Commission, Regulatory Guide, Directorate of Regulatory Standard, July 1974, Regulatory Guide 1.86

[4] Wikman P. Trivial risks and the new radiation protection system. Journal of Radiological Protection. 2004;**24**(1):3-11

[5] International Radiation Protection Association (IRPA). Statement on 'Reasonableness' in Optimization of Protection [Internet]. 2020. Available from: https://www.fs-ev.org/fileadmin/user_upload/02_Der_FS/IRPA/IRPA_Statement_on_Reasonableness_AS_consultation__May20.pdf

[6] Fumoto H. Radioactive waste disposal (III)—exemption and clearance in waste disposal. Radioisotopes. 2019;**68**:773-789

[7] Basic Safety Standards for Radiation Protection 1982 Edition. IAEA Safety Series No. 9. 1982

[8] Principles for the Exemption of Radiation Sources and Practices from Regulatory Control, IAEA safety series no. 89. 1988

[9] Organized Collections, Atomic Bomb Casualty Commission [Internet]. 1945-1982. National Academy of Science. Available from: http://www.nasonline.org/about-nas/history/archives/collections/abcc-1945-1982.html [Accessed: 01 August]

[10] Preston DL, Ron E, Tokuoka S, Funamoto S, Nishi N, Soda M, et al. Solid cancer incidence in atomic bomb survivors: 1958-1998. Radiation Research. 2007;**168**:1-64, © 2020 Radiation Research Society

[11] Sherbini S. Effects of Low Levels of Radiation, the LNT Model, and the HPS Position Statement, "Answer to Question #9694 Submitted to "Ask the Experts" [Internet]. 2020. US Health Physics Sciety, Specialists in Radiation Protection. Available from: http://hps.org/publicinformation/ate/q9694.html

[12] Sutou S. A message to Fukushima: Nothing to fear but fear itself. Genes and Environment. 2016;**38**:12 [Internet]. Available from: https://genesenvironment.biomedcentral.com/articles/10.1186/s41021-016-0039-7

[13] Silini G. Ethical issues in radiation protection-the 1992 Sievert lecture. Health Physics. 1992;**63**(2):139-148

[14] ICRP PUBLICATION 60: Recommendations of the International Commission on Radiological Protection. Annals of the ICRP. 1990;**21**(1-3):49

[15] Sources and Effects of Ionizing Radiation. UNSCEAR 2000 Report to the General Assembly, with Scientific Annexes, Volume I: Sources 2000

[16] Developments Since the 2013 UNSCEAR Report on the Levels and Effects of Radiation Exposure Due to the Nuclear Accident Following the Great East-Japan Earthquake and Tsunami, A 2017 white paper to guide the Scientific Committee's future program of work, 2017, UNSCEAR

[17] August 6, 1945: Statement by the president announcing the use of the A-Bomb at Hiroshima [Internet]. 1945

[18] Scope of Radiological Protection Control Measures. ICRP Publication 104. Annals of the ICRP. 2007;**37**(5):30

[19] Fumoto H. Radiation protection (lessons learned from the Fukushima accident). In: Advances in Health and Disease. New York: Nova Science Publishers; January 2021. ISBN: 978-1-53618-605-5 (To be published)

[20] Chan WL et al. Effects of Cobalt-60 exposure on health of Taiwan residents suggest new approach needed in radiation protection. Dose-Response. 2007;**5**(1):63-75

[21] Germany's Energy Transition—A Collective Project for the Future. German Ethics Commission for a Safe Energy Supply [Internet]. 2011. Available from: https://archiv.bundesregierung.de/resource/blob/656922/457334/784356871e5375b8bd74ba2a0bde22bf/2011-05-30-abschlussbericht-ethikkommission-en-data.pdf

[22] Cardarelli JJ et al. It is time to move beyond the linear no-threshold theory for low-dose radiation protection. Dose-Response. 2018;**16**(3):9. [Internet]. Available from: https://www.ncbi.nlm.nih.gov/pmc/articles/PMC6043938/

[23] JRC to Assess Nuclear Inclusion in EU Taxonomy. 2020. World Nuclear News (WNN), Nuclear Policies, July 6, 2020, [Internet]. Available from: https://world-nuclear-news.org/Articles/JRC-to-assess-nuclears-inclusion-in-EU-Taxonomy

Chapter 13

Medical Act and Negligence: Ethical Concerns

Julio Cesar Ballesteros Del Olmo

Abstract

To all doctors, Medical ethics must be in support of every medical action. Nowadays, ethics in medicine is an elective topic in college curricula, and therefore, unknown, forgotten or poorly learned in detriment of patient care and their wellbeing. Medical care lacking in ethics generates mistakes derived from lack of skill, negligence or recklessness. These are exacerbated by the lack of training and/or overconfidence, which at first glance can appear to be commonplace and even normal, and thus, resulting in medical malpractice. We must return to humanistic medicine. Combat medical mercantilism at the cost of the patients, and recover the social position that medicine has held with the utmost respect for centuries.

Keywords: medical ethics, medical act

1. Introduction

In memory of my unforgettable wife Dra. Laura Elena Mancilla RIP

The purpose of this document is for professors, general, specialists, subspecialists and medical trainers, to make a thorough insight of the grave deficiencies that the medical act has before a healthy or ill patient. We will start by defining the medical acts as every action executed by the doctor before the patient from the moment they seek to be treated.

The medical act must be surrounded by the medical ethics and this is based on the moral values and principles that define what is done right or not under the use of reason (ethics) and rule our conduct for the benefit of the patient (medical ethics). This principle called deontological are benevolence, equity, autonomy, confidentiality, respect, dignity, solidarity, honesty, loyalty, and justice and every doctor has the responsibility to apply them with humanism, knowledge, and experience at all times during the medical act. All these, taking into account that the ultimate purpose of the doctor is to provide quality in the medical attention with a humanistic sense (granting appropriate medical attention to the patient, pursuant to medical knowledge and applicable ethic principles, which allow the satisfaction of their medical needs and expectations), knowing that the medical responsibilities are aimed on the prevention, diagnosis, treatment, control, curing, palliation of the problem, or moral support when none of the above is possible in order to facilitate a dignified death. The patient's expectations value each one of the components of the medical act, that translate in to a professional, humane, and assertive treatment and also valuing the institutional capacity where this attention is given [1].

 IntechOpen

Thus, we can say that the medical act so practiced is a Good Practice or Lex Artis, and the medical act that does not comply with the requirements is called Malpractice and is identified frequently drawn from one or more unjustified errors, therefore it implies fault, and medical liability. Here we highlight negligence, incompetence, reckless incompetence, and willful misconduct. The justified or excusable errors are not deemed malpractice but the result of the continuous risk in the decision making even when they are adequate and that might arise from intolerances, allergies, over-infections, side and adverse effects, etc. The doctors are not infallible nor make willful mistakes.

Negligence: lack of the required precautions and attention; when in spite of having the knowledge, it is not applied and causes damage. Incompetence: Lack of skills and experience of technical and practical knowledge; acting without having the knowledge and causing damage. Reckless incompetence: taking unnecessary risks caused by the lack of knowledge. Willful misconduct: Scheme to cause damage. It is always sanctionable because it consciously breaches the law [2, 3].

When the actions become repetitive in the same place, by the same doctor, before the same patient, the need and importance of having mechanisms that allow to perform an evaluation process of the medical act arises, in order to identify if the clinical practice and the complete analysis rendered correct and complete diagnosis and treatment processes, based on the knowledge of the feelings of the patient and not only in the encyclopedic knowledge of medicine or the isolated analysis of the requested studies, additionally to the experience and empathy, if it was performed under continuous and adequate supervision, with explicit and precise instructions, and starting from the identification of the fact that there was a diagnosis and if it was correct, if the secondary effects and complications that might damage the patient were expressly explained, as well as the short and long term prognosis, and especially if every step of the medical act was taken with dignity and wellbeing.

This concern takes us to claim that the doctors in formation must retake the professional ethics as a way of personal life in order to give health and wellbeing, to their main objective, the patient. To achieve this, they need to exercise their profession with medical ethics, so tampered nowadays. Luckily there are many of us working towards that objective. But the road to the patient's wellbeing is filled with obstacles, or simply shortened because of mistakes arising from inexperience, negligence, or incompetence, covered by the lack of training, the excess confidence, or just for executing medical acts lacking ethics and humanism that may seem normal at plain sight, but giving rise to medical malpractice.

This is why, the specialized and sub-specialized doctors must reevaluate our actions before the patient; the medical malpractice has caused that in the past years, our roll as doctors has changed before the society, and our ethical behavior with lex artis, has been transferred to the personal and commercial ethics, the encyclopedic knowledge to diagnose with laboratory and cabinet studies more sophisticated and expensive and to give therapies for diseases without contemplating the patient as the center of everything and as a human being.

To exercise medicine with lex artis, we want to make the following observations and recommendations:

1. The doctors educate ourselves to prevent diseases and in its case, try to cure them, avoid complications, and premature death from the disease or from complications from such, favoring the recovery and rehabilitation for the family and social reintegration, always applying the best diagnostic and therapeutic strategies, with the least expense and the maximum benefit [4, 5] complying with the following responsibilities:

a) Acknowledging that their patient is a human being, such as them, that by falling ill, they are living with a crisis of pain, anguish, fear, confusion, uncertainty, and are asking for help; b) disposition to listen and act with empathy (asses in order to understand the disease and suffering of a patient "putting themselves in their shoes"); c) having sufficient medical and scientific knowledge and being an expert in the field; d) being aware of their deficiencies, abilities, and skills; e) accepting the need for training and updating; evaluating and contrasting their real skills with other doctors and in different educational and working sites; f) knowing how to talk and inform patients in the same communication channel and language, in a simple, direct, and explicit manner about every aspect related to their health or disease; g) asking in a timely manner for the help, opinion, or official valuation from their peers when we cannot make a diagnosis with certainty; h) highlighting the medical ethics and the human values as a way of life and work exercising respect, discipline, responsibility, timeliness, honorability, honesty, integrity, education, love, and passion; i) fostering the wellbeing; j) not trying to get rich at the expense of your patients; j) honoring the profession with your actions; k) educating and transforming your patients and alumnus; l) professing humanism in your daily life, towards excellence, exercising assertive medicine and not in a defensive way that damages the doctor, the patient, and the institutions; m) using the medical file as a working took to record all your medical actions, because when there are explicable or inexplicable mistakes; unconformities, and claims against the doctor might arise, in the midst of pain and resentment caused by the grieve of the family members, the instigation of lawyers, plus the media that makes a windfall from another's misfortune, the doctor is degraded to the level of a criminal by all the aforementioned and even the judges, that without any knowledge on the subject feel capable to comment and make medical conclusions and issue despicable sentences against us. A well-documented clinical file will then become a valuable tool in the defense before this scenario [6, 7].

2. In the climax of scientific development and technology applied in the medical field, the medical education at medicine schools prioritize the medical clinic subjects and minimize or exclude the hours of ethics and bioethics and in general, the psycho-social sciences from the curriculum design. The clinical rotations of the medicine students in the teaching hospitals are anxiously awaited and it is here when the young students are seduced and/or confirm their dreams of making a medical specialty, they are impressed by the medical personalities that are their teachers and that exercise their profession, with or without ethics, in the midst of the advanced diagnosis and treatment studies so that when they are done with their studies they can join medical residencies, which is often hard to achieve.

3. Now, let us focus in the public service in Mexico where millions of patients are treated, in most cases from the middle and lower class and the workload increases in the attention of emergencies, consults, and hospitalization. Let us suppose that 20 patients are treated in a family medicine consult, dedicating 6 minutes per patient for the information, exploration, issuance of laboratory, admission, and prescription requests. This time conditions a scares communication and does not allow for empathy, there are only brief questions and answers, without the patient feeling conformable or even being able to ask what they have and how the given treatments work, nor the doctor having the time or desire to do so. The work is deficient and fast, because there are other tasks to complete. Depending on the level of attention, the doctors are, most of the times, the residents being supervised or not by the attending doctor. And that is how they learn, without anyone telling them what is well done and what

is not. The great teaching hospitals, that are highly specialized, use advanced scientific and technological tools, new and expensive medicine and therapeutic schemes at the reach of a limited group of patients, those who have weird or complicated diseases, or those that are a public health issue (that even so, they exceed the capacity for attention of the units). The doctors exceeded by the patients in public medicine are the same ones that work in private medicine that will treat patients with more resources able to pay the exorbitant costs of their treatment whether in a doctor's office or in the hospital. The high quality and expensive medical attention is limited to the few patients that are trapped in social security (SS) in the various specialty clinics (most of the time for occupancy) or to the patients with high economic resources capable of paying for their treatment on a direct manner or by mayor medical expenses insurance policies that are another factor that intervenes in this mechanism [7].

4. And up to this point, the doctors have not yet received training in humanistic medicine, are still acting with the ethics that they have molded thought the years, piling up knowledge and personal experience, sometimes being a true reflection of their teachers that carry out the medical act on the path of personal ethics or convenience, in which they act convinced that what they are doing is right because it is convenient to them, leaving aside the humanist ethics in which this reflection is based upon for their personal benefit as well as for the patient's.

To continue with the problem, the public and private treatment define the patient as an object and not as an ill human being, and the health worker is thus, a public servant or a service renderer or provider (the medical attention, whatever it may be, in transformed into a service). A service is rendered to make a diagnosis of something and not of a disease in an ill person. They make a budget to fix your broken car and if you can afford it, you fix it, and if not, you leave it as is until it stops working. That is the exercise of medicine in a public and private level, splashed with great scientific and technological advances, the doctor slowly starts losing their doctor-patient relationship, and substitutes a good interrogation and exploration with lab and cabinet exams, that make the service and the attention more costly and delay the early diagnosis and the timely treatment. In private medicine, the doctors are encouraged to admit more patients and ask for more studies to keep the economy in hospitals, or to face the increase of the lease in their offices or being expelled from the medical group. The medicine is commercialized and many times it becomes fraudulent [8].

The malpractice has given rise to more indictments and medical detentions that damage the physical, moral, social, and professional integrity of the doctor, which warns us to review the environment of our medical actions, asses how we are failing and not only to protest against the judicialization of the wrongful medical act without explanation.

5. We will now submit to your consideration some cases (the first one unabridged for a better analysis) in which you can assess the medical action:

2. Case 1

Woman of 65 years with a background of abdominal pain in meso and epigastrium, mild to severe, inexplicable, during 20 years, associated with hypergastrinemia never above 500 picograms per milliliter, reason for which a sub total pancreatectomy is performed in a public institution, leaving behind only the head of the pancreas; no macroscopic tumor was removed nor were microscopic tumors detected. As of this moment, hypothyroid crisis, hypothyroidism, rheumatoid arthritis, detection of anti-thyroid antibodies, to mention the most important and those that required hospital services. The patient continues with the abdominal pain, and still in pain and without any advancements, she attends another health institute where she is diagnosed by endoscopic means with atrophic chronic gastritis and finding antibodies anti-parietal cells in the stomach, adding to the diagnosis from the same studies, gastric polyps and neuroendocrine tumors in different years, resolved by extirpation. The endoscopic exam was made every one or two years or as per the request of the patient because of the continuance of the symptomatology. Fast forward to three years ago, and the pain continues, adding halitosis, heaviness sensation, and postprandial abdominal bloating, abdominal distention, and mild to moderate gastroesophageal reflux (RGE). Eight months ago, we add bad digestion, nausea, and post prandial vomiting, so as per the insistence of the patient (hereinafter called ROMA) a new endoscopy is performed referring to a great amount of dense food residue in the stomach rendering the study impossible, programing a new study for the following week, prescribing a soft and liquid diet of 24 hours prior to the study. ROMA, at that time, had a month and a half tolerating only 30 to 40 ml of liquid diet four or five times a day according to her tolerance, suspended every time by intense nausea and constant moderate reflux. The endoscopic study is suspended once more due to a great amount of gastric residue arguing that ROMA has not been duly prepared, rescheduling it for the following week. For this reason, a private cabinet performs an ultrasound and an abdominal tomography that finds a very distended gastric chamber at the expense, presumably, of nutritional material, and of the water flow, and contrast medium, they cannot see the pylorus, but they can see an increase in the thickness of the gastric antrum wall, establishing the diagnosis of gastric obstruction or pyloric syndrome. With this results she goes to the institution for the programed study, with those results they perform a vacuum of the gastric content referring once again to a poor preparation for the performance of the study, reporting verbally and surely by the expert (the same one that has performed the studies in this time table), again an atrophic of chronic atrophic gastritis without any other associated or visualized pathology due to the fact that the endoscopy was not able to pass thought the pylorus, sending her for a gastroenterology assessment. The patient insists that a programmed and complete gastric lavage is performed, for an endoscopic study with ultrasound that has been programed and suspended twice, the doctor refuses and with a doctors' consensus by phone, the hospitalization is requested knowing that there are no available beds. It is important to highlight that the diagnosis of the consult never elaborated diagnosis of suspicion of cancer, nor multidisciplinary evaluation were performed in the many years of studies. The treatment for the discomfort of the patient was purely symptomatic the whole time, and in this 20 years of evolution, the abdomen was explored by the clinic only three times and in every semimanual or annual visit for the control of her specialist doctor, she was seen by different resident doctors that were not aware of the case, without the presence of the base doctor.

COMMENT.

ROMA with 20 years of evolution always interested in her health and wellbeing, always attending to doctors seeking attention, diagnosis, and treatment. It calls our attention that during this time after dozens of consults almost nobody was interested to perform a full physical exploration, correlate symptoms, physical findings, laboratory and cabinet studies, and on top of that it was impossible because the doctors were different, new, in training for being teaching institutions, inexperienced in the field of skills, and with scarce supervision. When there were inter-consulting services, each one attended the affected organ or system and of course, there were never multisystemic assessments or consults to help ROMA who was seen during 20 years without the doctor in charge of assessing her thinking to make analysis and scrupulous, clinical, lab, and cabinet correlations to go beyond the ignored hypergastrinemia, an immune chronic atrophic gastritis, and even less to comment that with someone of the same service or of the gastroenterology service in order to help the patient. Sadly, for the doctor of gastroenterology and other interconsultant services a case that was so painful, rude, and chronic did NOT stand out in a prestigious teaching institution, even when the patient seeks direct and urgent attention, in the end, loosing wellbeing and dignity on top of her heath before the collective indifference.

The medical act must be performed with professional and medical ethics, and here is evident, the lack of respect towards the patient, of medical consciousness, of empathy for the patient, of training, of supervision, total indifference to the suffering and needs of the patient, and what makes it worst, the medical action that must be credited with the application of knowledge, abilities, skills that give experience and wisdom to the service renderer, here we could see a lack of these abilities. Ignoring ROMA with loss of weight for an obstructive syndrome and the inability to eat, additionally to the cause of the problem, turns the medical act to medical negligence, for omission, carelessness, incompetence, arrogance, excess of confidence, actions that take the patient to unnecessary risks for her health or even death, as in this case, which turns these acts in a crime against health, in a medical conditioned crime due to a total lack of professional ethics [9].

In light of this, ROMA seeks alternative solutions and is hospitalized in the institutions where she was affiliated, at that time she had not been feeding, had lost 12 kg of weight, tolerating only scarce suspended liquids due to the sensation of satiety, nausea, and persistent reflux only reduced by abundant voluntary vomiting of abundant and fetid content. Once hospitalized and one week after her last endoscopy, a gastric lavage is performed for 24 hours as well as a pan endoscopy with ultrasound that informs of a mucous of gastric antrum very thickened and swollen almost 3 times the normal standard up to the pylorus, which is impermeable, with a macroscopic image very suggestive of linitis plastica. The trans ultrasound reports no invasion of extra gastric and extra abdominal organs. ROMA is programed to an exploring laparotomy, finding the stomach very increased in size, indurated almost completely, of antrum towards the pylorus, with invasion to the first portion of the duodenum, bile ducts and implantations in the epiploon. The surgeons refer that with carcinomatosis there is no indication of gastrectomy and only a gastro jejunum anastomosis is performed to promote feeding and closed, informing ROMA that she has a surgical diagnosis of gastric adenocarcinoma in state 4b type linitis plastica. The biopsy taken a few days before reports cells in sealing rim characteristic of the linitis plastica [10].

COMMENT.

The correct management function and the empathy of her doctors allow ROMA to be hospitalized in the second mentioned institution and subject to immediate testing, which allowed that four days after her admission a suspicious endoscopic diagnosis was made, and a surgery programed one week after her admission.

The deficient management work of the first hospital institution and the lack of medical ethics in her doctors reflects the mentioned mishaps and radically contrasts with the fact that within a week, the diagnostic appreciation of years with chronic atrophic gastritis changed abruptly to a linitis plastica [11, 12], that even though it is in fact strange and hard to diagnose when the patient is not studied, ROMA has been studied for 20 years and dozens of digestive tract endoscopies were performed, with clinical evidence of a gastric disease more complex than a gastritis and was ignored and rejected for studies in several opportunities when her manifestations became evident of gastric obstruction, inability to eat, and weight loss during the last two months to begin with, that should be cause for alarm for any self-respecting doctor, but here, the medical negligence and the great lack of professionalism and professional responsibility, became a constant in the negligent and reckless medical act and the lack of expertise of at least the endoscopies and gastroenterology services, that were aware of the grave problems of ROMA and finally agreed not to treat her as an emergency as it was merited, but instead to indicate hospitalization when they knew there were no beds available nor there would be in a near future and that would inevitably harm the patient.

The medical teams of the second institution acted promptly to try to solve a medical emergency and within a week they were operating the patient [13]. Unfortunately, the cancer was very advanced without expectations of medical and surgical treatment, except to favor ROMA to help her eat for 3 months, before the second obstruction. Great difference between the duties and action between one institution and the other one, as well as in their medical teams. Great difference in the professional exercise and the execution of the medical act with and without medical ethics.

Two months after the surgery, when the oral tolerance at home was of half servings, in fifths, with good tolerance, they assess, at the oncology hospital of the second institution, already immerse in the pandemic with great occupancy of patients in waiting rooms and medical offices, indicating oral chemotherapy (it is important to comment that there was no physical exams and a they documented a weight loss of 12 kg, with normal lab studies and a tomographic study that was poorly performed and very poorly interpreted). (It was said that for stratification she was not a candidate for any surgical medical treatment). The treatment causes the diminishment of oral tolerance, nausea, vomiting, loss of and extra 4 kg of weight in two weeks, so the patient suspends the treatment voluntarily and before the fear and great risk of contagion of coronavirus she stops attending the institution. As of that moment, the problems worsen, presenting low oral tolerance, with exclusive liquid diet and deterioration of her general condition, even though the lab tests are normal and there is no invasion to other extra-gastric and extra-abdominal organs in the new contrasted tomography performed in particular, although there was persistence of the gastric findings.

COMMENT.

Once again, the negligence (the convenience, need, and consequences of an alternative oral chemotherapy in a patient in terminal phase with oral tolerance were not deeply analyzed), the inexperience (lack of clinical capacity, experience, and knowledge led to a wrongful medical act) and the recklessness (indicating drugs without explanation and damaging to the patient) made the appearance in another group of oncology doctors during pandemic times. Not to even mention the high occupancy of patients in COVID times in the oncology hospital…

ROMA is still desperate because she cannot eat, and nobody is helping her. She consults two prestigious oncologists in a private hospital. One, after hearing the case by phone, orders to start triple chemotherapy without specifying drugs, effects, and when he was asked about this, he answered… What do you prefer, cancer or the effects of the chemotherapy? and by phone, requests for a budget. Another one, in a face to face visit and also without exploring or touching the abdomen recommends starting triple chemotherapy (both doctors are from the same public assistance institution). Afterwards, an oncology surgeon after knowing about the case and without exploring the patient, asks: Why did you come to see mee? What do you want me to do for you? Without further explanation he offers to perform a gastrostomy for drainage and a jejunostomy. Up to this point, none of the three doctors offered nutritional therapy in spite of the evident malnourishment of ROMA nor offered drug therapy for the pain that was on the rise. ROMA did not like like the lack of empathy and dehumanized attitude of the doctors and did not go back to them. She was conscious of her diagnosis and prognosis, as well as of these new complication of a possible intestinal tumoral invasion, because she could feel a big and painful tumor that increased in size in the upper part of the abdomen that was rapidly increasing. The clinical evidence of obstruction was corroborated in a different institute with a pan endoscopy and the placement of an orogastrojejunal trilumen tube, so they try to feed her for two weeks with specialized diets, without success, having a gastric loss always similar to the oral intake. She is reassessed by fluoroscopy, the placement of a tube that was angled and corrected, recommending not changing and continuing the feeding scheme.

COMMENT.

To the prior findings, a new cancer is added in the medical act, the commercial exercise of medicine, where it is more important to charge chemotherapy services without professional assessments, without regard of the clinical state of the patient, the state of malnourishment and diminishment of physical capacities, because there was no clinical exercise, than helping to alleviate the suffering and procuring the wellbeing in the disease of the patients. Something appalling at least in this school of oncology doctors that contaminate the good medical act of the majority. As reproachable is this as another oncology surgeon doctor, once aware of the case, asking ROMA; why did you come to me and what do you want me to do for you? Offensive questions to the intelligence and dignity of the patient that, seeking for help, finds the doctor asking her to tell him what to do. The lack of professional ethics with all its sad consequences. In the desperation and suffering the trips to attention centers continue every time more painful and filled with difficulties and ROME finds humanitarian and uninterested answer for help when a tube of trilumen is placed to favor her nourishment. The doctors still have souls. Unfortunately the intestine obstruction was causing harm making the whole feeding process and marking the immediate future of ROMA.

Immediately after, she is assessed by an oncology surgeon, one of the few that in this story is capable of being empathic with ROMA, treating her in a human manner and after a full evaluation with physical exploration including rectal tact where he corroborates intestinal pastrones and contemplating her terrible general and nutritional conditions, that was almost unable to move by herself and unable to speak on top of the advanced state of the tumor, indicates her to stoop seeking therapeutic palliative alternatives, that her and her family had done enough and all that was humanly possible and that she should be treated solely for algology and palliative medicine to mitigate her suffering and give dignity to her life, that she had to go home and not move anymore. It is important to say that ROMA was a retired doctor and that she seeks and asked for specialists in alternative medicine, private algologists, nutritionists since her discharge from the surgery and her husband tended after her at home during her last seven months of her life, he and family members searched for drugs and specialized nutritional supplements. All made more difficult by the ghost of the pandemic and the collapse of medical services [14], where the high occupancy of hospital services, distant or absent doctors, or those interested in the profit, that are only thinking of themselves not in the patient, new and unsalvageable administrative procedures, change of attention on the hospitals and transformation into COVID hospitals, inability to perform multidisciplinary assessments that were required seeking a complete focus of the problem, many months, maybe years, with a complicated and terminal cancer, situation that was never addressed by the consulting doctors.

The patient dies of starvation [15] in her home because of the inability to feed herself and due to the impossibility of hospitalization for the risk of infection or coronavirus and five months after her diagnosis, from a tumor invasion to the intestine with obstruction.

COMMENT.

To close this very sad chapter, we find ROMA completely weakened by the disease, hit by the medical acts lacking of ethics, and favoring only in a few cases like the latter where a group of doctors agreed to give ROMA a full assessment WITH PHYSICAL EXPLORATION, humanism and empathy, to convince her and her family that they had done everything that was humanly possible to help her, but that there were no probabilities for success. That the best was to rest at home aided by algology and palliative medicine to mitigate the suffering and wait for the end of her life. With this it is important to analyze a little bit the role of her husband, also a doctor, dedicated to complying every aspect of his wife's wishes. Here in that dedication to procure certainty in the diagnosis and palliating treatments aimed to mitigate her pain and suffering [16], the family member encounters the difficult situation of distinguish the very thin line that divides the therapeutic efforts from therapeutic cruelty, and even worse in case of a close family member. It was only this last assessment that made him see this thin line in order not to cross it.

As we can see in this type of rapid analysis, it is necessary to reconstruct the medical education in the social fields, ethics, and bioethics aimed to provide help, solution, and peace of mind to the patient. The social compromise of the doctor has been forgotten in favor of their economic benefit and not few times the work overload hinders the rational medical act and in other teaching institutions for doctors dedicated to providing encyclopedic education graduating doctors that diagnose lab and cabinet exams more and more sophisticated, when they are taught well, forgetting that in front and behind those studies there is a patient demanding for help for their sufferings. Another chapter to envision in full is the training of the doctors in medical schools where they are prepared to face a medical specialty in teaching hospitals, forgetting about the greater demand of attention at the fist level of attention and forgetting specially the ethical formation of the doctors. We need to remember that our formation with moral principles, social values at home and in our family consolidate our ethical actions as individuals and we consolidate this with the formation in schools and hospitals. As in pediatrics THE KIDS LEARN WHAT THEY SEE… the adult doctors in formation also add the bad or good medical example to their professional actions in the future. If the teachers (and all the teachers mentioned herein) act with no professionalism and with personal ethics codes and not dedicated to impact in the wellbeing of the patient, the learning alumnus will learn the same mistakes and are and will become, the doctors that asses us and literally those we will face, like ROMA, when we fall ill.

3. Case 2

Male, 70 years old with background of systemic arterial hypertension of some years of evolution, apparently controlled, and smoking 10 cigarettes daily since he was young. The control drugs are unknown. He arrives from his work for lunch complaining of a terrible headache. He stands up suddenly saying he cannot stand the pain and loses consciousness.

He is taken to a private hospital received in the emergency room with a cardiopulmonary arrest. He receives CPR for 10 minutes, with the recovery of heart frequency but not automated breathing, keeping him intubated with ventilation assistance. With the suspicion of ICP and a deep secondary comma, they perform a CAT and pulmonary puncture and a diagnostic of grave subarachnoid hemorrhaging [17] WITH VENTRICULOMEGALY, and severe intercranial hypertension. 6 hours later he is declared brain dead. Even so, the family members were asked for authorization to perform an evacuation of the hemorrhaging without any explanations given and is subject to surgery with primary decompressive craniectomy. Inexplicably, the patient is pronounced dead after the surgery. The costs associated with the surgery were included in the invoice of the wife, also without further explanation [18].

COMMENT.

The patient arrives with cardiopulmonary arrest, secondary to a grave vascular cerebral probably hemorrhagic event. From 35 to 50% of the cases die before reaching the hospital. Once revived, a diagnosis is made of a subcarinal grave hemorrhaging with ventricular dilation and intercranial hypertension and braindead at 6 hours of live. The first critique of the case arises because a lumbar puncture was performed with the risk of worsening the bleeding of an extremely grave patient and by the conditions of his admission, with a very bad prognosis, with the hospital resource of a tomography (negligence, inexperience, and recklessness as medical acts committed at one moment) and the worst part, having already diagnosed with encephalic death, all medial heroic treatments should have been suspended and especially surgical, which lack therapeutic basis in a patient in those conditions (lack of medial professionalism, lack of medical ethics with unjustified therapeutic cruelty for the patient, absolute negligence and commercialization of the medicine for charging procedures in a patient declared with encephalic death which is an irreversible condition and therefore is not a candidate to be kept under vital support (unless he is an organ donor) and even less to be subject to surgical medical treatment).

4. Case 3

The NICU of a hospital receives a concentration phone call regarding the arrival to a second tier hospital, of a patient pregnant with triplets, preeclampsia, and premature labor of 27 weeks of gestational age, the phone call is to request 3 places in the NICU that is at the time, at its maximum occupancy, without expectations of discharges in a week nor internal movements to free spaces. It is suggested to start specific treatment to the mother and transfer her urgently to an obstetrics and gynecological hospital with the capacity to provide assistance to the mother and to the newborns. Unfortunately the specific treatment is stared, she is not transferred, and the birth occurs one week later, rendering 3 premature newborns with the need of neonatal intensive care unit assistance because of their prematurity and respiratory distress syndrome on top of complications for not reviving the adequate attention for having full occupancy and for not transferring them to a third tier attention or high specialty hospital and the newborns start dying with a difference of days before the impotence of the doctors and despair of the parents.

COMMENT.

The correct thing would have been to transfer the patient at risk of a premature labor in a timely manner to a unit with the adequate infrastructure for the attention of her problem and of the triplets at birth. Not doing so caused the death of the neonates that required assistance in an intensive care unit due to their base problems for being premature and the inherent complications to the inadequate treatment. In this case, there was medial institution liability, negligence for not doing what was supposed to, recklessness by taking decisions that put in high risk the lives of the patients clearly exemplifying a malpractice. What is evident in the case of the newborns is an absolute lack of prenatal control by the obstetrician to evaluate the state of mother's health or illness, of the product or products, and its channeling to specialized centers in case of finding pathologies that would rendered it necessary and consequently and the lack of communication between the obstetrician and the doctor who would attend the babies at birth. If the latter would have existed, together they could have made a diagnostic-therapeutic strategy to favor the adequate evolution of the pregnancy and a better birth. There are many problems that impact the pre-natal and post-natal periods; if the mother is under the legal age, if there is malnourishment, obesity, diabetes, hypertension, cardiac or renal disease, multiple pregnancy, etc., prematurity, pulmonary, diaphragm, cardiac, and nervous system disease in the product, malformation or genopathies, etc. are problems that should be detected by the doctors who treat pregnant women, that are not necessarily obstetricians, and they should inform is to the doctor that receives the product, that is not necessarily a pediatrician or neonatologist [19, 20].

5. Final comment

Without the intention to demonize the doctors of these cases and in the understanding that we are only analyzing the medical action over the information given, it is clear that an emergency must be immediately treated to reduce suffering, pain, stabilizing, to make diagnosis, and decide the best therapeutic conduct pursuant to the expertise and skills of the doctor and the hospital infrastructure, and to the benefit of the patient in order to improve, cure, mitigate, or offer palliative care as the case may be.

There is doubt in the three mentioned cases whether the doctors that intervened were able, prepared, experimented, organized, with knowledge of the protocols, ethic, and if they made the best decisions for the benefit of the patient, but if the place where they are found does not have the resources and infrastructure required for treatment, there is institutional negligence. We can only state for a latter reading, that the reader makes and additional exercise assessing with their own judgment each of the cases and think what they would have done in each one to make a lex artis.

In the cited cases there are details that move the doubt on whether the medical act was the appropriate one and if, in ourselves, a bad decision is possible, even more in the patients and family members that are misinformed and miscounseled [21].

Lets go back to the humanistic medicine, lets fight medical commercialism at the cost of the patients and take back the well-respected social standing of centuries ago.

Author details

Julio Cesar Ballesteros Del Olmo
Academia Mexicana de Pediatría, Mexico City, Mexico

*Address all correspondence to: jc56bo@gmail.com

References

[1] Aguirre GH. Principios éticos en la práctica médica. Cir Ciruj 2004; 72: 503-510Volumen 72, No. 6, noviembre-diciembre 2004

[2] http://www.conamed.gob.mx/gobmx/boletin/pdf/boletin5/acto_medico.pdf

[3] Aguirre GH, Zavala-VJ, Hernández TF, Fajardo DG. Calidad de la atención médica y seguridad del paciente quirúrgico. Error médico, mala práctica y responsabilidad profesional Cir Cir 2010;78:456-462

[4] *Humanismo y medicina;* Ruy Pérez Tamayo http://www.ejournal.unam.mx/rfm/no53-6/RFM053000607.pdf

[5] *Nuestra realidad frente a la ética humanista* https://www.elpais.cr/2015/11/30/nuestra-realidad-frente-a-la-etica-humanista/

[6] *Formación Humanista y modo de actuación del médico. Estrategia para su integración* http://www.revcmpinar.sld.cu/index.php/publicaciones/article/view/2239

[7] *Ética y bioética; medicina y derecho. Encuentro por la salud* https://www.derecho.unam.mx/direccioneditorial/assets/etica_bioetica.pdf

[8] *La responsabilidad del médico como servidor público* https://archivos.juridicas.unam.mx/www/bjv/libros/5/2499/27.pdf

[9] https://www.gastrointestinalatlas.com/espanol/cancer_gastrico_iii_.html

[10] Belén Mendoza de Molano, MD,1 Rocío López, MD,2 Mauricio Palau, MD,2 Shirley Andrews, MD. Linitis plástica. Presentación como gastropatía hipertrófica Linitis plastica. Presentation as hypertrophic gastropathy. Rev Gastr Vol 24, N 1 indv http://www.scielo.org.co/pdf/rcg/v24n1/v24n1a11.pdf

[11] Adán Merino L., Martín Arranz E., Luca de Tena F., Gómez Senent S., Alonso Gamarra E., Olveira A. et al . Diagnóstico atípico de linitis plástica gástrica. Rev. esp. enferm. dig. [Internet]. 2009 Ene [citado 2020 Oct 17] ; 101(1): 79-80. Disponible en: http://scielo.isciii.es/scielo.php?script=sci_arttext&pid=S1130-01082009000100016&lng=es.)

[12] Ikoma N., Agnes A., Chen H. et al. Linitis plastica, a distinct type og gastric camcer. J Gatrpintest Surg 24, 1018-1025 (2020) https://doi.org/10.1007/s11605-019-04422-7

[13] Blackham AU, Swords DS, Levine EA, Fino NF, Squires MH, Poultsides G, Fields RC, Bloomston M, Weber SM, Pawlik TM, Jin LX, Spolverato G, Schmidt C, Worhunsky D, Cho CS, Maithel SK, Votanopoulos KI. Ann. Linitis plastica a contraindication for surgical Resection:A multi institution study of the US Gastric Cancer Collaborative Surg Oncol. 2016 Apr;23(4):1203-11. doi: 10.1245/s10434-015-4947-8. Epub 2015 Nov 3.PMID: 26530447

[14] https://economia.nexos.com.mx/?p=2922

[15] Sosa-Sánchez,R, Sánchez-Lara K, Motola-Kuba D, Green-Renner D. . Síndrome de anorexia-caquexia en el paciente oncológico. Gac Méd Méx Vol. 144 No. 5, 2008 https://www.anmm.org.mx/GMM/2008/n5/64_vol_144_n5.pdf

[16] Cuidados Paliativos. Clinicas oncológicas de Iberoamerica. Guillén, RN. Ed. Pydesa, 2019

[17] Kairys N, M Das J, Garg M. Acute Subarachnoid Hemorrhage (SAH) [Updated 2020 Aug 10]. In: StatPearls

[Internet]. Treasure Island (FL): StatPearls Publishing; 2020 Jan-. Available from: https://www.ncbi.nlm.nih.gov/books/NBK518975/

[18] Michael T. Lawton, M.D., and G. Edward Vates, M.D., Ph.D Subarachnoid Hemorrhage N Engl J Med 2017; 377:257-266

[19] https://www.who.int/es/news-room/fact-sheets/detail/newborns-reducing-mortality

[20] Pérez Zamudio Rosalinda, López Terrones Carlos Rafael, Rodríguez Barboza Arturo. Morbilidad y mortalidad del recién nacido prematuro en el Hospital General de Irapuato. Bol. Med. Hosp. Infant. Mex. [revista en la Internet]. 2013 Ago http://www.scielo.org.mx/scielo.php?script=sci_arttext&pid=S1665-11462013000400005&lng=es.

[21] *Mala práctica* http://www.conamed.gob.mx/comisiones_estatales/coesamed_nayarit/publicaciones/pdf/mala_practica.pdf

Chapter 14

Value-Based Healthcare

Patrick Rech Ramos

Abstract

Value-based healthcare is a new health-care model in which what is important is value to the patient. Value is a broad term, but in essence, it is the best outcome for the patient per dollar spent. To provide value to the patient, the medical practice should be centered around conditions and care cycles and the results must be measured. We now know that the model we have right now, the fee-for-service model, is not linked to quality of the patient. All around the world, many hospitals and clinics are making the transition to this value-based model. To provide the best for the patient, we must have the best medical evidence to follow. In the following chapter, we will cover a few aspects of value-based healthcare, its reimbursement model, the integrated practice units, and the information technology necessary to implement it.

Keywords: value-based, healthcare, value, accountable care, cost-efficiency

1. Introduction

Value-based healthcare (VBHC) is a term coined by Harvard Professor Michael Porter. Along with Elizabeth Teisberg, he published his book in 2006 entitled *Redefining Health Care Creating Value-Based Competition on Results* [1]. They proposed that healthcare should be restructured and focused on competition and improved outcomes for patients.

Some level of competition is important to drive improvement forward. In other fields of expertise, competition is what drives knowledge forward and thus improve value to its consumers, such as in technology. In health, this competition also occurs today but is dysfunctional and does not equate to value to the patient.

Value is defined, according to Dr. Porter, around the costumer and that is achieving the best outcome at the lowest cost, in other words better health per dollar spent. Conrad defines health as maximum health benefit at minimum cost [2]. The shift from today's model and the value-based model is a change that must be physician led and focused around three principles: 1—the goal is value, 2—medical practice should be organized around medical conditions and care cycle, and 3—results must be measured [3].

Moving to a value-based structure is challenging but feasible and the best way to contain costs is to improve outcomes [4], but containing costs alone will not solve the problem. The focus on value is key to a sustainable health-care system [5]. Achieving and maintaining good health is less costly than dealing with poor health, according to Dr. Porter [4].

Not only physicians but the industry itself is moving toward a value-based system. For example, in orthopedics, we have value-based implants [6]. To cut costs of sterilization and sales representatives, they are manufacturing single-use kits.

There are some barriers to the use of these implants such as the surgeon's conflict of interest with the industry [6], but they can be overcome.

Right now, we have a fee-for-service model for reimbursement that over the past several years is shifting toward this value-based model that attempts to link quality and value to payment [7]. The difficulty in implementing it is to quantify quality and value. Professional societies are trying to develop different programs to attempt to define what high value is.

Tools that quantify if we are achieving our goals are needed. In VBHC, we need quality measures that quantify health-care processes, outcomes, patient's experiences, and organizational systems to evaluate the effectiveness of delivered care as it benefits the patient [8]. Value and good outcome may differ from person to person and from condition to condition. It is hard to build a single tool that can be used for every condition.

But how does this model fit in the real world and how can we make the transition to this value-based model keeping in mind we need to improve value to patients? That's what we are trying to answer in this chapter. It is a rather simple question but with a complex answer. A few hospitals in the United States and around the world are adhering to this type of healthcare based on value to the patient [9]. We are going to review a few of them and how they implemented it.

Value-based healthcare may be considered a merge between evidence-based medicine, patient-centered care, and cost-effectiveness [10], even though in essence they are not the same thing.

2. The goal is value for patients

Today's healthcare is not necessarily structured that way. Hospitals want to increase revenue, health plans want to cut costs, and physicians want to increase revenue to their practices. Those practices not necessarily mean better outcome or results for the patient. Patients only want good outcomes with less office visits, less procedures, and less tests [3]. A more individualized practice is needed to meet all these goals.

Many argue that genetic testing is a possibility in the near future [11], but that raises many other questions. The majority of physicians are not trained to interpret the results of a genetic test and that may lead to wrong interpretations and harmful treatments. When that is done correctly, by a specialist, that raises the concern that sometimes an asymptomatic patient or one that did not developed the disease, whether they need treatment or not.

The concept of value remains misunderstood. It is not supposed to be confused with cost reduction, although it encompasses it. Value should be defined around the patient and what they see as a good result, and the creation of value should be rewarded. Value depends on results not volume of services, and the two should not be confused [12].

The cost related to value is the total cost of care cycle, not only the cost of a single procedure or surgery as it is today. Often we need to spend more money in some services to reduce the need for others, which in the end will reduce the total cost of care. The outcome is condition specific, and no single outcome captures the results of care [12].

This value-based model strengthens the role of primary care. There are four features of primary care as stated by Starfield in his 2005 paper: 1—first-contact access, 2—long-term person-focused care, 3—comprehensive care for most health needs, and 4—coordinated care [13]. In primary care, value should be defined for similar groups with similar needs. Primary care and preventive medicine should be divided by need, for example, healthy children, single chronic disease, and so on [12]. This will be addressed further along the text.

The structure we have today makes it difficult to measure value, and most providers fail to do so. Some argue that measurement is necessary but not sufficient to improve quality. One of the barriers to improve quality and value to the patient is the lack of a uniform, simple, and reliable measurement. This difficulty is being addressed by The International Consortium for Health Outcomes Measurement (ICHOM), as we will see next.

Outcomes must be reported publicly to benefit patients and providers. These public reports will further accelerate innovation by motivating their peers to improve their own results. The costs for achieving value to the patient must be measured around the patient, not specialty or around departments. Measuring cost around an entire cycle of care will reduce costs through reallocation of servicer, elimination of others, and better use of the local capacity.

The change in the reimbursement model from a volume-based to a value-based model will allow a reform in payment. It will reward value by providing bundled payment covering the full cycle of care, covering periods of months to 1 year, or longer, according to the condition treated. We will cover this topic further along.

The payment must fit five conditions: payment covers the overall care required to treat the condition, payment is contingent on delivering good outcomes, payment is adjusted for risk, payment provides a fair profit for effective and efficient care, and providers are not responsible for unrelated care or catastrophic cases [14].

3. Medical practice should be organized around medical conditions and care cycle

The organization we have today is by specialty, so a patient who has a condition that needs the effort of different specialties will bounce around from office to office to get his treatment. The reform should be made that patients only go to one place and have a team ready to address their different problems related to the initial condition in the same visit. Organizing around medical conditions and care cycle will be a major change for physicians but a great improvement for patients [3].

Effective care should be centered around a medical condition. That will need the effort of multiple physicians and other health professionals. This organization is known as integrated practice unit (IPU). The IPU is formed by physicians and nonphysicians who provide the full cycle of care for the patient. We will review them further along the text.

The scope of services should be accounted for concentrating volume in fewer locations, choosing the right location for each service line, and integrating care across locations.

Defining the scope of service is to reduce or eliminate service lines where value cannot be achieved. Another possibility is to create partnerships or affiliations with services that you have eliminated because of the lack of possibility of creating value for patients [15].

The concentration of volume in fewer locations is to create a consumer-oriented healthcare. Volume matters for value. The more you treat a disease and the more you learn, the better your treatment will be and more value will be created for the patient. This can be very difficult for organizations to achieve [15].

To choose the right location for each service line is of high value for patients. Less complex conditions should be moved away from high-value facilities to low-cost facilities. It's important to match complexity and the skills needed to the right location. That will optimize cost and productivity [15].

The integration of care across locations is the final component for health system integrations. This concentration of services around different locations must be tied together to improve patient's experience across the sites. All have been directed by IPUs and their physician managers [15].

Value for the patient comes from the effect of the entire set of activities and not only from a single specialty, and the value is greater when all of the four changes above are made.

The MD Anderson Cancer Center in Texas is one of the places where this set up was made. They are organizing around cancer type, and all relevant specialties needed are found in every one of those centers [1].

This approach changes the way physicians manage their practices and their patients. In primary care, for example, they will participate in a number of care cycle teams and they will focus, maybe, in disease diagnosis [3]. The complete cycle of care includes many areas and may take months or even years for the cycle to come to an end. Surgery is one part of care and physical therapy another.

It is thought that increasing value for the patient and the patient feeling well taken care of will reduce the number of malpractice suits. When you have more outcomes measured and a good data collection system, if you get sued you have better data that you can use to defend yourself.

4. Results must be measured

There cannot be an improvement in value for patients without measuring the results. The outcomes for every medical condition and the cost for achieving it need to be measured. Good measures are vital, and they enable professional insight and the development of expertise [16].

This is easier said than done because it may not be so straightforward to measure value or outcome. They can mean different things to different people, and unifying that is a challenge. Many medical associations all over the world are trying to do just that, some with relative success.

One thing we need to recognize is that health consists of physical, mental, and social health. All three must be in order to consider someone healthy and that need to be taken in consideration when measuring results for patients [17] and when a measurement tool is being done. To measure results by improvement on the initial condition alone is not good enough and should not be done.

The results should be measured by condition and care cycle, not specialty or even intervention. It should cover the full cycle of care until after care is completed and taken in consideration the social and mental status. According to Dr. Porter, the outcomes fall into three tiers. Tier 1 involves the health status achieved. Tier 2 outcome relates to the nature of the care cycle and recovery. Tier 3 outcomes relate to the sustainability of health [15]. If all tiers of outcome work well, costs will go down and productivity will go up.

If we want the value-based model to be successful, we need to measure outcomes. If we measure a minimum sufficient set of outcomes for every major medical condition and then standardize them nationally, we are one step closer to this model's success, but that has proven to be difficult.

First, quality is not defined as improvement in outcomes by today's standards. Second, the measurements that have been done are done by specialty societies but the aim is to treat the patient around a care cycle, not by a single specialty or a single procedure. Third, outcome measurements have focused on clinical status rather than functional outcome, which is the patient goal after all, to improve quality of life. And finally, every organization and even physician have their own set of measurements and outcomes, and that leads to inconsistencies in definitions and results. A regional, national, and global standardization is needed, but that is hard to achieve [18].

The International Consortium for Health Outcomes Measurement (ICHOM) has convened groups of experts on specific diseases to set a minimum standard set outcome and risk factors using a structured process [18]. Once this is done, it should be fairly easy to spread round the country and around the world.

One important thing for this to work is the implementation of information technology. The development of software that can automatically collect and aggregate the data for future analysis, such as electronic medical records.

It is believed that in the near future, this is something that will be implemented all over the world with good results for everyone involved in healthcare, especially with excellent outcomes for the patients.

5. Integrated practice units

An integrated practice unit (IPU) is a multispecialty team that collaborates to provide the best outcome to the patient at the lowest cost. These IPUs are encouraged to compete among themselves for the best possible outcome at the lowest cost during the cycle of care. The IPU will treat not only the disease but also all related conditions of the patient.

The team is responsible for the patient's full cycle of care. That encompasses outpatient, inpatient, rehabilitation, and supporting services such as nutrition, social work, and others. The team is also accountable for the outcomes and costs.

Usually with IPUs, we have faster treatment, better outcomes, and lower costs. All that are achieved by the amount of patients they are able to see.

Since the IPU focuses on disease, it is not clear how a patient with multiple diseases at the same time, and not necessarily correlated, will be conducted. Does he have to seek multiple IPUs to treat each of his diseases or only the one? Some say that the need to go to multiple IPUs may cause almost the same problem we have in today's system.

The West German Headache Center can be considered an IPU. It includes neurologists, physical therapists, and psychologists who work together to treat every patient. The patient sees all experts they need in a single visit. If diagnostic imaging is needed, it is obtained from a nearby partner provider [19].

Care delivered in an IPU should be structured. Just the fact that everybody is in the same place does not mean it works well and is integrated. The creation of evidence-based guidelines will incrementally improve value to patients.

One important thing for an IPU is volume. Volume is needed to achieve better results and improve value to patients. The more you study and the more you treat a disease, the better you get at it. Experience is a key point for the deliverance of value. With that you can incorporate more parts of the cycle of care in your facility.

The creation of an IPU can be challenging. A good example of how to make it work is as follows. The Navy launched in Jacksonville at their hospital a value-based program. They selected four of the most common condition to be the starting point [20]. A physician and a nurse were selected to lead each of the four IPUs that were created, and then they recruited other physicians, physical therapists, nurses, and others to be on the IPU. The teams received training on VBHC by external experts and the entire hospital too. Evidence-based treatment and outcomes were defined for later examination; the location, structure, and schedule were also defined by the team. The IPUs met weekly to monthly to discuss patients and treatments. When a treatment was not working, the team would come to an agreement to change it [20]. Three out of the four IPUs created were successful.

Another example of an IPU is at the Dell Medical School at the University of Texas. The musculoskeletal group implemented an IPU team. They followed these steps. First, they choose a condition, symptom, or patient segment to focus. They choose lower extremity joint pain. Next, they set the standards to meet for the patient to be able to go back to primary care. The next step is to define the clinical and nonclinical staff of the IPU, such as the IPU multidisciplinary team and the physical location of the IPU, for example, the building they are located. For their lower extremity joint pain IPU, all patients were initially evaluated by a mid-level orthopedic provider and if surgery was as option they would consult with the orthopedic surgeon and address any questions of the patient. All decisions were discussed with the patients as a shared-decision making. Data collection and feedback is an important step in an IPU since those measurements will be used to address the value of care. The final step is to identify opportunities to improve value to the patient, increasing the overall health and maintaining the patient engaged in care [21].

This is the basic structure to initiate an IPU at your local hospital to get started. At first, we can select a few specific conditions, the most common ones. Later, when you have the first data collected and analyzed, if they are successful, others IPUs can be created. For the data collection and patient information to be readily available, we need the implementation of information technology, such as electronic data records.

We will cover this topic of the collection of data next.

6. Primary care

Primary care is essential for healthcare. Primary care physicians are hard to find, and when patients do, they feel frustrated with the ability of primary care to meet their needs. The problem is that primary care needs to be organized to deliver and demonstrate measured value [22].

Primary care needs to be deconstructed. Instead of one single set of services, it is actually a group of services delivered for multiple subgroups of patients [22]. Like VBHC is organized around conditions so should value-based primary care. It will be needed to transform care into subgroups of patients with new ways of measuring outcome and costs, new payment models, and new approaches to integrate primary care with specialty care [22].

The problem with primary care is that the patients are heterogeneous. The diversity of needs these different patients create is the challenge to implement value-based primary care. It is impractical to measure outcomes achieved relative to costs for such diverse patients [22]. There are five elements to shift primary care to a value-based model [22].

The first element is basing primary care on patients' needs. It is to group patients by their needs. It is designed to create value to patients. The "needs" include types of services and effective methods for patients to access care [22].

The second element is integrating delivery models by subgroup. Once the subgroups are defined, we can move over to the second element. A few questions must be answered. First, the team should be composed of the physician and other personnel according to the subgroup and their needs. Second, the facilities should also be organized around the subgroup and their needs, and they can be arranged to each day of the week to receive a different group of patients. Third, providers must function as teams, a leader must be recognized, and the team must meet regularly to address the patients' needs [22].

The third element is measuring value for each patient's subgroup. Identification of the outcome that matters to patients is key; also, the measurement of the total cost should be done, including those costs outside primary care. All of the care processes must be mapped by subgroups; then, the resources needed can be identified and the costs ascertained.

The fourth element is aligning payment with value. The payment system should be redesigned to a time-based bundled payment or a payment for a total package of services for a defined primary care subgroup during a specified time period. Additional fee-for-service payment could be available for patient's acute care need [22].

The final element is integrating subgroup teams and specialty care. Some patients will need coordination between primary care and specialty secondary and tertiary care. Healthy children and adult may have all their needs met by primary care. Chronic conditions will need to be integrated with specialty care according to their needs [22].

This concept of organizing care around subgroups may seem different than the purpose of primary care but this approach is something that will make primary care more efficient, integrative, and holistic [22]. Electronic data record systems are needed in primary care also. All the participants of the teams must have access to it, and it must be integrated with secondary and tertiary care units and their IPUs. We will revise this topic on information technology up next in further detail.

7. Information technology

All over the world the interest in VBHC is growing. With this growing interest and rapid acceptance of both patient and providers, it is important to have the right tools to record and analyze patient's data toward a value-based model. That is why the implementation of a value-enhancing information technology system, such as a patient electronic data record, is so important.

It is critical for the implementation of value-based healthcare to be successful such as the use of electronic data record. The completion of data and reduction of the potential loss of data, by not keeping patient paper records, are critical for the correct measurement of outcomes [23].

Some electronic records today are very good for keeping data but make it hard to export those records for later analysis. There are six elements that are key for a value-enhancing IT platform for the IPUs [15].

First, the platform must be centered on the patient. The system needs to follow the patients across the services and through time for the full cycle of care. Data are aggregated around the patients not locations. So, all parts of the team have access to the same and complete records, instead of the physicians having access only to his notes or other physicians' notes, he is capable of accessing the records from the nurse staff, physical therapists, and so on [15].

Second, it needs to use common data definitions. The data fields related to diagnose, medical history, and other aspects of care are standard according to the condition being treated so everyone can understand what it means and it is easy to export when needed across the entire system [15].

Third, it encompasses all types of patient data. Notes, images, laboratory tests, and many other are stored in the same place and in a standard format. Like said before, everybody has access to everybody's notes and to the complete patient record. Access is not limited to the IPU team leader [15].

Fourth, the access is available to all parties involved in care. That means that the data collected have to be available to the patients and any referring physicians. The best information technology system possible is the one in that the patients can schedule appointments, refill their prescription, and communicate with their physicians and to the rest of the IPU team, in a simple and easy way. It also should be made easy to access some types of information needed for the evaluation of the care given to the patient [15].

Fifth, every medical condition should have its own template. This set of templates makes it easy and efficient for the IPU teams to retrieve the data they need in order to execute procedures and measure the patient's outcomes and risk factors and the costs of the full cycle of treatment [15].

And finally, the system must be easy to extract information. In a value-enhancing system, the data to measure outcomes, track costs, and control the patient risk factors must be easy to extract. They should also allow the patient to report on his/her own outcomes, so that clinicians can make better decisions [15].

The Cleveland Clinic is a good example of an institution that followed all those steps when adopting a value-enhancing data system [24].

8. Reimbursement

The reimbursement changes in a value-based model. Instead of fee-for-service like in a volume-based model, the reimbursement occurs after the full cycle of care. It is essential to have this payment reform. Physicians paid in a fee-for-service tend to provide more care compared with salaried physicians [25]. Also, the fee-for-service payment method is not necessarily aligned with value to the patient.

Payment per activities encourages more procedure done, maintains fragmentation, and discourages prevention, which does not stimulate high-quality care to the patient [26]. According to the authors, high-value care should limit per capita cost, improve patient experience, and improve population health [27].

Emphasis of VBHC is developing and implementing a bundled-payment known as pay-for-performance (P4P). This payment method focuses on aspects of value that can be measured using indicators of quality [27]. Cattel and Eijkenaar in their 2019 article give a comprehensive view on a new payment initiative that combines two elements: 1—global base payments and 2—explicit quality incentives [27].

The rationale of their initiative is that in essence, the global payment is a bundled payment, with the bundle being constructed at a higher level than at the level of conditions or treatments. The second component, the quality incentives, is sort of a P4P payment that rewards measurable aspects of value [27].

Some aspects of value cannot be explicitly measured such as well-coordinated care or many health outcomes are difficult or impossible to measure. While important, they cannot be explicitly accounted for in the payment contract [28]. Designing the base payment as a global payment facilitates cost-consciousness and well-coordinated care across the full cycle of care, with a focus on the patient instead of on separate conditions [27].

Global base payment transfers the risk from payer to provider which may cause a few problems such as diminishing quality or attempting to underprovide expensive services. These concerns have been addressed by Frakt and Meyers [29], and they can be addressed by risk-sharing arrangements in global payment and explicit quality incentives.

The components of the global base payment are, to a multidisciplinary provider group, for a cohesive set of care activities to a predefined population, fixed for a defined period of time, risk-adjusted, and with risk-mitigating measures. The components of the explicit quality incentives need to have a method of linking payment to quality, with quality measures and with quality incentive structure [26].

This payment initiative described above is a little different than that proposed by Dr. Porter. In his initial model, reimbursement would be done after the complete cycle of care and would include all services and medications, and treating inpatients and outpatient's services together. This model would reward true value and incentivate innovation among physicians [3]. Bundled reimbursement allows for all the system to benefit from value improvement [9].

Today, reimbursement takes place for discrete services not for the entire cycle of care. This works against value, according to Dr. Porter [9]. Value is created by the entire care cycle, not the parts. A change in the payment method is required for the VBHC to work. In essence independently of what reimbursement model you use, for value-based healthcare to work, reform is needed. A fee-for-service model, which is the prevailing way of reimbursement today, does not work in a value-based health-care model.

9. Comparison

The fee-for-service is the prevailing model of healthcare in the US and around the world. The patient pays for a medical service, such as visits, tests, and surgical procedures. In theory, the physician charges to cover their costs and for a profit, the patient knows through itemized bills what they are paying for and they can compare the prices with other providers. This competition will drive prices down.

The value-based health-care model has a pay-for-performance reimbursement system. In primary care, for example, the patient pays a monthly, quarterly, or annual retainer fee. This fixed price is regardless how many visits or test the patient requires. As long as the patient is satisfied to continue this plan, physicians will get paid.

10. Limitations and obstacles

The limitations of the value-based health-care model are that it must be led by physicians and that can pose as a problem. If physicians sense that this new model can limit their gains with reimbursement, they may be inclined not to follow through with the necessary steps to make it work.

Physicians are also worried they have little to no time with the IPU team, lack of transparency with the providers, and find it hard to meet quality expectations. Some physicians are not implementing a value-based healthcare because they fear it is too risky with no real assurances.

Other physicians say that this model is beyond the scope of their practices. Should an internist be concerned about organizing someone's efforts to quit smoking? And so, this only adds on to the physicians' responsibilities and with their work load.

Another problem raised is that some fear that tying the physician salaries directly to outcomes might encourage them to refuse to treat the sickest patients who are more likely not to get better.

11. Conclusion

There is a strategic agenda for creating value-based health-care system. It should encompass what we have seen so far in this chapter: organize into IPUs, measure outcome and costs for every patient, move to value-based reimbursement models and bundled payments, integrate and coordinate care in multi-site care delivery systems, expand across geography, and build an enabling information technology platform [30].

When we have achieved all these goals, we will have reached a VBHC system. The road getting there is rough but rewarding. This shift in health-care design from volume to value is already happening all over the world in all specialties, some with success and some not, but the first step in the right direction must be taken.

It is possible to achieve better results at lower costs as we have seen, and at the same time, creating value to patients. This topic needs to be further studied, but all the components to make it work all already in place and ready to be put to the test.

Conflict of interest

The author declares no conflict of interest.

Author details

Patrick Rech Ramos
Santa Monica Hospital, Erechim, Brazil

*Address all correspondence to: prramos@terra.com.br

References

[1] Porter ME, Teisberg E. Redefining Health Care: Creating Value-Based Competition on Results. Harvard Business School Press; 2006

[2] Conrad DA. The theory of value-based payment incentives and their application to health care. Health Services Research. 2015;**50**:2057-2089

[3] Porter ME, Teisberg EO. How physicians can change the future of health care. Journal of the American Medical Association. 2007;**297**(10):1103-1111. DOI: 10.1001/jama.297.10.1103

[4] Porter ME. A strategy for health care reform – Toward a value-based system. The New England Journal of Medicine. 2009;**361**:2

[5] Bozic KJ, Wright JG. Value-based healthcare and orthopaedic surgery. Clinical Orthopaedics and Related Research. 2012;**470**:1004-1005

[6] Lybrand KE, Althausen PL. The role of value-based implants in orthopaedic trauma. Orthopedic Clinics of North America. 2018;**49**:437-443. DOI: 10.1016/j.ocl.2018.05.005

[7] Ray JC, Kusumoto F. The transition to value-based care. Journal of Interventional Cardiac Electrophysiology. 2016;**1**:61-68. DOI: 10.1007/s10840-016-0166-x

[8] DeBaun MR, Chen MJ, Bishop JA, Gardner MJ, Kamal RN. Orthopaedic trauma quality measures for value-based health care delivery: A systematic review. Journal of Orthopaedic Trauma. 2019;**33**(2):104-110

[9] Porter ME. Value-based health care delivery. Annals of Surgery. 2008;**28**(4):503-509

[10] Sampaio SGSM, Motta LB, Caldas CP. Value-based medicine and palliative care: How do they converge? Expert Review of Pharmacoeconomics & Outcomes Research. 2019;**5**:509-515. DOI: 10.1080/14737167.2019.1651645

[11] Vilhelmsson A. Value-based health care delivery, preventive medicine and the medicalization of public health. Cureus. 2017;**9**(3):e1063. DOI: 10.7759/cureus.1063

[12] Porter ME. What is value In health care? The New England Journal of Medicine. 2010;**363**:26

[13] Starfield B, Shi L, Macinko J, et al. Contribution of primary care to health systems and health. The Milbank Quarterly. 2005;**83**:457-502. DOI: 10.1111/j.1468-0009.2005.00409.x

[14] Porter ME, Kaplan RS. How to Pay for health Care. Available from: https://hbr.org/2016/07/how-to-pay-for-health-care [Accessed: 02 June 2020]

[15] Porter ME, Lee TH. The Strategy That Will Fix Health Care. Harvard Business Review; 2013. Available from: https://hbr.org/2013/10/the-strategy-that-will-fix-health-care

[16] Berwick DM, James B, Coye MJ. Connections between quality measurement and improvement. Medical Care. 2003;**41**(suppl):I30-I38

[17] Putera I. Redefining health: Implication for value-based healthcare reform. Cureus. 2017;**9**(3):e1067. DOI: 10.7759/cureus.1067

[18] Porter ME, Larsson S, Lee TH. Standardizing patient outcomes measurement. The New England Journal of Medicine. 2016;**374**:6. DOI: 10.1056/NEJMp1511701

[19] Porter ME, Guth C, Dannemiller E, et al. The West German Headache

forms of competition for access to local public assistance services, and, finally, for generating overload and scarcity in local health systems [1].

Secondly, because bioethics aims to analyze and understand the morality of the actions of moral agents on moral patients. It is understood that the transformations related to the process of implantation, operation and uninstallation of productive processes in the territories start from the decision of a certain agent, the State, and have consequences for the moral patients (those who suffer from the effects of the decision) represented by the residents, workers and professionals working in local assistance policies.

These are not only susceptible and vulnerable to possible consequences resulting from the State's action, but concretely and vulnerable. Thus, if the impacts on health and the environment can, in principle, affect anyone in the areas of influence of the enterprises, the negative consequences are concentrated, in fact, on specific individuals who work in the production process, those who live at the same time. Around and health professionals who are responsible for health care [2]. Thirdly, because, in its origin, the word ethos means "den" or "abode" and has a semantic proximity to oikos or "house."

Understood as thematization of ethos [3] ethics has in its spectrum of concern and performance the purpose of protecting susceptible and vulnerable subjects [4]; therefore, it has a relationship with health protection in the environmental licensing process. There is, therefore, a legitimate object, but little treated in the field of bioethics. Thus, the objective is to indicate tools of bioethics that contribute to the implication of health protection in Brazilian environmental licensing.

2. Method

An integrated review of scientific literature was carried out which consists of constructing an analysis of the literature regarding discussions on research methods and results, as well as reflections on future studies [5].

For this, the following steps were established: (1) structuring the research question—which tools of bioethics contribute to the implication of health protection in environmental licensing; (2) search for evidence in the databases SciELO (Scientific Eletronic Library Online), VHL—Virtual Health Library and Redalyc—Network of Scientific Journals of Latin America and the Caribbean, Spain and Portugal, through the search feature (licensing OR licensure OR concesión de licencias) AND (bioethics OR bioethics OR bioethics) AND (public health OR public Health OR public health); and (3) application of inclusion criteria—indexed articles, published between 1990 and 2016, in Portuguese, English and Spanish; relevant content for research in abstracts; and exclusion criteria—incomplete articles, articles whose content did not meet the research design.

With the perspective of justifying the relevance and legitimacy of bioethics to support the relationship between the morality involved in the environmental licensing process and the consequences for the health of individuals, populations and ecosystems, the deconstructive method is used, as presented by Derrida, which seeks to make evident in the text that which sought to command it from outside [6].

3. Result

In the initial search process, 71 articles were identified. When applying the inclusion and exclusion criteria, nine were selected. Chart 1 summarizes the integrated review of the scientific literature with articles located in the SciELO, Lilacs and Redalyc databases, in the period 1990–2016, according to the database, title, objectives and results related to elements of protection bioethics which contribute to the implication of health protection in the licensing of large enterprises in Brazil.

In an article published in 1995, Schramm reviews the relevance of a natural ethics that considers the complexity of the health field and considers that human intervention on the environment has power over life.

3.1 Principle of quality of life

In view of this, the principle of quality of life becomes a fundamental tool in issues related to being together, equity, justice, and general well-being. It questions the supremacy of science, proposing dialog for the transformation of scientific knowledge into common sense committed to the norms and values of societies and for the translation of common sense into questions for scientific investigation [7].

In 1997, Silva and Schramm [8] analyzed the environmental problem in the context of scientific rationality, in which the conflict between the relationship between man and the natural environment is evident and gives rise to social movements that denounce the environmental impacts produced by the highly techno-industrial model polluter, consumer of natural resources and generator of global biosphere disorder. It highlights the need for an ethics of solidarity involved with dialog, regulation, action, inclusion, with the recognition of conflict, with co-responsibility in the face of the advancement of technoscience.

3.2 Bioethics and public health

Schramm and Kottow [9] characterize the moral problems in public health and consider that principlism a particular current, originating in the United States of America, which provided a bioethical model for biomedical practices, whose core are the principles of autonomy, beneficence, non-maleficence and justice—it is not suitable for this field because it does not effectively fulfill the principle of justice in which interventions must promote the reduction of inequality.

Thus, they propose the principle of protection, which would be more suitable for bioethical purposes in public health, in which protection should be directed to the subjects who actually need it, through the implementation of public policies, morally correct and effective from the point of view of technicians. In 2002, an article tried to characterize the development of bioethics and its potential to deal with problems related to research with human beings. It presents lato sensu bioethics as a planetary ethics that is concerned with responsibility for the damage produced by human action on the environment [2]. Pontes and Schramm [10] studied the bioethics of protection and the role of the State with regard to unequal access to drinking water as a public health problem.

3.3 Bioethics and accountability

The authors consider that bioethics contributes to the State's accountability as a strategic protective agent in the construction of a just and equitable society, committed to the protection of the health of its members, as well as to the promotion of its legitimate personal development projects. Schramm [1] carried out an analysis of the problem of applied ethics, bioethics and environmental ethics. Identifying that the common denominator between them is in the reference of each to ethics and ethos, as well as by the common methods to construct their specific objects;

Year	Base	Title	Purpose	Results
1995	SciELO	The third margin of health: "natural" ethics	Discuss the relevance of a complex natural ethics for the health field escaping the main disjunctions of the subject/object, public/private, value in itself/value in itself.	Points out the centrality of ethical reasons in the field of health that imply the need to outline a practical-discursive universe of ethics capable of articulating technoscientific knowledge with ethical principles of justice, equity and general well-being in the context of available resources and priorities each concrete situation. This universe includes reflections on duties towards the environment; the rights of present generations linked to future generations. Health ethics is conceived as a field of links and possibilities between the bioecological dimension of the life of individuals and populations in a territory and its socio-cultural dimension.
1997	SciELO	The ecological question: between science and ideology/utopia of an era [8].	Analyze important elements in the formation of technoscientific culture—techno-industrial model—and developments in the field of Ecological Science and related social movements.	It establishes the need for an ethics that enables dialogs on ecological issues in which politics takes a strategic place for making democratic decisions in the context of advancing technoscience. Proposes an ethics of solidarity based on an open dialog of pluralist and interdisciplinary confrontation; regulatory ethics; in pragmatism; in the non-exclusion of feeling—the affective expression of judgment—of the set of elements that cooperate in making ethical decisions; in the ethics of ambivalence, in the sense that this is a choice, not a logical conclusion, or a mechanical result; evolutionary ethics and reversibility of principles; in the ethics of co-responsibility; integrate efforts to overcome conflicts, become aware of responsibilities so that action can be taken accordingly.
2001	SciELO	Bioethical principles in public health: limitations and proposals [9].	Characterize the specificity of moral problems in public health and analyze the applicability of principlism as a standard for resolving conflicts in this field.	It is considered that although relevant to clinical bioethics, principlism is not applicable to public health dilemmas, since it is based on the morality of doctor-patient relationships. They propose to link Jonas's ontological concern and the transcendental Levinas, through the protection principle that would be more adequate to the purposes of a public health ethics, allowing to clearly identify the objectives and the actors involved in the implementation of morally correct and pragmatically effective public policies.

Year	Base	Title	Purpose	Results
2002	Lilacs	Bioethics, its development and importance for Life and Health Sciences [2].	To characterize the development of bioethics and its usefulness to face the problems that arise in research involving human beings, taking into account both the disciplinary and methodological requirements of bioethics and some needs of ethics.	It establishes that bioethics is part of a context in which it coexists with the conflicting structures of human reality and attempts to build convergences of solutions, considered equidistant. Lato sensu bioethics corresponds to a planetary ethics concerned with responsibility for the harmful effects that can result from human acts on individual human life, social and cultural life, the biosphere and the ecosphere. Thus, the concept of lato sensu bioethics encompasses and is adapted to complex situations, being a theoretical and practical tool for understanding (analyzing) conflicts; converge the proposed solutions (prescribe better solutions); protect individuals and populations (provide means of protection). Bioethics is the thematization of ethos, taking into account human practices that can have irreversible effects on other humans, living beings in general and the environment.
2004	SciELO	Protection bioethics and the role of the State: moral problems in unequal access to drinking water [10].	Examine unequal access to drinking water as a health issue, analyzing the moral implications, primary needs, situations of fragility and threat of population groups and responsibilities for water supply.	Bioethics contributes to the State's accountability for the provision of sanitation services, and, in particular, the provision of good quality water. From a view of access to water as a right, privatization policies in favor of public policies that aim to correct situations of social injustice, protect the health of populations and, in particular, of population groups constantly threatened, are promoted, promoting conditions for a better quality of life.
2009	SciELO	Applied ethics, bioethics and environmental ethics, possible relationships: the case of global Bioethics [1].	Analyze the issue of applied ethics, bioethics and environmental ethics, identifying the possibility of integration in research.	It shows that, in the Age of Globalization, applied ethics, bioethics and environmental ethics are in fact intertwined. This implies conceptual, methodological, disciplinary, interdisciplinary and transdisciplinary tools used by the three forms of thematization of the ethos under examination in order to be able to account, at the same time, for the identities and differences between these three areas of knowledge and Ethics practices

Year	Base	Title	Purpose	Results
2010	Redalyc	Bioethics as a form of resistance to biopolitics and biopower [4].	Deconstruct the concepts of biopolitics and biopower and create conditions for a correct performance of bioethics, understood both as an analytical and normative tool for the morality of biopolitics and biopower.	It assumes that the concepts of biopolitics and biopower are used inconsistently, which affects their power of intelligibility to understand the profound transformations of contemporary society. Discusses the proposals for biopolitical democracy and democratic biopolitics, showing the need for bioethical control of biopolitics as a practical application in the form of resistance and democratic dissent in relation to morally questionable effects, resulting from biopolitical practices and the inappropriate uses of such concepts to carry them out.
2012	Redalyc	Elements for a bioethical analysis of recent urban transformations in Rio de Janeiro from the perspective of globalization [11].	Identify elements for a bioethical analysis of conflicts related to urban transformations in the city of Rio de Janeiro related to major world events—Olympics and World Cup—inscribed in the phenomenon of globalization and the social consequences of the removal of inhabitants from slums and occupations.	They carry general aspects of the phenomenon of globalization and its background—the civilizing process -, as well as its urban implications. As an analytical focus for the bioethics produced in Brazil—in particular the bioethics of protection and the bioethics of intervention—they propose the resistance processes present in urban conflicts over housing and urban housing. They approach bioethics of protection and bioethics of intervention, in its theoretical aspects, and criticizes the relevance they give to the role of the State as the sole agent of transformation, indicating that the analytical focus of bioethics turns to the autonomous and self-managing processes of social movements.
2014	Redalyc	Dialectic between liberalism, state paternalism and biopolitics. Conceptual analysis, bioethical and democratic implications [12].	Analyze the concepts of liberalism, paternalism, biopolitics, bioethics, separately, and then relate them dialectically to each other.	The thesis is defended that the terms, as a whole, have a dialectical relationship, since liberalism would take the place of thesis and paternalism, that of antithesis, whose synthesis would be represented by the moment of biopolitics, which would in turn constitute a new thesis, initiating a new dialectical process in which the place of antithesis would be represented by bioethics, both of which would converge to a new synthesis, represented by the empowerment of citizens, constitutive of democratic societies, or that are intended to be such.

Table 1.
Articles used in the SciELO, lilacs and Redalyc databases, from 1990 to 2016, according to the database, title, objectives and results related to elements of protection bioethics that contribute to the implications of health protection in the licensing of large enterprises in Brazil.

that is: (a) the description and understanding (in the double sense of "representing" and "presenting") of the conflicts that exist in the ethos; and (b) the prescription and banning of human behavior.

Assumpção and Schramm [11] studied the urban transformations in the city of Rio de Janeiro related to major world events—Olympics and World Cup—in the light of the elements for a bioethical analysis. It is an important contribution in pointing the State as an agent that produces vulnerability in the context of economic globalization and the civilizing process that produces gentrification under the discourse of environmental revitalization, expelling people from their places of life and work to make way to the interests of a globalized elite.

These authors show that there is an ethic of resistance on the part of human groups vulnerable by the State, and that bioethics has elements that contribute to the strengthening of the struggle of these groups that come together to interact in social movements. For this reason, bioethics needs to return, more and more, its analytical focus to processes of production of subjectivity, autonomy and resistance of these movements.

3.4 Bioethics and biopolitics

Schramm [12] analyzed the concepts of liberalism, paternalism, biopolitics and bioethics, establishing dialectical relations among themselves and opening spaces for forms of resistance to threats to the quality of life of people and populations resulting from questionable actions. In this study, bioethics is rightly seen as a form of resistance that includes the analysis of macroproblems and collective conflicts through the previous theoretical deconstruction of categories, an ethical criticism and a concrete political opposition to an unjustified annexation of bioethics to biopolitics, when, in fact, it is possible to consider bioethics as a form of resistance to biopolitics. It is a liberating alternative to biopolitical practices as it mediates normative issues involved in the relations of organic life (zoe), practical life (bios) and these with politics (polis), enabling the empowerment of citizens (**Table 1**).

4. Discussion

The selected and described articles address theoretical, conceptual and practical aspects related to elements of protection bioethics that contribute to the implication of health protection in environmental licensing.

4.1 Bioethical problems addressed

Although they do not attempt to analyze, specifically, environmental licensing, it is evident that this problem is implicated with the bioethical problems addressed, which are discussed below. Global capitalism, local development and the environmental issue Initially, it is worth considering that the implantation of large enterprises in Brazil must be understood in relation to economic globalization, inscribed in the aspects of technological and biotechnological advances that intensify social relations in a global dimension, as well as in conflicts and, in principle, in the possibilities of finding points of convergence [11].

Similarly, the oil refining process is located in the environmental problem because of its local impacts that become global, such as, for example, the production of greenhouse gases, both by the emission of industrial pollutants, and by the use mass of automobiles.

4.2 Power relations

Therefore, it is inserted in the complex circuit of global capitalism in which the expansion of the infrastructure to make productive processes feasible represents a perspective for the cross-borderization of politics and the economy [11]. Shedding light on the process of economic domination, Assumpção and Schramm [11] contribute to the reflection on decision-making processes in the context of globalization, a phenomenon that is not constituted through horizontal agreements between those involved but is achieved through vertical and hierarchical power relations and/or effective conflicts.

These power relations refer to the global and national elites that constitute a supranational power or an empire in which there is a transfer of sovereignty from nation states to a higher entity marked by great tension between a place institutional and the series of global instruments used by capital, as well as by a network or set of multiple power relations that cross, characterize and constitute the social body [13, 14]. In order to establish these power relations, the local economic development discourse useful to the incorporating power of global capitalism is used, which co-opts the political agents to adapt to the global market logic [11].

4.3 Economic globalization and conflicts

This cooptation process is necessary for homogenization that is preceded by adaptation to local power, history and diversity. Adaptations occur as these three fields become commodities, which, as such, meet economic globalization. For Santos [15], globalization is a vast field of conflicts and imposes itself as a hegemonic field, acting on the consensus of its most influential members. Such a consensus gives it domination and legitimates it as the only possible or most appropriate, consolidating itself from the simultaneous denial and affirmation of the consensus.

Santos [15] considers globalization, in general, perverse to increase local inequalities abysmally. In view of this complex scenario of interrelationships between the implementation of enterprises, globalization and local economic development, one must analyze human acts and the significant irreversible effects on the biosphere [1]. It is also important to give globalization an ethics directly focused on the long-term survival of the human species.

This will happen for the protection of human dignity and for the preservation and restoration of a healthy environment [1]. In this way, the relationship between ethos and oikos can be considered, taking into account all the problems about the effects of biopolitical and biopower devices, and their moral assessment and political consideration [4].

4.4 Environmental licensing and bioethical implications

Environmental licensing implies technoscientific responsibilities on the part of analysts environmental factors (when setting the parameters of the terms of reference) and the specialists (consultants) who carry out the environmental impact studies, and political and social responsibility for the decision to approve the introduction of risk situations in the territories where ecosystems are located, people living in the territories [16, 17].

The results show that environmental licensing is part of the field of concerns of bioethics as it is an intervention on a territory in which, in general, people and other living beings live, which are generally affected by the transformations produced.

Protection bioethics has been considered the ethics applied to human actions related to vital phenomena and processes through concepts, arguments and norms that value and ethically legitimize human acts whose effects deeply and irreversibly affect, in a real or potential way, the systems vital, being a crucial issue in environmental licensing with the perspective of protecting collective health and ecosystems [1].

The conception of bioethics best suited to this context is that of protection adopted by Pontes and Schramm [10], as it is comprehensive to account for the vast spectrum of human action on the living world and which can affect human beings positively or negatively, living beings and the delicate autopoietic balances that characterize the environment.

4.5 Risk situations

The complexity involving risk situations, health impacts and the production of new ways of becoming ill and dying, due to the introduction of polluting production processes, requires a critical understanding of the consequences of an action, answering substantial philosophical questions. Regarding the nature of ethics, the value of life , including the consequences of public policies, in particular those of health [16–18].

In accordance with this perspective, it can be said that environmental licensing is a strategic moment for the protection of public health. As it is a time of conflict, it requires an open dialog on the consequences of the actions of environmental analysts. and specialists (moral agents) who contribute to the authorization of the transformations produced by the productive processes on the lives of populations, living beings and ecosystems (moral patients) [2, 18].

The concern with morality in environmental licensing is similar to that of research in human beings. However, market interests exert greater pressure on moral agents, increasing suspicions of conflicts of interest. An example of this is the hiring of consultants for the preparation of EIA by the entrepreneurs themselves [19].

There is a suspicion that those interested in the implantation will certainly not produce evidence against themselves, indicating the real environmental impacts. In the case of public undertakings, the most interested party has been the Brazilian State itself, which has the responsibility to protect populations and groups threatened, including, by market interests.

The concept of a protective state presupposes that it is committed to the requirement of health justice, such as the principle of protection, which must be exercised in order to cover the basic needs for the construction of a fair social order and to protect the quality of life of the populations [10].

Protection is the perspective of environmental licensing; however, carrying out actions that promote better quality of life depends on the quality of the EIA. These should indicate the real impacts and the respective compensatory mitigating measures to be developed [18]. In this case, the protection principle recommended by Schramm and Kottow [12] applies.

4.6 Principle of responsibility

This should be understood as a specification of the principle of responsibility, as the most appropriate to address moral problems related to public health. In this way, the bioethics of protection presents itself as an ethics of social responsibility on which the State is based to assume sanitary obligations towards human populations considered in their real contexts, which are, at the same time, natural, cultural, social and eco-environmental [1, 10].

In this perspective, the protection principle requires that it be clearly specified what should be protected, who should protect what and for whom the protection is directed, becoming, therefore, operational. In particular, the population groups to be protected for their specific needs should be made aware of protective measures; otherwise, they can only be perceived as paternalistic and/or arbitrary, thus making them ineffective [12].

In addition, the information produced by environmental analysts and specialist consultants should not be reduced to the intricacies of research carried out for the preparation of the EIA, with language known only by professionals working in the paradoxically closed universe of institutions as foundations for supporting universities., environmental agencies and private environmental management companies. For these reasons, the fact that there are specialist researchers in the consultancies with the supposed exemption, determination to do good, integrity of character and scientific rigor, does not guarantee ethics or exempt any scientist from suspicion [2].

An example has been the absence of important aspects of health protection in the EIA in Brazil, which implies suspecting the non-identification of impacts, with a view to favoring the authorization of implantation, disregarding the production of a health risk situation [16–18].

An alternative in the sense of qualifying the EIAs would be the integrated environmental licensing, in which the environmental agencies would count on the participation of reference institutions (universities, research centers, institutes) through technical opinions for the elaboration of the term of reference, as well as well as the request for consent, after a critical analysis of the EIA by collegiate bodies with democratic participation, such as health and environmental councils. It is noteworthy that the ethics committees linked to the National Health Council are virtuous places in which cognitive, normative and protective tools of bioethics are used.

The evaluation system constituted by the research ethics committees of the National Research Ethics Commission, inspired by secular bioethics, represents itself as a legitimate and prima facie tool effective in contributing to the ethnicity of environmental licensing with regard to public health. The conceptual tools to be used for the approach of environmental licensing should not be those of the principled model, based on the four principles of "non-maleficence," "beneficence," "autonomy," and "justice" because they present themselves as inadequate in the treatment problems that occur in collective contexts, such as public health or global health [8, 10].

The recognition of conflicts is essential because all social practice inevitably falls within the dialectic between conflicts and cooperation that shapes historical societies. For this reason, bioethics serves as practical knowledge that aims precisely to account for the moral implications, seeking to understand, explain the reality of conflicts and trying to establish convergences to obtain a kind of harmony [3].

4.7 Social movements of resistance

In view of the conflict of interests, in which the Brazilian State is most interested in expanding infrastructure through major development works, and in which social movements of resistance to undertakings in local communities are manifested, here we defend, that protection bioethics, in its broad sense dimension, represents a virtuous form of resistance, just as it does with biopolitical practices that subject ethical questioning to supposed pragmatic needs of political realism, considered more concrete, effective and legitimate in its management of bodies, populations and life in general, sometimes with the cynicism of a public interest to justify morally unjust practices [4].

An example of how to apply a resistance bioethics is the critical analysis of EIA, in socio-anthropological aspects, through the operation of deconstructing the concepts that materialize the contents and their reconstruction in the light of protection bioethics, producing mitigating measures and compensatory measures aimed at providing resistance to harmful effects on health and the environment, reconstructing forms of resistance in the name of what cannot be subject to deconstruction: justice.

Deconstruction is an analytical and interpretative method of moral conflicts inscribed in biopolitics, but also a tool that justifies bioethical practices that question biopolitics and biopower [4]. In this sense, it is possible to produce tools of resistance that contribute to the strengthening of social control through the cooptation and manipulation of the State, as in the case of public hearings provided for in environmental licensing [19].

The perspective is that public hearings are, in fact, devices of power displaced and returned to the common use of democratic participation and the production of open dialog for social justice.

4.8 Bioethics, resistance to biopolitics and biopower

According to Schramm [12] bioethics represents, in addition to questioning and criticism, a resistance to biopolitics, and its reductionism to the biological. According to the author, there are interrelationships that include the interests of the economy and public management. Bioethics enables a practical synthesis in the process of "empowerment" or "liberation," understood as an existential result of the concrete exercise of citizenship, represented by participatory democracy.

When deciding on the implementation of a productive process, which transforms the dimensions of life by authorizing the construction of works by mere political decision without taking into account the lives of human beings, other living beings and the entire ecosystem, this alienation is a form of manifestation of biopolitics about life.

And so, it is up to bioethics to establish itself as a form of resistance to biopolitics and biopower; that is, an alternative, mediating for the empowerment of citizens [12].

5. Conclusion

Supported by an integrative literature review, those related to protection were identified among the elements of bioethics, which involve other ethical dimensions: responsibility, solidarity and resistance.

The deconstruction method used as an analytical and interpretative method of conflicts contributes to the establishment of health protection means in the environmental licensing process in Brazil.

In this sense, relations have been established that characterize environmental licensing as linked to a morality that implies complex transformations in the territories where populations and other living beings in ecosystems live, and which fall within the field of lato sensu protection bioethics, being pertinent to adoption of practices aimed at improving the quality of life and empowering citizens, in the face of threats of introducing a situation of harmfulness to health and irreversible environmental damage.

The topic is not exhausted, requiring new studies—theoretical and empirical—on other aspects of bioethics, such as health protection tools in situations of environmental licensing of large enterprises.

The limitations of the method are recognized, mainly regarding the search resource using the term licensing, however the issue involved is implicated with bioethical problems—environmental ethics, global bioethics, natural ethics, biopower, biopolitics, globalization, social movements.

It is concluded that, in the context of the environmental licensing process of large enterprises in Brazil, bioethics can contribute as a theoretical and practical tool to mediate the existing moral conflicts, carrying out the detailed description of conflicts and dilemmas, the criticism, the justification and the proposal of morally acceptable measures for the protection of life.

Acknowledgements

Fermin Roland Schramm and Lia Giraldo da Silva Augusto, my admired philosophers and bioethicists.

Conflict of interest

The author declares that there are not conflict of interest.

Author details

José Marcos da Silva
Federal University of Pernambuco, Recife, Brazil

*Address all correspondence to: jose.marcoss@ufpe.br

References

[1] Schramm FR. Applied ethics, bioethics and environmental ethics, possible relationships: The case of global bioethics. Collective Health Notebooks. 2009;**17**(3):511-630

[2] Schramm FR. Bioethics, its development and importance for life and health sciences. Revista Brasileira de Cancerologia. 2002;**48**(4):609-615

[3] Maliandi R. Ethics: Conceptos y Problema. Buenos Aires: Biblos; 2004. p. 158

[4] Schramm FR. Bioethics as a form of resistance to biopolitics and biopower. Revista Bioética. 2010;**18**(3):519-535

[5] Souza MT, Silva MD, Carvalho R. Integrative review: What is it and how to do it? Einstein. 2010;**8**(1):102-106

[6] Derrida J. Positions. Belo Horizonte: Autêntica; 2001. p. 12

[7] Schramm FR. The third margin of health: "Natural" ethics. History, Sciences, Health-Maguinhos. 1996, 1995;**1**(2):54-68

[8] Silva ER, Schramm FR. The ecological question: Between science and the ideology/utopia of an era. Public Health Notebooks. 1997;**13**(3):355-382

[9] Schramm FR, Kottow M. Bioethical principles in public health: Limitations and proposals. Public Health Notebooks. 2001;**17**(4):949-956

[10] Pontes CAA, Schramm FR. Bioethics of protection and role of the state: Moral problems in unequal access to drinking water. Public Health Notebooks. 2004;**20**(5):1319-1327

[11] Assumpção LA, Schramm FR. Elements for a bioethical analysis of recent urban transformations in Rio de Janeiro from the perspective of globalization. Revista Bioética. 2012;**20**(1):106-118

[12] Schramm FR. Dialectic between liberalism, state paternalism and biopolitics. Conceptual analysis, bioethical and democratic implications. Revista Bioética. 2014;**22**(1):10-17

[13] Negri A. Five Lessons on Empire. Rio de Janeiro: DP&A; 2003. p. 28

[14] Negri A. Back. Rio de Janeiro: Record; 2006. p. 99

[15] Santos BS. The globalization process. In: Santos BS, editor. A Globalização e as Ciências Sociais. Cortez; 2005. pp. 25-94

[16] Santos M. For another Globalization: From Single Thought to Universal Consciousness. Rio de Janeiro: Record; 2000

[17] Silva JM, Santo MO, Gurgel IDG, Augusto LGS. Sustainable development and worker health in the environmental impact studies of refineries in Brazil. Saúde e Sociedade. 2013;**22**(3):687-700

[18] Silva JM, Gurgel AM, Gurgel IDG, Augusto LGS. The interrelationship between health, work and environment in the licensing of the northeast refinery. Tempus: Minutes of Collective Health. 2009;**4**(4):72-83

[19] Silva JM, Gurgel IDG, Augusto LGS. Health, knowledge ecology and environmental impact studies of refineries in Brazil. Interface—Communication, Health and Education. 2016;**20**(56):111-122

Chapter 16

Climate Change: A Forced Choice Ethical Paradigm

Daniel Londyn Menkes

Abstract

Notwithstanding the political debates in the media, climate change presents a unique set of ethical challenges faced by all the planet's inhabitants. To understand the current challenges facing humanity, it is important to retrace the evolution of human society as this underlies the ethical foundations that internalize a group's beliefs and norms. Humans have modified the environment on a global scale that is unsustainable that has resulted in climate change-a disease process with dire implications. Understanding the root causes of a disease process is the best means of devising a treatment plan. Climate change solutions must be syntonic with a bio-psychosocial model that addresses culture and belief systems. The six main ethical theories-utilitarianism, egoism, deontology, virtue, divine command, and relativism all have their inherent flaws. Beauchamp and Childress concatenated these constructs into the four main bioethical principles of autonomy, beneficence, non-malfeasance, and justice. Of these, autonomy is least applicable to climate change as decisions made by a subgroup of one species will have an impact on all terrestrial lifeforms both present and future. Humanity must accept the reality of climate change and effect solutions based on these four principles. Failure to act will lead to catastrophic climate changes that may lead to the sixth mass extinction. Effective climate change solutions must embrace an integrative approach by supporting leaders who will embrace science and will advocate for universal human rights.

Keywords: climate change, global warming, ethics

1. Introduction

"Global warming" and "climate change" are not synonymous as global warming refers to an increase in the average atmospheric temperature whereas climate change describes the downstream impact of a global temperature increase [1]. Thus, the term climate change will be used preferentially in this manuscript. It should also be stated that weather and climate are not synonymous as weather refers to short term atmospheric fluctuations whereas climate refers to what can be expected over a longer-term period. In this context, an occasional 90 °F or 32 °C summer day in Fairbanks, Alaska would not be an unusual event for its weather whereas 30 consecutive days at this temperature in the autumn might portend climate change. Thus, climate change evolves slowly and is often only discernible in retrospect. A comprehensive understanding of this process requires an interdisciplinary knowledge of mathematics, biology, chemistry, and physics. One must understand the underlying issues to devise viable solutions.

IntechOpen

Problem solving requires several steps including identification that a problem exists, discerning the facts, framing the problem, and then proposing solutions. Errors may occur in any step of this process. With respect to climate change, there is no universal agreement that a problem exists. Irrespective of the preponderance of scientific evidence, there continues to be debate on climate change's existence as well as its root causes. Anurag Shurie once remarked that "A half-truth is even more dangerous than a lie. A lie, you can detect at some stage, but a half truth is sure to mislead you for long" [2]. Scientifically speaking, there are no half-truths in that a statement is factually correct or it is not. Most "half-truths" are opinions that reflect a different interpretation of the available data. To paraphrase the late United States Senator Daniel Patrick Moynihan, everyone is entitled to their own opinions but not their own facts. Solutions may only be achieved when all stakeholders agree on the facts that frame the issue as well as the certainty to which these facts have been established. In the absence of such agreements, solutions are unlikely to be achieved. If climate change is to be addressed, then there must be a commitment to accepting reality while understanding that there will always be some level of uncertainty. The next step is to frame the problem such that potential solutions will become apparent. Ideally, the best decision is made based on a risk to benefit analysis with the understanding that the failure to act is also a decision. Given that most of humanity appears to be mired in the fact gathering stage, the author will address the current facts in evidence.

2. The salient facts of climate change

The intergovernmental panel on climate change stated that the scientific evidence for warming of the climate system is unequivocal [3]. The Potsdam Institute for Climate Impact Research stated that "business as usual" climate change trends portend a warming of 4 °C or 7.9 °F this century's end [4]. Notwithstanding the political debate, an overwhelming number of scientists concur that the current rise in global temperature is from human activity [5]. Those who have an opposing viewpoint correctly state that the earth has experienced numerous warming and cooling cycles. One previous warming event that occurred 252 million years ago, likely from volcanic activity, resulted in an average global temperature rise of 10 °C, which induced an extinction of 75% of terrestrial life and 95% of all marine life [6]. Dinosaurs arose to fill this void for the next 165 million years wherein being large bodied and cold-blooded was advantageous. This provides an example of Darwinian selection in which those organisms best suited to their natural environment survived to reproduce and dominate that niche, which Darwin termed "survival of the fittest" [7]. The age of the dinosaurs ended with an asteroid impact 65 million years ago that extinguished 75% of all lifeforms on earth including all land-based lifeforms weighing more than 25 kilograms or 55 pounds [8]. Mammals arose to fill this gap, some which were capable of altering their environments such as beavers, elephants, and humans. However, none of these mammals made a significant impact on the atmosphere until the Industrial Revolution, which consumed fossil fuels for energy. The current Anthropocene era has ushered in significant changes including global warming, habitat loss, changes in atmospheric composition and incipient mass extinctions. Prior history indicates that novel lifeforms will adapt to these changes. However, this is the first time in geological history that climate change did not occur from natural phenomena. Humanity has altered the biosphere, yet humanity also can avoid the most profound effects of self-induced climate change with a commitment to action.

Climate change induced by global warming may seem complex initially as it involves an understanding of chemistry and physics, specifically the laws of thermodynamics. The first law of thermodynamics states that energy may neither be created nor destroyed such that one can only change its form. For example, consider using a coal stove that combusts hydrocarbons with oxygen to create heat and light. Implicit in this law is that the conversion is imperfect such that some of the energy is lost as heat. Thus, burning hydrocarbons results in the net addition of heat to the atmosphere some of which cannot be radiated into space. Hydrocarbon combustion on a massive global scale will increase the amount of heat retained in earth's atmosphere thus raising the average global temperature. However, heat production itself is not the only issue as the covalent bonds holding the greenhouse gases (GG) together are capable of absorbing infrared radiation without breaking chemical bonds. The net result is that the excess GG radiate heat (infrared) energy back to earth at a later time much in the way that a glass container may be put in an oven, removed intact, and then cooled down to food serving temperature by releasing the excess heat into the environment [9]. The downstream effects of this additional heat production and heat trapping are explored in subsequent sections.

Another issue facing humanity is that of our current, unsustainable rate of resource consumption. Presently, humanity requires the annual equivalent of 1.6 Earths or 20 months to provide the resources we require and absorb our waste, which is the environmental equivalent of "deficit spending" [10]. This rate of resource consumption and waste generation is simply unsustainable. The United States Army Field Manual 21–76 states, "Remember that nature and the elements are neither your friend nor your enemy-they are actually disinterested. Instead, it is your determination to live and your ability to make nature work for you that are the deciding factors" [11]. The current willful disregard for nature portends the sixth mass extinction, the Anthropocene extinction, which will result from humans' alteration of the environment [12]. One prediction is that, unless drastic action is taken, the earth will experience catastrophic climatic and negative socioeconomic changes within the next 30 years [13]. This author further stated that a solution requires "political change producing policy change" [14]. If such changes are to be realized, then an understanding of the science underlying climate change must be understood within the framework of human evolution, the rise of political systems and the ethics that arose from these political developments.

3. The science of climate change

The tendency of CO_2 and other gases to trap heat is often summarized as the "greenhouse effect" in which solar radiation penetrates the earth's atmosphere but only 30 percent is reflected into space [15]. This acts as a warming blanket such that the earth's average temperature supports terrestrial life. Without naturally occurring GG, Earth's average temperature would be near 0 °F (or – 18 °C) instead of the much warmer 59 °F (15 °C) that currently exists [16]. The five main gases that have significant global warming potential [GWP] are CO_2, methane (CH_4), nitrous oxide (NO_2), fluorinated gases and water vapor [17]. These are potent trappers of infrared wavelength energy as the covalent chemical bonds between them are relatively weak such that the molecule remains intact despite adding energy. Nonetheless, these gases are not equivalent in their GWP as this is dependent on their concentration, their ability to trap infrared wavelengths, (i.e. heat), and their atmospheric functional lifespan. GWP uses CO_2 as a benchmark when calculating 100-year relative effects. For example, methane is 84-fold more effective in trapping heat but it persists in the atmosphere for only slightly more than a decade reducing its 100 year impact to 28 times that of CO_2 [18]. Notably, CO_2 released today will persist in the atmosphere for 300–1000 years such that it will have a much greater overall temporal impact. In comparison, nitrous oxide persists in the atmosphere for more than a century such that its short term and long-term effects are identical at 28 fold the impact of CO_2 [19]. Fluorinated gases are even more potent as R-22, the most common refrigerant currently in use, has a 100-year GWP of 1,810, which is almost 2,000 times the potency of CO_2, such that one pound of R-22 is nearly as potent as a ton of CO_2 [20]. This same reference stated that releasing one 30-lb tank of R-22 into the atmosphere is nearly equivalent to the CO_2 emitted by driving 7 additional cars each year, (source data available at CARB's Cool California Calculator). Notwithstanding these effects, water vapor currently exerts the greatest greenhouse effect at this time given its higher concentration in the atmosphere. While the effects of water vapor are relatively short-lived, the amount of water vapor in the atmosphere will increase as warmer atmospheres correlate with greater degrees of humidity [21]. The positive feedback generated by adding additional water vapor into the atmosphere is such that a 1 °C temperature increase from excess CO_2 production has a net effect of increasing atmospheric warming by 2 °C [22]. Ice (solid water) also has an impact on climate change as snow and ice reflect a greater degree of incoming sunlight than does water known as the "albedo effect" [23]. Polar ice cap shrinkage leads to reduced reflectivity or reduced albedo. Conversion of polar ice in the Northern Hemisphere to liquid water leads to a relatively darker ocean surface, which facilitates the absorption of additional heat from the sun, thus melting large masses of ice in the ocean [24]. Melting ice will lead to a rise in sea levels coupled with more frequent storm surges leading to more frequent and intense flooding [25]. Sea level rise will result in less habitable land for terrestrial based life forms as was noted during the past interglacial period when the earth's average temperature was 1 to 2 °C warmer and sea levels were 4 to 6 meters or circa 13–20 feet higher [26]. Even with limiting global warming to 1–2 °C, many of our coastal cities will be submerged.

The most recent extreme period of relative warming in Earth's history was that of the Paleocene-Eocene Thermal Maximum (PETM) about 55–56 million years ago when the earth's mean temperature rose by 5–8 °C (9–14 °F) to an average temperature of 22.8 °C or 73 °F [27]. These authors further stated that concurrent paleoclimate data from fossilized phytoplankton and ocean sediments recorded a massive release of CO_2 into the atmosphere, at least doubling or possibly even quadrupling the background CO_2 concentrations. The net result was that crocodilians and palm trees thrived at the polar regions. Thus, the geological record is clear that failure to address climate change and permitting average temperatures to rise will lead to lands unsuitable for large scale agriculture and a drastic reduction in the biodiversity of the planet. Therefore, the moral imperative is to mitigate the probability of this result. The United Nations' International Panel on Climate Change (IPCC) reported that global temperatures will likely rise to 1.5 degrees Celsius above pre-industrial levels in the time interval of 2030 and 2052 if GG induced warming continues at the current rate [28]. The Paris Agreement, in which all countries agreed to cooperate in order to limit average global temperature increases to between 1.5 and 2 degrees Celsius above pre-industrial levels [29]. Even if this goal is reached, climate change will have a significant impact on global ecology. Restated, it is not debatable if climate change will occur but rather that of the rapidity and the severity of the ongoing climate change.

Climate change is already having impacts. A review article stated that there is a 97% consensus within published climate research that is robust and consistent with other surveys of climate scientists and peer-reviewed studies [30]. Global weather patterns are changing such that there is less precipitation in the Western United States and greater precipitation in the Midwest [31]. Other climate change effects may be summarized as melting polar ice caps and glaciers leading to coastal flooding, loss of biodiversity, and a redistribution of species such that some are becoming extinct while others, (e.g. invasive pests and disease carrying vectors), are expanding their range [32]. All these data indicate that the earth is unwell and is suffering from the disease of global warming that has already induced permanent changes to the global environment and portends greater degrees of climate change.

4. Medical model of disease

One approach to understanding climate change is that of the medical model of disease. This model assesses how several risk factors and causative triggers interact to produce a "disease" characterized by specific pathology that presents with a combination of symptoms and signs that help establish a diagnosis and suggest potential treatments [33]. This has been greatly enhanced using technology such as laboratory testing, imaging studies and genetic analyses. Agusti opines that this approach is more applicable to acute disease rather than chronic disease as chronic diseases tend to induce secondary effects and produce additional comorbidities leading to ever increasing adverse impacts on the afflicted organism. The chronic disease model integrates more risk factors and triggers that interact, (*e.g.* aging and smoking), which induce damage to several organ systems concurrently, (*e.g.* cardiovascular and respiratory) [34]. Type two diabetes mellitus, a disease that results in elevated blood sugar, provides an example. This disease occurs from inadequate insulin production and/or insulin resistance that often leads to multiorgan dysfunction. These comorbidities include large vessel complications such as stroke and cardiovascular disease as well as microvascular complications such as nephropathy, retinopathy, and polyneuropathy. The net result is a significantly reduced quality of life, wherein the person survives but in a compromised state. To avoid these complications, the disease process must be identified, confirmed as a diagnosis, and treated as soon as practical. In the early stages of type two diabetes, the body can tolerate some excess glucose, but there is a threshold beyond which compensatory mechanisms fail and hyperglycemia begins to exert its deleterious effects. In this analogy, global temperature rise is equivalent to increasing blood sugar levels that must be identified and reduced before long-term complications arise. Restated, it is better to recognize and treat an asymptomatic person with an elevated blood sugar with oral medications than to begin aggressive high dose insulin when that person is admitted to an intensive care unit with a hyperglycemic coma. While an asymptomatic person may deny their disease's existence despite laboratory testing, it is only a matter of time before the consequences are self-evident such that denial is no longer possible. However, a person's willingness to accept a medical diagnosis, seek treatment and adhere to that treatment is highly variable between individuals and reflect cultural background, education, and acceptance of scientific principles.

George Engel MD described the biopsychosocial model of disease in which social and psychological factors have a significant impact on disease development and management [35]. For example, the Pima people in the Sonoran Desert region have one of the highest rates of diabetes on the planet [36]. Their ancestors had to adapt to an environment where nutrients were scare such that evolution favored the survival of those individuals who could extract the most calories from limited food sources. Human evolutionary biology has prioritized calorie dense foods as their consumption favored survival and subsequent reproduction such that humans will consume these foods preferentially [37]. The Pima people, who did not live in an area conducive to intense agriculture, consumed whatever calories were available to endure episodes of relative famine, especially calorie dense foods such as animal fat. This ancient adaptive strategy became maladaptive once these people adopted a Western diet high in processed sugar and saturated fat that induced an exponential increase in the incidence of diabetes mellitus. For the Pima people, education strategies designed to limit the consumption of high calorie foods is more likely to be successful than insisting that they return to their ancestral diet. In this analogy, the modern world is unlikely to be willing to return to a pre-industrial state. Few people would be willing to eschew modern conveniences such as electricity, indoor climate control, and internal combustion engine modes of transportation. Therefore, any viable solution to climate change must integrate the realities of the Western standard of living. Just as it would not be feasible for the Pima people to resume their Pre-Colombian lifestyle, it would not be feasible to return to a pre-industrial civilization.

5. Origins of human civilization

As humans evolved from primates, brain development eventually resulted in language and the ability for critical thinking. Our evolutionary history likely began as small bands or tribes of hunter-gatherers in which the individual was required to subordinate his or her immediate desires to that of the tribe's overall benefit. For example, a hunter who successfully killed an animal would benefit from its sole consumption, but that hunter's long-term survival would be threatened if the consequences were expulsion from that group. This struggle between critical thinking that resides in the cortex that accepts delayed gratification and emotional behavior residing in the limbic system favoring instant gratification has been traditionally viewed as the struggle between good and evil [38]. Therefore, it became important to define "good" or ethical behavior from "bad" or unethical behavior. Behaviors that enhanced tribal survival were more likely viewed as "good," "ethical" or "normative" thus necessitating the suppression of contrary behaviors. "Proper behavior" included deference to the tribe and the forces of nature, as natural phenomena were understood as the vicissitudes of arbitrary spirits or deities. Pre-agricultural societies such as the San people of the Kalahari Desert tended to have an egalitarian culture and a belief system that can be generally characterized as a struggle between good and evil. In such societies, subordination of the individual to the needs of the group was more likely to insure survival than individual efforts. Gender parity likely existed in which women performed the child rearing and gathering while men provided meat derived from hunting that was not invariably successful. This arrangement optimized child survival given the relatively prolonged time required for human development and dependency on others for survival. Given that such tribes were nomadic, few material possessions would be accumulated such that there were fewer disparities between the richest and poorest members of a tribe. However, this remains speculative as this occurred before recorded history.

The agricultural revolution altered this dynamic in that permanent city states arose in which food production was more reliable and a nomadic lifestyle was no longer required. Hierarchies arose in which there was a stricter division of labor, greater wealth accumulation and a need to defend the city state from neighboring tribes. This created a need for a warrior caste governed by a monarch and supported by a religious order designed to enforce the monarch's will and placate temperamental deities. This political, religious, and military aristocracy would value its members and its offspring over others. The subjugation of women likely followed as it was important for an aristocratic male to ensure that his offspring were his own such that his offspring would inherit his accumulated wealth and social position. Thus, rigid rules regarding women's roles and their sexual behavior were strictly enforced. Since few labor-saving devices existed, the society's survival depended on manual labor. A shortfall in labor was likely met through indentured servants or slaves. Slavery has been widespread throughout recorded human history and did not start to diminish until after the start of the Industrial Revolution and the invention of labor-saving devices. Slaves were supplied by persons unable to pay their debts, birth into a slave family, child abandonment, war, or punishment for a crime [39]. The monarch's authority was absolute and bolstered by a religious order that would threaten divine retribution for failure to comply. In exchange, the subjects of the city-state would be protected against outsiders and the vicissitudes of temperamental deities. Expulsion from the city-state could lead to a reduced probability of survival or death at the hands of a hostile tribe. Given these alternatives and the pressure of adhere to societal norms, most subjects acquiesced to this reality. Adherence to authority, temporal and divine, was integrated into a belief system, expressed as morality, and codified into law. Although greater democratic participation has evolved, humanity still functions within a hierarchy ruled by some combination of politicians, religious leaders and the nobility.

The philosopher Thomas Hobbes articulated these concepts when he advocated for a strong centralized authority that would prevent the expression of baser instincts in which people would stop at nothing to further their own interests including theft and murder [40]. He famously opined that existence outside a society without a rigid authority would be "nasty, brutish and short." As the son of a clergyman, Hobbes was likely familiar with the Bible and its emphasis on obedience to higher authorities. One particular passage from Genesis 1:26 has been translated as follows, "Let us make mankind in our image, in our likeness, so that they may rule over the fish in the sea and the birds in the sky, over the livestock and all the wild animals, and over all the creatures that move along the ground" [41]. One interpretation of this passage is that humans are the "in group" and that nature is available for exploitation. A more sanguine interpretation is that humans were commanded to be stewards of the earth's resources. Nonetheless, with few exceptions, modern history has been one of nature's exploitation out of proportion to conservation. If the worst outcomes of climate change are to be avoided, then it will be necessary to develop a belief system or moral code that respects nature and embraces science while accepting the reality of the political and religious foundations of modern society.

6. Ethical constructs

The Merriam Webster dictionary defines ethics as "the principles of conduct governing an individual or a group" [42]. Thus, ethics may be viewed as behavioral guidelines designed to enhance the survival of the individual and the group to which that person belongs. Implicit in this definition is that there are those within the group and those that are external to the group. Thus, the size of the "group" may range from one individual to all life forms on the planet. It is how those external to the group are treated that sets the stage for conflict. While conflict is inherent in any instance of resource scarcity, it is how conflict is resolved that determines the outcome. Many leaders invoke morality to bolster their position in such conflicts. Nonetheless, conflict resolution can be achieved by five different methods: avoidance, competition, accommodation, collaboration, and compromise [43]. These potential solutions to climate change are discussed in the penultimate section of this chapter.

Four major ethical theories discussed in the literature may be summarized as utilitarianism, deontology, virtue, and relativism [44]. Utilitarianism attempts to maximize benefit and minimize harm to all stakeholders involved. In such a paradigm, decisions are made without consideration of the costs involved. Utilitarianism would advocate the same standard of living across the globe irrespective of the economic impact. Deontology focuses on rules that distinguish "right from wrong." Deontology tends to be rather rigid in that it focuses on adhering to rules without appreciating nuance [45]. Immanuel Kant promulgated this approach in which people are morally obligated to act in accordance with a certain set of principles and rules independent of the outcome [46]. Although many religious leaders are deontologists as they promulgate adherence to divine authority, belief in a deity is not required. One variant of deontology, Natural Law, opines that there is an order to human behavior that can be deduced independent of religious or secular authorities [47]. The contrast between deontology and utilitarianism would be apparent in a situation in which a homeowner is harboring 30 refugees illegally and is confronted by the police. Utilitarianism would dictate that the owner should lie as this would protect 30 people whereas a deontologist would insist that the owner should follow the law of the land and tell the truth even if this adversely affected the refugees.

The other two major theories are virtue and relativism. Virtue is an ethical framework that evaluates a person's overall character as opposed to their actions. When questionable behavior is observed, the virtue theory requires that the person's past actions and temperament be taking into consideration when evaluating the act. For example, if a person is known as a mild mannered, temperate and a pillar of the community who embezzles money then the act needs to be evaluated in the context of prior behavior. Virtue based theory would recommend greater leniency for this person as opposed to someone who had a reputation as a scofflaw. Relativism opines that moral obligations and beliefs tend to be based on the environment and that acts need to be judged within that context. Thus, a relativist would not categorically condemn cannibalism as this may be an accepted practice in some cultures.

Beauchamp and Childress discuss a different framework in which they promulgate the four ethical pillars of autonomy, beneficence, non-malfeasance, and justice [48]. They opine that these must be taken into consideration when faced with a moral dilemma. Autonomy expresses the concept that an affected individual has the right to make decisions that directly impact them. For example, autonomy dictates that a person should be able to act in accordance with their religious beliefs and refuse a blood transfusion even if this decision could result in death. Beneficence implies that a decision should always be based on achieving a good outcome whereas non-malfeasance is a requirement to minimize harm as epitomized in the Latin expression *primum non nocere*. Justice implies that one is obligated to treat all stakeholders fairly. In health care ethics, this can be subdivided into three categories: fair distribution of scarce resources (distributive justice), respect for people's rights (rights-based justice) and respect for morally acceptable laws (legal justice) [49]. Alperovitch et al. described an alternative view in which there are two elements of justice, namely equality and equity [50].

It should be evident from this brief discussion that each of these ethical theories have advantages and disadvantages such that no one theory is always superior when faced with a moral dilemma. While the four ethical pillars allow for autonomy, an apocryphal quote often attributed to United States Justice Oliver Wendell Holmes states, "Your right to swing a punch ends at the bridge of my nose." The earth is a closed system such that an individual cannot act in isolation as one individual's consumption of resources will have climate impacts. Moreover, the right to assert "climate denial" is untenable given the overwhelming scientific evidence of climate change. Nonetheless, autonomy should not be excluded altogether. One such approach is a carbon tax in which a person or a corporation can use more than their fair share of fossil fuels but would pay a premium to do so [51]. A person would be able to make choices although some of these choices would be economically prohibitive.

7. Climate change solutions

Can humanity adapt its social norms and integrate science into a solution? The late cosmologist, Professor Stephen Hawking opined that, "There is a fundamental difference between religion, which is based on authority, [and] science, which is based on observation and reason. Science will win because it works" [52]. However, religion and a belief in supernatural phenomena predates recorded human history. It is unlikely that humans would reject millions of years of belief in supernatural phenomena and suddenly embrace science. Moreover, even humans who possess a high degree of scientific literacy do not make decisions solely on scientific principles as political and religious backgrounds factor into these decisions such that those persons with more hierarchical and individualistic worldviews rated climate risk significantly lower [53]. Therefore, it is important for scientists to align with religious and political leaders in order to meet the challenge of climate change. The challenge is to convince humans of diverse political backgrounds, cultures, and religions to overcome tribalism, accept that climate change is a crisis and act in accordance with scientific principles to address its most deleterious effects. From a scientific perspective the options are GG removal from the system, decreased production, and sequestration.

Returning to the diabetes analogy, excess blood sugar can be addressed by excreting it from the body, decreasing glucose production (e.g. consume fewer calories) and sequestration in which it is stored in an unusable form in the body. Indeed, a comprehensive treatment for Type II diabetes usually involves weight loss from decreased calorie consumption, medications to store glucose within the body and, in some cases, medicines designed to facilitate glucose excretion. In this analogy, the earth's disease is a rising average temperature due to excess GG production, predominantly CO_2. As an overview, the main solutions are removal of atmospheric CO_2, (e.g. send it into outer space), sequestration, and reduced production of GG.

Removal involves sending greenhouses gases out of earth's orbit never to return whereas sequestration involves converting atmospheric GG into a different form, (e.g. pumping underground or storing in a liquid or solid form). While theoretically possible, pumping GG out of the atmosphere would require building a pipe in the form of a space elevator up to 53,000 km, (circa 33,550 miles), an altitude wherein these GG would be at escape velocity [54]. Using rockets to remove GG is impractical given the economic costs and relatively limited payloads in addition to the possibility that rocket launches might actually result in a net addition of GG to the environment [55]. Thus, removal is not practical as it is cost prohibitive, technologically challenging and may be counterproductive.

CO_2 sequestration involves capturing and storing atmospheric carbon dioxide of which there are two main methods: geologic and biologic [56]. This government source states that, "Geologic carbon sequestration is the process of storing CO_2 in underground geologic formations. The CO_2 is usually pressurized until it becomes a liquid, and then it is injected into porous rock formations in geologic basins. This method of carbon storage is also sometimes a part of enhanced oil recovery, otherwise known as tertiary recovery, because it is typically used later in the life of a producing oil well. In enhanced oil recovery, the liquid CO_2 is injected into the oil-bearing formation in order to reduce the viscosity of the oil and allow it to flow more easily to the oil well" [57]. The United States Geological Survey estimated that 2,400 to 3,700 metric gigatons of CO_2 could be stored by this method [58]. Nonetheless, this method has its limitations in that during calendar year 2017, the United States produced 5.1 metric giga tons of energy-related carbon dioxide, while the global emissions of energy-related carbon dioxide totaled 32.5 metric gigatons [59]. Using the lower estimate, the United States could store approximately 74 years of global CO_2 emissions at 2017 production levels. Objections to this method include the possibility of inducing seismic activity and contaminating drinking water although the United States' Environmental Protection Agency has proposed mitigation strategies [60]. There are other geologic CO_2 sequestration methods that may be used to sequester CO_2, but they are beyond the scope of this discussion. Nonetheless, such a strategy would only be temporary such that longer term strategies are required.

Biological carbon sequestration is the storage of CO_2 in vegetation such as grasslands or forests, as well as in soils and oceans [61]. Animals, including insects, also contribute to the planetary biomass. The oceans absorb 30% of annual CO_2 emissions, which has mitigated the full effect of GG emissions at the expense of acidifying the oceans from an historic pH of 8.2 to a current pH of 8.1 [62]. This increased ocean acidity impairs the ability of shell-forming marine life to survive including some of the microscopic plankton that forms the base of the marine food chain. Coral reefs comprise less than 1% of the ocean floor yet support over 25% of all known marine species and provide food to over one billion people [63]. Increasing ocean temperatures and acidification portends an increasing probability of this fragile ecosystem's collapse [64]. Biological sequestration is also problematic as a recent publication indicated a greater number of biological consumers than producers [65]. These processes could eventually lead to the collapse of the food chain and mass starvation. Therefore, increasing biomass will likely be only a small part of any climate change solution.

Of these, the most effective approach is to reduce the production of GG as part of an integrated strategy as proposed by Project Drawdown. Project Drawdown's mission statement is to help the world reach "Drawdown"-"the the point in the future when levels of greenhouse gases in the atmosphere stop climbing and start to steadily decline, thereby stopping catastrophic climate change — as quickly, safely, and equitably as possible" [66]. While an integrated approach using all modalities available is logical, any viable solution must minimize the impact of future climate change by a significant reduction in the production of GG. Notably, the wealthiest 10% of the planet produces nearly half of GG emissions whereas the people in the lowest half of global income produce only 10% of these gases [67]. In 2014, the top CO_2 emitters comprising 70% of all emissions were China (30% of total), the United States (15% of total), the European Union (9% of total), India (7% of total), the Russian Federation (5% of total), and Japan (4% of total) [68]. Thus, the United States and China were responsible for nearly half of all CO_2 emissions. While the entire earth is vulnerable to climate change, the greatest impacts are likely to occur in countries that were not major contributors to GG production. The countries most vulnerable to climate change often have the greatest degree of population growth and relatively lower educational levels in their populace. These socioeconomic disadvantages are risk factors for extremism that increases the probability of violence and reduces the probability of collaboration [69].

8. Ethical solutions for climate change

It should be clear that climate change is underway and that failure to act will have catastrophic consequences. While each ethical approach has inherent advantages and disadvantages, utilitarianism and deontology are not viable ethical constructs for this issue. Economic realities are such that it is highly unlikely that a universal standard of living acceptable to all humanity can be achieved in the current political climate. Deontology is unlikely to be a successful strategy as humanity does not subscribe to one religion much less one set of "divinely" or "naturally" inspired moral principles. Given the general lack of scientific literacy and current anti-science movements, it is unlikely that humanity will embrace science over belief systems that predate recorded history. Humanity must accept acknowledge its history, recognize that there is an ongoing disease, reject maladaptive behaviors and embrace a new paradigm that will enhance our survival as a species. As the late Japanese athlete Morihei Ueshiba stated, "Each and every master, regardless of the era or the place, heard the call and attained harmony with heaven and earth. There are many paths leading to the top of Mount Fuji, but there is only one summit – love" [70]. Humanity must embrace one concept of relativism that that there are many belief systems but only one summit, which is that of mitigating climate change. Humanity needs to focus on this common goal for failure to act may endanger our future survival. We can opt to exercise individual autonomy and elect political leaders who are committed to embracing science and addressing the current climate crisis. These political leaders should partner with religious authorities to encourage all their constituencies to act in a manner congruent with mitigating climate change for the betterment of all.

Humanity must also reduce its per capita resource consumption so that our use of the earth's resources does not exceed its regenerative capacity. This can be realized either by reducing the amount of resources consumed per person or population reduction with the same average rate of resource consumption. While the concepts of beneficence and justice would assess these approaches as equivalent, non-malfeasance would favor a voluntary population reduction as this would not require lessening the standard of living for some to achieve global economic parity. Population reduction can be achieved through education and gender equality as educated women tend to have fewer children [71]. This article also cited Project Drawdown, which listed potential solutions to mitigating climate change, including an estimate that educating girls and securing women's voluntary right to high-quality family planning together could reduce atmospheric carbon dioxide by 85 gigatons, making this one of the most powerful solutions to climate change [72]. Thus, to be concordant with Beauchamp's and Childress' four ethical pillars, gender equality must be achieved. At the present time, no country has achieved gender equality although the top 10 counties include those in Northwestern Europe, New Zealand, the Philippines, and Nicaragua [73]. Implicit in women's rights and gender equality is the worldwide elimination of child labor and slavery, which adversely affects both genders but disproportionately affects females [74].

As stated in the introduction, autonomy is of less importance than the other ethical principles when global solutions are involved yet autonomy remains important. The author CS Lewis remarked that "Integrity is doing the right thing even when no one is watching" [75]. We are all empowered to make choices everyday including decisions regarding recycling, public versus private transportation, resource consumption and family planning. Minding the science underlying climate change and the other three ethical principles should guide us in making proper individual choices that minimize our impact on climate change.

9. Climate change conflict resolution

As previously stated, conflict resolution can be achieved by five different methods; avoidance, competition, accommodation, collaboration, and compromise.

The scientific evidence is such that avoidance is not possible. Climate change is underway such that a rise in global temperature will occur even if all GG production ceased immediately. Competition does not necessarily imply an adversarial approach as a friendly competition among nations to reduce GG production would be beneficial. A contrary approach is one in which warfare is used to compete over scarce resources, which is an event that would likely add more GG and result in greater degrees of environmental degradation. Accommodation to the new normal might be possible for wealthier nations but is probably not viable for those nations most vulnerable to climate change. Of these options, collaboration and compromise provide the best pathways forward for climate change once most of humanity recognizes the scope of the problem. Adherence to the four ethical pillars will favor solutions based on collaboration and compromise. Given human nature, the author hopes for worldwide collaboration and compromise but fears that failure to act in the early stages of this disease will lead to global warfare and unprecedented destruction.

10. Conclusions

Climate change is underway. It was induced by human activity that commenced with the Industrial Revolution. The medical disease model frames climate change as the downstream effects of a rise in global temperatures caused by humanity's overconsumption of resources and the production of GG. If the earth is viewed as the human body, then humanity can reduce the amount of toxin production or behave like a metastatic cancer consuming all desired resources, which will eventually result in the organism's death and the demise of the cancer as well. Those who created the problem are obligated to address it. Humanity must acknowledge the current climate crisis, agree on a factual framework, and identify viable solutions. Irrespective of the ethical framework utilized, treatment of this disease process requires the application of the four pillars of autonomy, beneficence, non-malfeasance, and justice. Thus, everyone must act to reduce individual resource consumption and the production of GG. It is unlikely that these goals will be realized without gender equality that will attenuate human population growth and its associated rate of resource consumption. Project Drawdown has proposed a roadmap for addressing climate change. It remains to be seen if humanity will rise to this challenge.

Author details

Daniel Londyn Menkes
Department of Neurology, Oakland University William Beaumont School of Medicine, Royal Oak, Michigan, USA

*Address all correspondence to: daniel.menkes@beaumont.edu

References

[1] National Aeronautics and Space Administration. Overview: Weather, Global Warming and Climate Change. Retrieved from https://climate.nasa.gov/resources/global-warming-vs-climate-change/

[2] Shourie A. Spirit Button. Retrieved from: https://www.goodreads.com/author/quotes/15197943.Anurag_Shourie

[3] Intergovernmental Panel on Climate Change. Retrieved from https://www.ipcc.ch/

[4] Potsdam Institute for Climate Impact Research (PIK). "Four-degrees briefing for the World Bank: The risks of a future without climate policy." ScienceDaily. ScienceDaily, 19 November 2012. www.sciencedaily.com/releases/2012/11/121119104842.htm

[5] National Aeronautics and Space Administration [NASA]. Scientific Consensus: Earth's Climate is Warming. Retrieved from https://climate.nasa.gov/scientific-consensus/

[6] Whitcomb I. Has the Earth ever been this hot before? Retrieved from https://www.livescience.com/65927-has-earth-been-this-hot-before.html

[7] Darwin C. On the Origin of Species by Means of Natural Selection or the Preservation of Favoured Races in the Struggle for Life. New York: D. Appleton and Company, 1859.

[8] Natural History Museum. How An Asteroid Ended the Age of the Dinosaurs. Retrieved from https://www.nhm.ac.uk/discover/how-an-asteroid-caused-extinction-of-dinosaurs.html

[9] TheAmplituhedron. What is the relationship between covalent bonding and the greenhouse effect? Retrieved from What is the relation between covalent bonding and the greenhouse effect? (theamplituhedron.com)

[10] Global Footprint Network. Ecological Footprint. Retrieved from https://www.footprintnetwork.org/our-work/ecological-footprint/#:~:text=Today%20humanity%20uses%20the%20equivalent,we%20use%20in%20a%20year.

[11] The United States Army Survival Manual. Department of the Army Field Manual 21-76. Berkley, CA: Ulysses Press, 2009.

[12] Kolbert E. The Sixth Extinction: An Unnatural History. New York: Henry Holt and Company, 2014.

[13] Wallace-Wells D. The Uninhabitable Earth: Life After Warming. New York: Tom Duggan Books, 2019.

[14] Young R, Mitchell J. Humans Have 30 Years To Stave Off Climate Catastrophe, 'Uninhabitable Earth' Author Says. Retrieved from https://www.wbur.org/hereandnow/2019/05/13/climate-change-uninhabitable-earth-david-wallace-wells

[15] National Center for Atmospheric Research. Retrieved from https://scied.ucar.edu/longcontent/greenhouse-effect

[16] Greenhouse Gases. Redefining the role of CO2. Retrieved from https://www.giss.nasa.gov/research/briefs/ma_01/#:~:text=Without%20naturally%20occurring%20greenhouse%20gases,naturally%20over%20geological%20time%20scales.

[17] Greenhouse Effect 101. Natural Resources Defense Council. Retrieved from https://www.nrdc.org/stories/greenhouse-effect-101

[18] CO2 equivalents. Climate change connections. Retrieved from https://climatechangeconnection.org/emissions/co2-equivalents/

[19] CO2 equivalents. Climate Change Connection. Retrieved from https://climatechangeconnection.org/emissions/co2-equivalents/

[20] What is Global Warming Potential? Retrieved from https://ww2.arb.ca.gov/resources/documents/high-gwp-refrigerants#:~:text=Global%20Warming%20Potential%2C%20or%20GWP,destructive%20a%20climate%20pollutant%20is.&text=The%20most%20common%20refrigerant%20today,a%20ton%20of%20carbon%20dioxide.

[21] Hansen K. Water Vapor Confirmed as Major Player in Climate Change. Retrieved from https://www.nasa.gov/topics/earth/features/vapor_warming.html#:~:text=%E2%80%9CThis%20new%20data%20set%20shows,atmospheric%20humidity%2C%E2%80%9D%20Dessler%20said.&text=And%20since%20water%20vapor%20is,the%20warming%20from%20carbon%20dioxide.%22

[22] Frank J. Explaining how the water vapor greenhouse effects works. Skeptical Science. Retrieved from https://skepticalscience.com/water-vapor-greenhouse-gas.htm

[23] Albedo. Retrieved from https://climate.ncsu.edu/edu/Albedo

[24] The Climate Institute. Oceans and sea level rise. (n.d.) Retrieved from http://www.climate.org/topics/sea-level/index.html

[25] Tebaldi, C., Strauss, BH, Zervas, CE (2012). Modelling sea level rise impacts on storm surges along US coasts. Environmental Research Letters, 7(1):014032.

[26] Lambeck K. Dutton A. Ice Volume and Sea Level during the last Interglacial. Science 2012; 337:216-219.

[27] Michon S, Lindsey R. What's the Hottest that Earth's ever Been? Retrieved from https://www.climate.gov/news-features/climate-qa/whats-hottest-earths-ever-been

[28] United Nations International Panel on Climate Change. Retrieved from https://www.ipcc.ch/

[29] Paris Agreement international treaty [2015]. Retrieved from https://www.britannica.com/topic/Paris-Agreement-2015

[30] Cook J, Oreskes N, Doran PT et al. Consensus on consensus: a synthesis of consensus estimates on human-caused global warming (2016). Environ. Res. Lett. 11 048002

[31] Climate change impacts. National Oceanic and Atmosphere Administration. Retrieved from https://www.noaa.gov/education/resource-collections/climate/climate-change-impacts

[32] National Geographic. Effects of Global Warming. Retrieved from https://www.nationalgeographic.com/environment/global-warming/global-warming-effects/#close

[33] Agusti A. The disease model: implications for clinical practice. European Respiratory Journal 2018 51. Retrieved from https://erj.ersjournals.com/content/51/4/1800188

[34] Tinetti ME, Fried T. The end of the disease era. Am J Med 2004; **116**: 179-185.

[35] Engel, GL. "The need for a new medical model: a challenge for biomedicine" (PDF). *Science* 1977; 196: 129-36.

[36] Schulz LO, Bennett PH, Ravussin E at al. Effects of Traditional and Western Environments on Prevalence of Type 2 Diabetes in Pima Indians in Mexico and the U.S. Diabetes Care 2006; 29(8): 1866-1871.

[37] DeVries R, Morquecho-Campos P, deVet E et al. Human spatial memory implicitly prioritizes high-calorie foods. Retrieved from https://www.nature.com/articles/s41598-020-72570-x

[38] Siyabona Africa. San. Retrieved from http://www.krugerpark.co.za/africa_bushmen.html

[39] Restavek Freedom. The History of Slavery. Retrieved from https://restavekfreedom.org/2018/09/11/the-history-of-slavery/

[40] Hobbes T. Leviathan. Retrieved from https://www.gutenberg.org/files/3207/3207-h/3207-h.htm

[41] Old Testament. Genesis 1:26. Retrieved from https://biblehub.com/genesis/1-26.htm

[42] "Ethics." Merriam-Webster Dictionary. Retrieved from https://www.merriam-webster.com/dictionary/ethics

[43] The Participation Company. 5 Conflict Resolution Strategies that We All Use. Retrieved from https://theparticipationcompany.com/2016/06/5-conflict-resolution-strategies/

[44] Hamilton V. Types of Ethical Theories. Retrieved from https://www.theclassroom.com/types-of-ethical-theories-12083413.html

[45] Ethics Unwrapped. Deontology. Retrieved from https://ethicsunwrapped.utexas.edu/glossary/deontology

[46] Seven pillars Institute for Global Finance and Ethics. Kantian Duty Based Ethics. Retrieved from https://sevenpillarsinstitute.org/kantian-duty-based-deontological-ethics/

[47] Thought company. Natural Law: Definition and Application. Retrieved from https://www.thoughtco.com/natural-law-definition-4776056

[48] Beauchamp TL, Childress JF. Principles of Biomedical Ethics (7th Edition). New York: Oxford University Press, 2013.

[49] Gillon R. Medical ethics: four principles plus attention to scope. BMJ 1994; 309: 184-188.

[50] Alzheimer Europe. Justice. Retrieved from https://www.alzheimer-europe.org/Ethics/Definitions-and-approaches/The-four-common-bioethical-principles/Justice

[51] Amadeo K. Carbon Tax, Its Purpose and How It Works. Retrieved from https://www.thebalance.com/carbon-tax-definition-how-it-works-4158043

[52] Hawking S. Retrieved from https://abcnews.go.com/WN/Technology/stephen-hawking-religion-science-win/story?id=10830164#:~:text=When%20Sawyer%20asked%20if%20there,will%20win%20because%20it%20works.%22

[53] Lombrozo T. Scientific Literacy: It's Not (Just) About The Facts. Retrieved from https://www.npr.org/sections/13.7/2015/09/14/440213603/scientific-literacy-it-s-not-just-about-the-facts

[54] Could we get rid of CO2 if we pumped it through a pipe into space? Retrieved from https://www.askamathematician.com/2019/05/q-could-we-get-rid-of-co2-if-we-pumped-it-through-a-pipe-into-space/

[55] Ross M, Vedda J. Op-ed-Time to clear the air about launch pollution.

Retrieved from https://spacenews.com/op-ed-time-to-clear-the-air-about-launch-pollution/

[56] What is carbon dioxide sequestration? Retrieved from https://www.usgs.gov/faqs/what-carbon-sequestration?qt-news_science_products=0#qt-news_science_products

[57] What's the difference between geologic and biologic carbon sequestration? Retrieved from https://www.usgs.gov/faqs/what-s-difference-between-geologic-and-biologic-carbon-sequestration?qt-news_science_products=0#qt-news_science_products

[58] Has the USGS made any Biologic Carbon Sequestration assessments? Retrieved from https://www.usgs.gov/faqs/has-usgs-made-any-biologic-carbon-sequestration-assessments?qt-news_science_products=0#qt-news_science_products

[59] How much carbon dioxide can the United States store via geologic sequestration? Retrieved from https://www.usgs.gov/faqs/how-much-carbon-dioxide-can-united-states-store-geologic-sequestration?qt-news_science_products=0#qt-news_science_products

[60] Carbon Dioxide Capture and Sequestration: Storage Safety and Security. Retrieved from https://archive.epa.gov/epa/climatechange/carbon-dioxide-capture-and-sequestration-storage-safety-and-security.html

[61] What is Biological Sequestration? Retrieved from https://climatechange.ucdavis.edu/science/carbon-sequestration/biological/

[62] Garcia M. Global Warming and Ocean Acidification Accelerate. Retrieved from https://www.counterpunch.org/2020/07/21/global-warming-and-ocean-acidification-accelerate/

[63] Scripps Institution of Oceanography. Value of Corals. Retrieved from https://sioweb.ucsd.edu/projects/coralreefsystems/about-coral-reefs/value-of-corals/

[64] Anthony KRN, Kline DI, Diaz-Pulido G, Dove S, Hoegh-Guldberg O. Ocean acidification causes bleaching and productivity loss in coral reef builders. PNAS 2008; 105:17442-17446.

[65] Bar-On YM, Phillips R, Milo R. The biomass distribution on earth. Retrieved from https://www.pnas.org/content/pnas/115/25/6506.full.pdf

[66] Project Drawdown. Retrieved from https://drawdown.org/

[67] Garcia M. Global Warming and Ocean Acidification Accelerate. Retrieved from https://www.counterpunch.org/2020/07/21/global-warming-and-ocean-acidification-accelerate/

[68] United States Environmental Protection Agency. Global Greenhouse Gas Emissions Data. Retrieved from https://www.epa.gov/ghgemissions/global-greenhouse-gas-emissions-data

[69] Smith AG. Risk Factors and Indicators Associated With Radicalization to Terrorism in the United States: What Research Sponsored by the National Institute of Justice Tells Us. Retrieved from Risk Factors and Indicators Associated With Radicalization to Terrorism in the United States: What Research Sponsored by the National Institute of Justice Tells Us whitepaper (ncjrs.gov)

[70] Ueshiba M. Retrieved from Morihei Ueshiba quote: There are many paths leading to the top of Mount... (azquotes.com)

[71] Women's rights. Retrieved from https://populationmatters.org/the-facts/womens-rights

[72] Project Drawdown. Retrieved from https://www.drawdown.org/solutions/table-of-solutions

[73] World Economic Forum. The Global Gender Gap Report. Retrieved from http://www3.weforum.org/docs/WEF_GenderGap_Report_2013.pdf

[74] International Labour Organization. Gender and Child Labour in Agriculture. Retrieved from https://www.ilo.org/ipec/areas/Agriculture/WCMS_172261/lang--en/index.htm

[75] Lewis CS. Retrieved from https://www.goalcast.com/2018/03/26/15-c-s-lewis-quotes/c-s-lewis-quote1/#:~:text=C.S.-,Lewis%20Quote%3A%20Integrity%20is%20doing%20the%20right%20thing,when%20no%20one%20is%20watching.

Chapter 17

The Ethical Desirability of Geoengineering: Challenges to Justice

Augustine Pamplany

Abstract

Geoengineering or climate engineering is defined as a deliberate and intentional intervention into the earth system to combat dangerous climate change. Solar Radiation Management (SRM) and Carbon Dioxide Removal (CDR) are two dominant approaches in geoengineering. From an ethical point of view, both these approaches pose serious challenges to justice from the intergenerational, distributive and procedural point of view. Intergenerational equity and the risk-transfer to future generations suggest major challenges to justice in geoengineering. Abdicating our responsibility is a form of injustice to future generations. Unequal distribution of cost and benefits and benefits and harms is a major challenge to distributive justice in SRM. Paying compensation to those harmed by SRM is presented as a way out of ethical deliberations. But there are serious challenges with regard to compensation for SRM, such as, who ought to pay the compensation, who are the beneficiaries and how much to pay. Participation across vulnerable sections alongside indigenous people and their central involvement remains a concern of procedural justice. Food justice is at stake as the adverse impact of SRM on agriculture and food production is considered to be a major challenge.

Keywords: geoengineerig, climate engineering, justice, equity, intergenerational justice, distributive justice, procedural justice

1. Introduction

It is a general convention that developments in technology beginning with the Industrial revolution has been largely responsible for the unabated exploitation of the earth which in turn has produced the dangerous climate change. Ironically, the awareness of the dangers of climate change has attributed a rectificatory mission to technology, whereby technology itself emerges as a potential option to combat climate change. The technology under reference here is geoengineering, also called climate engineering. This technology is still at its conceptual levels. However, if developed and deployed, geoengineering will carry unprecedented levels of planetary outreach as it is to be deployed in the open and non-encapsulated system of the earth. Serious recourse to geoengineering as a possible response to climate change began with the paper by Paul Crutzen [1] in *Climatic Change*. IPCC's assessment report in October 2014 had references to geoengineering. Solar Radiation Management (SRM) and Carbon Dioxide Removal (CDR) are the two major schemes of technologies under geoengineering. SRM aims at the reduction in the amount of sunlight that reach the earth by deploying

2.2 The issue of compensation

Compensating the harms as a condition for ensuring justice is often proposed in geoengineering discussions [4, 7, 13, 15, 26–34]. Preston [10] underscores the provision for compensation to the most affected in the likely scenario of the poor becoming poorer in the aftermath of geoengineering deployment. Even in that regard, the challenges to justice are not adequately addressed. Study shows that SAG coupled with compensation would not be justified, as such a deal would significantly shoot up the cost of SAG [11].

The proposal of compensating for harm is not that smooth a solution as it appears to be. It invites a series of questions. What is the baseline to decide on the definition of harm and compensation? Howe to adjust compensation to the parties who have caused the harms? What will be the moral responsibility of individual nations to various consequences? How to identify the losers and gainers in the absence of clear baselines and standards? [28]. The very case of Canada and Uganda may be taken as an example of the complexities highlighted here. If there is reduction in global temperature due to SRM, Canada's agricultural yield will decline significantly and conversely Uganda's reduction in agricultural production will be due to the decline in precipitation. It can be seen that both these reductions are of different moral standing calling for different standards for calculating compensation. This motivates Bunzl [28] to propose differentiated moral assessment of the harms caused by SRM. He concludes, "[...] it is unfair for some to be worse off than others through no fault of their own among equally deserving people, it follows that it is also unfair for some to be better off than others though no more deserving. But in that case, those who are better off under such circumstances can have no complaint if they lose their better-off status" ([28], p. 73).

Similarly, there are also dormant paradoxes in the seemingly sound ethical assumption of compensation [35]. That there is a possibility for compensating harm cannot be considered as a licence or justification to inflict harm. The general ethical practice of penalising the parties who caused the harm to pay the compensation will make any sense if only the benefits of geoengineering is greater than the costs it incurs. As of now, there is no conclusive evidence to suggest that the benefits will outweigh the harm. Accordingly, the issue of compensation carries an inherent contradiction. Thus there is no justification for the "infliction of all manner of costs onto some purely for the benefit of others,... without any discussion of matters such as rights, justice and responsibility" ([35], p. 7).

Marion Hourdequin [36] thinks that there is every chance of climate injustice being exacerbated if the governance, research and deployment are confined to a very few powerful hands. Monopoly of research and deployment is least compatible with justice. She thinks that only the ideals of solidarity and relationship at the societal and technological levels can ensure justice in this context. Hourdequin [37], shares an optimism that a collective response can ensure distributive justice in the context of technological intervention. Hourdequin [36, 38, 39] has highlighted several major nuances of the issues of justice in geoengineering. McLaren is of the opinion that present risk managerial approach to justice is insufficient in the geoengineering context and we need a "relational, care-based imaginary of the future" ([40], p. 2).

2.3 Distributive justice and food justice

Recently, concerns have also been raised over the dangerous impact of SRM on cultivation, and food production. The consequences of SRM for food justice is to be significantly correlated with the issues of distributive justice in geoengineering [41].

Due to complex relationality between geoengineering and food production, it is normatively obligatory to ensure sufficient and sustainable production of nutritional food before advancing with geoengineering [42].

2.4 Care and virtue ethics perspectives

Concerns about distributive justice in SRM are raised also from the viewpoints of virtue ethics and care ethics. From the perspectives of virtue and care ethics, the assessment is that the principle of fairness will not be respected in the SRM scenario [43, 44].

It appears that from the justice point of view, even researching geoengineering could be like opening a Pandora's Box. It is unequivocally agreed by the parties in the debate that greater research is essential for addressing the issue of distributive justice. With the present range of research that are confined mostly to computer simulation, there can be no definitive judgement on the challenges to distributive justice in geoengineering. Unsurprisingly, the dominant approach in the literature on justice in geoengineering is to see geoengineering as a serious challenge to distributive justice from whichever form of geoengineering, mostly stratospheric aerosol injection. This is not to overlook the nominal voices that argue that geoengineering would present itself as providing positive opportunities for global distributive justice and equity [45, 46].

It could be noted that there are no adequate context-specific studies on the impact of geoengineering on justice. Unfortunately, the debate on distributive justice is extremely polarised towards the analysis of SRM technologies with less attention paid to the distribution of the harms or benefits of CDR approaches. Though the issue of climate justice in relation to anthropogenic climate change is extensively researched (E.g., [47]), most of those researches fall short of addressing the challenges to justice from geoengineering.

3. Intergenerational justice in geoengineering

Geological history shows that there is a global impact for any local climatic intervention. The temporal impact of such interventions cannot also be confined to a particular period. This fact is of particular importance in geoengineering as it is self-evident that the impact of the climatic interventions by this generation will not be confined to this generation. The future generations are naturally brought into the debate on the ethics of geoengineering. This is how intergenerational justice is of decisive value in the geoengineering debate. While distributive justice is challenged by the spatial factors resulting from geoengineering, intergenerational justice is challenged by the temporal imbalances.

The proprieties of distributing harms and benefits between the present and future generations is the focus of intergenerational justice. It assumes that natural resources are not to be entitled unlimitedly to any particular generation. As custodians of natural resources, each generation has to fulfil its obligations to the future generations. It involves the safe custody and preservation of the natural resources for the sustenance of the future generations. This is the reason why intergenerational justice forms a major component in any theory of ethics. It is a happy state of affairs that due importance is given to this principle in international treaties and conventions. There can be no fair treatment of justice in geoengineering without adequately appropriating the challenges to intergenerational justice.

As we discuss below, the contested issues of intergenerational justice in geoengineering revolve around the concerns over the problem of sudden termination of

SAG, questions concerning the agencies of pollution, the challenge of moral hazard caused by the technical interventions, the danger of treating the symptom over the cause, and the present generation transferring the risk to future generations.

3.1 Responsibility of the current generation

The paradoxical issue in intergenerational justice in geoengineering is that future generations are forced to bear the brunt the harms caused by the unnatural ways followed by the current generation. The policies and practices of the present generation concerning development and the consumption of natural resources are largely instrumental in creating a situation of having to geoengineer. However, the effects of geoengineering by this generation will be transferred to the future generations [27, 28, 35, 48–56]. This implies that this generation will reap the benefits by transferring the risks and harms to the future generations. This is often termed as the risk-transfer argument [54] or responsibility abdication objection [57]. A fair practice in this regard would be the polluter-pays principle. This principle, formulated by Betz and Casean [27] assumes that those who caused the dangerous climate changes should also pay for it. Abdicating our responsibility for the dangerous climate change imply that the present generation lets itself off from its offences.

3.2 Moral hazard

One of the major arguments against geoengineering is the challenge of moral hazard – the fear that geoengineering may water down the efforts at mitigation. Royal society coins the phrase "get out of jail free" ([58], p. 276) to mean the same. The ramifications of moral hazard are extensively discussed in the geoengineering debate. Moral hazard is often coined in the insurance context meaning that the security offered by the insurance coverage may trigger the confidence of the insured to venture into riskier activities. Similarly, the true or false hope in the technical solution by geoengineering may alleviate the efforts at mitigation. The assumption that there is a solution to an imminent problem will defer the aggressive measures that may otherwise be warranted in such a scenario. The luxurious life-style of the present generation is largely responsible for the ecological havocs and conservative solutions like change of life-style is called for to fix it. Now, as championed by certain proponents, if geoengineering is economically so feasible, the psychological impetus for a conservative solution naturally withers away. Such a lose commitment to mitigation by this generation means a heavy penalty upon the future generation for something which they are least responsible for.

The possible postulation of a hope of solution leads to avoiding the moral obligations towards climate change by the present generation. As for Gardiner [51] geoengineering is an evasive loophole found by the present generation to skip its moral obligations. As for the present generation, the problem of climate change is less apparent and imposing owing to factors like geographical dispersion of the various and diverse agents and effects of climate change and the pertinent scientific uncertainties about it. These are justificatory weapons of the present generation against its moral obligations. Gardiner [51], p. 408, thinks that climate change is such a problem that "provides each generation with the cover under which it can seem to be taking the problem seriously ... when really it is simply exploiting its temporal position." The vices of the present generation include moral corruption – subversion of the moral discourse to one's own favour – and passing the buck to the future generations. Researching and pursuing geoengineering is an acknowledgement that the present generation has "failed to take on the challenge facing us, and instead have succumbed to moral corruption. Indeed, the decision to geo-engineer might reveal just how far we are prepared to go to avoid confronting climate change directly, and this may constitute a tarnishing, even blighting, evil" ([51], p. 408).

3.3 The termination problem

Termination problem is the possible danger of global temperature bouncing back rapidly if SAG is suddenly terminated. Scientific estimations suggest that if SAG is terminated, there is the possibility of global temperature shooting up faster than the pre-geoengineering phase. This scenario imposes serious restrictions on the choices of the future generation to combat climate change. Most ethicists consider the problem of sudden termination to be the most challenging issue from the point of view of intergenerational justice.

If SRM is discontinued for unforeseen reasons, the worst case scenario is that it could result in the extinction of several species including humans. Svoboda et al. [11] used the theoretical model of Dworkin [21], Rawls [22], Sen [23], and Wigley [24], to assess the issue of intergenerational justice in the likely scenario of sudden termination. They found that in all these models there is a serious violation of intergenerational justice. According to Svoboda et al., "... intergenerational justice requires the present generation to ensure that future generations have access to food, water, shelter, and education.... any generation that implements SAG ...accepts the risk that it might later be discontinued, but the subjects of this risk are the future generations who would suffer the harmful effects if SAG should be discontinued abruptly" (2011, p. 173).

Apart from sudden termination, the long-term deployment of SRM also add to miseries of the future generations. There are scientific estimations predicting that a continuous deployment of around 500 to 1000 years may be required to contain the global warming. It means that the values and priorities of the future generations will be significantly conditioned by the existential challenge of SRM [11].

There are serious methodological limitations in estimating the issues of intergenerational justice in geoengineering. For instance, in the given scientific scenario, it is not clear how many future generations will be impacted by geoengineering and it is impossible to determine whether a future climatic impact is due to geoengineering or due to natural reasons. The identity and population of the future generations are also unknown. Accordingly, scientific uncertainties with regard to geoengineering poses serious hazards in assessing the full scale and length of the concerns with intergenerational justice in geoengineering.

3.4 The governance challenge in intergenerational justice

At this juncture it could also be asked if there are any positive factors in SRM towards facilitating intergenerational justice. After all there are voices claiming that SRM would promote equity as it is capable of avoiding the tragedy of the commons by doing away with the various forms of injustice caused by anthropogenic climate changes. It is also argued that the present generation empowers the future generation to contain the dangers of climate change by SRM [45, 46]. There are arguments that SRM would shield the future generations from otherwise future catastrophe. This is termed as the buying-time argument implying that SRM allows sufficient time for this generation and future generations to combat climate change. Thus proper governance mechanism would ensure intergenerational justice.

This observation, thought seemingly positive, is loaded with major practical challenges. The study by Burns [48] and Svoboda et al. [11] show that even in such scenarios SRM will be incompatible with intergenerational justice. Given the nature of the present international treaties on climate and environment, no law or convention is capable of absorbing the possible complexities posed by SRM ensuring a consensus on the deployment of SRM in a manner compatible with intergenerational justice. Treaties such as UNFCCC (United Nations Framework Convention on Climate Change), ENMOD (United Nations Convention on the Prohibition of Military or Any Other Hostile Use of Environmental Modification Techniques), and CBD (United Nations Convention on Biological Diversity) are not framed for geoengineering and as such they enjoy no comprehensive governance over it. Burns [48] opines that even if UNFCCC may claim certain authority, the lack of political determination will not ensure the just deployment of SRM. Since the very need for SRM is caused by the lack of political will, it cannot be assumed that the same would be present in ensuring justice in its deployment. Although ENMOD permits interventions with environment for peaceful purposes, the limited number of signatories to it does not give a credible mandate for ENMOD over SRM. Similarly, though CBD may be invoked in SRM or CDR, the terms of the CBD have no binding force on the parties as they are only recommendations to parties. The absence of proper governance mechanism seems to confirm that there is no way of deploying geoengineering in a manner compatible with intergenerational justice though some authors tend to think so.

As the critique so far had been around the challenges of SRM to intergenerational justice, one might be inclined to consider CDR to be compatible with intergenerational justice. It is clear that CDR does not invoke concrete problems like sudden termination. At the same time CDR is not freed from the possible moral hazard that it may cause. The moral hazard issue of alleviating the aggressive commitment to mitigation is equally present in CDR projects too. Besides, the required sustained deployment of CDR techniques "would deny them (future generations) the full panoply of options that the principle of intergenerational equity demands" ([48], p. 218). It may be granted positively that on a comparative scale, CDR schemes are not as challenging as the SRM schemes in regard to intergenerational justice. It should be noted alongside, that despite the reduced challenges to justice from CDR, the almost exclusive focus in the debate is on the SRM techniques with very little research being done along the CDR line. A full-blown commitment to the issues of intergenerational justice would require that this strategy needs serious rectification. Burns' [48] formulation that SRM "sows the seeds of a major peril for future generations" ([48], p. 209) may sum up the gist of the discussion on intergenerational justice in the context of geoengineering.

4. Procedural justice in geoengineering

Perhaps what is most rewarding at this stage of the debate on justice in geoengineering is the discussion on procedural justice. It is to the merit of the ethicists that the challenges of procedural justice have been brought to the forefront at the deliberative level itself. As it stands, the discussion on the choice of technologies can be significantly influenced by the concerns with justice.

An untested technology in search of its ethical normativity, but confronted with looming uncertainties about side-effects, will warrant a clear articulation on the procedures towards policy decisions on the choice of technologies, governance mechanism, field tests, etc. Accordingly, the leading issues pertaining to procedural justice coined today are participation and consent, moratorium on field tests, evaluation of the results of technology, security threats stemming from the deployment of technology, etc.

4.1 Consent and participation

The principle of informed consent emphasized by ethicists for the safety of the research subjects is a universal norm in research ethics. It ensures that the subject who voluntarily partake in a research is adequately informed about the risks involved in such a participation and the subject's consent is obtained only after imparting sufficient information. The issue of consent presents itself as the leading contender challenging procedural justice in geoengineering [4, 5, 10, 50, 56, 59]. Preston [10] has rightly identified informed consent to be a formidable challenge in geoengineering at the level of research and deployment. "If the problems of participation and consent first arise in the context of research, there is no doubt whatsoever that their reappearance in the context of implementation is one of the biggest ethical challenges geoengineering faces. As an engineering project promising global impacts, some form of consent—at least from the representatives of those affected—would appear to be a non-negotiable requirement of just procedure" ([10], p. 29).

While informed consent is essentially significant for geoengineering researches, obtaining such a consent is extremely problematic given the complexities involved in geoengineering. The conventional models of informed consent are no longer useful in the geoengineering context. The principle of informed consent meets challenges such as identifying the victims of the research and deployment, the huge number of population who will be affected by the technology, the difficulties of representative consent, etc. The conventional practice of obtaining representative consent look impractical in a technology with global impact.

The solution proposed by Morrow et al. [9], upholding the principle of respect as a motivation towards ensuring consent for geoengineering, does uphold the values of procedural justice. Morrow et al. [9] suggest that "[...] the scientific community secure the global public's consent, voiced through their governmental representatives, before beginning any empirical research [on geoengineering]" ([9], p. 1). This norm does prevent the public from having to accept a policy to which they have given no consent.

4.2 Unilateral deployment and issue of consensus

Another problematic that is anticipated in regard to procedural justice is the issue of a single nation most hit by the dangers of climate change unilaterally deciding to deploy geoengineering in a desperate situation. Even in this regard, procedural challenges cannot be ignored as the impact of the deployed technology is not limited to the nation under consideration. Informed consent cannot be assumed even in a such a desperate scenario [60].

The scope for the unilateral deployment is a central challenge to procedural justice in SAG [27, 45, 61–65]. The leading approach among ethicists is to caution against unilateral deployment.

Assessing procedural justice in geoengineering against the theoretical frame of Rawls does not give nod to research and development. From the Rawlsian point of view of procedural justice, in the present state of affairs with geoengineering, there is no deliberation let alone agreement among all stakeholders and those who would be affected by it. Such a consensus is central to the Rawlsian procedural justice. As such the projected fear about unilateral deployment of SAG should not occur in the Rawlsian context. Analysis by Svoboda et al. [11] showed that these conditions cannot be met in unilateral deployment, particularly as there is no governance mechanism for appeal against SAG.

Denouncing unilateral deployment does not imply that SAG itself is procedurally unjust. Svoboda et al. [11] has opened another unique stream of thought along procedural justice arguing that the unilateral SAG does not make geoengineering in itself procedurally unjust.

4.3 Non-ideal theory of justice

The proponents have recently introduced the notion of non-ideal theory of justice into the geoengineering debate. As the world and its structures are never ideal as it ought to be, it is important to have a realistic approach to geoengineering. It requires considering what matters for justice in circumstances where there is only partial compliance. Accordingly, a non-ideal approach should be taken towards SRM for SRM is a typical instance of the on-ideal theory of justice [4, 13, 66].

Procedural issues in geoengineering should be driven by non-ideal considerations as well. Conversely, recourse to clinical theory, a subset of non-ideal theory, which holds that "politically feasible institutions or policies that would address existing... injustice without violating certain kinds of moral permissibility constraints" ([66], p. 85) is also made in the discussions on procedural justice in the current context.

4.4 Public engagement

Yet another recommendation made towards developing geoengineering researches procedurally just is to treat geoengineering as a public good. The Oxford geoengineering group has proposed the idea of considering geoengineering as a global public good [67, 68]. Treating geoengineering as a public good would imply public participation in decision making process, ensuring transparency and disclosure of research methods, independent assessment of the impacts and developing proper governance mechanisms before deployment. A modified version of the Oxford principles was also endorsed by the Asilomar geoengineering conference in March 2010. Preston [10] observes that "Oxford principles are notable for stipulating that geoengineering should not be driven by profit-raising questions..." ([10], p. 28). Despite the popularity of the Oxford principles among geoengineering ethicists, it has not gone without critical scrutiny. Gardiner has expressed strong reservations against treating geoengineering as a public good as that alone would not suffice for geoengineering to be procedurally just [60]. According to Gardiner, Oxford principles fail to meet the conditions of non-excludability and fairness. Alternately, he proposed the tollgate principle with greater emphasis on fairness, legitimacy and respect [5].

Despite its vital importance, empirical studies showed that justice concerns still remain an under-recognised factor in the response of the public towards geoengineering [69, 70]. Some models of public engagement include upstream public engagement [71], and supermajority rule [72]. The importance of involving the public in decision making process was emphasised by the Royal Society [58] too.

4.5 Principles of beneficence and minimization

Principles of beneficence and minimization [9] are also coined as normative principles in the geoengineering debate to make research and development of geoengineering procedurally just. Principle of beneficence coupled with justice warrants that there should be a "favourable risk–benefit ratio and a fair distribution of risks and anticipated benefits [...]." As the long time span of geoengineering does not permit achieving a favourable risk–benefit ratio, they also advocate the minimisation principle. As the term itself suggests, this principle suggests keeping the extent and intensity of the research and field tests to the minimum. The purpose of minimum intervention is to avoid as much risks as possible. In the absence of "risk-knowledge calculus" [9] informed by scientific input on the risks and benefits, a maximin approach can be normatively helpful. As per the maximin approach population that are most vulnerable to risks and least likely to benefit deserves special attention.

As already discussed in this paper, ethical deliberations in geoengineering are operating against a lot looming uncertainties. Accordingly, the precautionary principle, a tool towards making decisions under uncertainties, finds it natural inroads into the geoengineering debate [73]. Although precautionary principle could provide some useful tips to make it procedurally just, the debate scenario does not provide a consensual opinion on the interpretations of the precautionary principle in the geoengineering debate. Some strong variants of the precautionary principle call for a total ban or moratorium on researches on geoengineering. The weak version emphasises the focuses on avoiding harm in matters of choices under uncertainty and hence an uncompromising approach to harm would be the norm for geoengineering researches too.

The possibility of the research and development being skewed towards military intentions is a major issue that demands proper procedural protocols [74]. The chequered history climate modifications is loaded with such misuse of technology as in the case of Vietnam War. The prevalent terrorist challenges pose maximum procedural caution against the technology being hijacked by ill-intentioned groups [71]. Guarding against such possible aberrations is a necessary condition for advancing procedural justice in geoengineering.

5. Conclusion

This paper tried to analyse the ethical desirability of geoengineering from the point of view justice. The analysis suggests that geoengineering, particularly SAG, conceived in its present format carries serious and almost irreparable damages to justice in its three major variants of distributive, intergenerational and procedural justice. Although the present analysis may seem to go heavily against geoengineering, it could be noted that the ethical desirability of geoengineering is not exclusively confined to the issues of justice. As such the motive here is not to reject geoengineering altogether, rather to motivate the proponents of geoengineering to meet the conditions of justice before researching, developing and deploying geoengineering.

Author details

Augustine Pamplany[1,2]

1 Institute of Science and Religion, Little Flower Seminary, Aluva, India

2 Little Flower Institute of Philosophy and Religion, Aluva, India

*Address all correspondence to: apamplany@gmail.com

References

[1] Crutzen P. Albedo enhancement by stratospheric sulfur injections: A contribution to resolve a policy dilemma? Climatic Change. 2006;**77**:211-220

[2] Adelman S. Geoengineering: Rights, risks and ethics. Journal of Human Rights and the Environment. 2017;**8**(1):119-138

[3] Bunzl, M. (2007). "Geoengineering and the Problem of Comparative Judgements of Harm," *American Geophysical Union* (2007). Available at http://adsabs.harvard.edu/abs/2007AGUFMGC52A.12B. Accessed on February 23, 2015.

[4] Callies D. *Climate Engineering: A Normative Perspective*. Rowman & Littlefield; 2019

[5] Gardiner S, Fragnière A. Geoengineering, political legitimacy and justice. Ethics, Policy and Environment. 2018;**28**(3):265-269

[6] Jones N. Safeguarding against environmental injustice: 1.5 c scenarios, negative emissions, and unintended consequences. Nature. 2018;**631**:635-636

[7] Lambini C. Internalising solar radiation management technological externalities: An ethical review on the design of economic instruments. Advances in Climate Change Research. 2016;**1,** 7:109-112

[8] McLaren D. Whose climate and whose ethics? Conceptions of justice in solar geoengineering modelling. Energy Research & Social Science. 2018;**44**:209-221

[9] Morrow, D., Kopp, R. & Oppenheimer, M. (2009). Toward ethical norms and institutions for climate engineering research. *Environmental Research Letters,* 4(4), 045106. 1-11.

[10] Preston C. Ethics and geoengineering: Reviewing the moral issues raised by solar radiation management and carbon dioxide removal. WIREs Climate Change. 2013;**4**:23-37

[11] Svoboda T, Keller K, Goes M, Tuana N. Sulfate aerosol geoengineering: The question of justice. Public Affairs Quarterly. 2011;**25**:157-179

[12] Svoboda T. Solar radiation management and comparative justice. In: Preston CJ, editor. *Climate Justice and Geoengineering: Ethics and Policy in the Atmospheric Anthropocene*. London: Rowman and Littlefield; 2016. pp. 3-14

[13] Svoboda T. *The Ethics of Climate Engineering: Solar Radiation Management and Non-ideal Justice*. New York: Routledge; 2017

[14] Tuana N, Sriver R, Svoboda T, Olson R, Irvine P, Haqq-Misra J, et al. Towards integrated ethical and scientific analysis of geoengineering: A research agenda. Ethics, Policy & Environment. 2012;**15**(2):136-157

[15] Wong P. Maintenance required: The ethics of geoengineering and post-implementation scenarios. Ethics, Policy & Environment. 2014a;**17**(2):186-191

[16] Wong, P. (2014b). Distributive Justice, Geoengineering and Risks. http://www.geoengineering-governance-research.org/perch/resources/workingpaper17wongdistributivejustice.pdf. Accessed 23 April 2020.

[17] Ray A. Alternative responses to climate change: An inquiry into geoengineering. Stanford Journal of Public Policy. 2010;**1**(1):35-50

[18] Schneider S. Geoengineering: Could we or should we make it

work? Philosophical Transactions of the Royal Society A Mathematical Physical and Engineering Sciences. 2008;**366**(1882):3843-3862

[19] Bunzl, M. (2008). An ethical assessment of geoengineering. *Bulletin of the Atomic Scientists,* 64 (2), 18. Available at https://www.researchgate.net/publication/241458892_Geoengineering_and_the_Problem_of_Comparative_Judgments_of_Harm. Accessed October 10, 2020.

[20] Jamieson D. Ethics and intentional climate change. Climatic Change. 1996;**33**:323-336

[21] Dworkin, R. "What is equality? Part 1: Equality of welfare," *Philosophy and Public Affairs* 10 (1981a): 185-246; "what is equality? Part 2: Equality of resources", Philosophy and Public Affairs 10 (1981): 283-345.

[22] Rawls, John. 1999. *Theory of Justice* (Cambridge, Massachusetts: Harvard University Press, First Published in 1971).

[23] Sen A. Equality of what? In: Sen A, editor. *Choice, Welfare and Measurement*. Cambridge: Cambridge University Press; 1982

[24] Wigley TML. A combined mitigation/geoengineering approach to climate stabilization. Science. 2006;**314**:452-454

[25] Carr W, Preston C. Skewed vulnerabilities and moral corruption in global perspectives on climate engineering. Environmental Values. 2017;**26**(6):757-777

[26] Barrett, S. (2008). The incredible economics of geoengineering. *Environmental and Resource Economics* 39. https://link.springer.com/article/10.1007%2Fs10640-007-9174-8?LI=true. 39, 45-54. Accessed 12 March 2020.

[27] Betz G, Casean S. *Ethical Aspects of Climate Engineering*. Karlsruhe: KIT Scientific Publishing; 2012

[28] Bunzl, M. (2011). Geoengineering harms and compensation. *Stanford Journal of Law, Science and Policy*. https://www-cdn.law.stanford.edu/wp-content/uploads/2018/05/bunzl.pdf. Accessed 12 March 2020.

[29] Gardiner, S. (2007). Is geoengineering the "lesser evil"? http://environmentalresearchweb.org/cws/article/opinion/27600. Accessed 20 February 2020.

[30] Heyward C. Benefiting from climate geoengineering and corresponding remedial duties: The case of unforeseeable harms. Journal of Applied Philosophy. 2014;**31**(4):405-419

[31] Heyward C. 21 normative issues of geoengineering technologies. In: Letcher TM, editor. *Managing Global Warming – An Interface of Technology and Human Issues*. London: Academic Press; 2019. pp. 639-657

[32] Lawford-Smith H, Currie A. Accelerating the carbon cycle: The ethics of enhanced weathering. Biology Letters. 2017;**13**:1-6

[33] Svoboda T, Irvine P. Ethical and technical challenges in compensating for harm due to solar radiation management geoengineering. Ethics, Policy & Environment. 2014;**17**(2):157-174

[34] Svoboda T, Irvine P. Response to commentaries on 'ethical and technical challenges in compensating for harm due to solar radiation management geoengineering'. Ethics, Policy & Environment. 2015;**18**(1):103-105

[35] Gardiner, S. (2013a). The desperation argument for geoengineering. January, *Symposium:* 28-33. https://www.cambridge.org/core/journals/

ps-political-science-and-politics/article/desperation-argument-for-geoengineering/23D9326AEA5756D07C05DA7B24140A86. Accessed 12 March

[36] Hourdequin M. Climate change, climate engineering, and the 'global poor': What does justice require? Ethics, Policy & Environment. 2018;*21*(3):270-288

[37] Hourdequin M. Geoengineering, solidarity, and moral risk. In: Preston CJ, editor. *Engineering the Climate: The Ethics of Solar Radiation Management*. Lanham, MD: Lexington Press; 2012. pp. 15-32

[38] Hourdequin M. The ethics of geoengineering. The Philosophers' Magazine. 2015;(71):44-50

[39] Hourdequin M. Geoengineering justice: The role of recognition. Science, Technology, & Human Values. 2019;*44*(3):448-477

[40] McLaren, D. (2017). *Mirror, mirror: Fairness and justice in climate geoengineering*. Ph.D. Thesis, Lancaster University. https://eprints.lancs.ac.uk/id/eprint/89109/. Accessed October 25, 2020.

[41] Dékány A. Climate justice and geoengineering: Ethics and policy in the atmospheric Anthropocene. Hungarian Geographical Bulletin. 2018;**67**(2):191-193

[42] Kortetmäki T, Oksanen M. Food systems and climate engineering: A plate full of risks or promises? In: Preston CJ, editor. *Climate Justice and Geoengineering*. *London*: Rowman & Littlefield; 2016. pp. *121-135*

[43] Preston C, Carr W. Recognitional justice, climate engineering, and the care approach. Ethics, Policy & Environment. 2018;*21*(3):308-323

[44] Roeser S, Taebi B, Doorn N. Geoengineering the climate and ethical challenges: What we can learn from moral emotions and art. Critical Review of International Social and Political Philosophy. 2019:1-18 https://doi.org/10.1080/13698230.2020.1694225

[45] Michaelson J. Geoengineering: A climate change Manhattan project. Stanford Environmental Law Journal. 1998;**17**:1-53 https://papers.ssrn.com/sol3/papers.cfm?abstract_id=2147499

[46] Svoboda T, Irvine P, Callies D, Sugiyama M. The potential for climate engineering with stratospheric sulfate aerosol injections to reduce climate injustice. Journal of Global Ethics. 2019;*14*(3):353-368

[47] Shue H. The unavoidability of justice. In: Hurrell A, Kingsbury B, editors. *The International Politics of the Environment*. Oxford, UK: Oxford University Press; 1992. pp. 373-397

[48] Burns W. Climate geoengineering: Solar radiation management and its implications for intergenerational equity. In: Burns W, Strauss A, editors. *Climate Change Geoengineering – Philosophical Perspectives, Legal Issues and Governance Frameworks*. Cambridge: Cambridge University Press; 2013. pp. 200-220

[49] Caney S. Just emissions. Philosophy & Public Affairs. 2012;**40**:255-300

[50] Flegal J, Gupta A. Evoking equity as a rationale for solar geoengineering research? Scrutinizing emerging expert visions of equity. Int. Environ. Agreem. Politics Law Econ. 2018;**18**:45-61

[51] Gardiner S. A perfect moral storm: Climate change, intergenerational ethics, and the problem of moral corruption. Environmental Values. 2006;**15**:397-413

[52] Gardiner S. Is 'arming the future' with geoengineering really the lesser evil? Some doubts about the ethics of

intentionally manipulating the climate system. In: Gardiner S, Caney S, Jamieson D, Shue H, editors. *Climate Ethics*. Oxford, New York: Oxford University Press; 2010. pp. 284-313

[53] Gardiner S. *A Perfect Moral Storm: The Ethical Tragedy of Climate Change*. Oxford University Press; 2011

[54] Goes M, Tuana N, Keller K. The economics (or lack thereof) of aerosol geoengineering. Climatic Change. 2011;**109**:719-744

[55] Hale B. The world that would have been: Moral hazard arguments against geoengineering. In: Preston CJ, editor. *Engineering the Climate: The Ethics of Solar Radiation Management*. New York: Lexington Books; 2012. pp. 113-132

[56] McLaren D. Framing out justice: The post-politics of climate engineering discourses. In: C. J. Preston, *Climate Justice and Geoengineering: Ethics and Policy in the Atmospheric Anthopocene*. London: Rowman and Littlefield; 2016. pp. 139-160

[57] Hale B. The world that would have been: Moral Hazard arguments against geoengineering. In: Preston CJ, editor. Engineering *the Climate: The Ethics of Solar Radiation Management*. New York: Lexington Books; 2014. pp. 113-132

[58] Royal Society. (2009). Geoengineering the climate: Science, governance and uncertainty. www.Royalsociety.org. Accessed 13 March 2019.

[59] Whyte K. Indigenous peoples, solar radiation management, and consent. In: Preston CJ, editor. *Engineering the Climate: The Ethics of Solar Radiation Management*. Lanham, MD: Lexington Press; 2012. pp. 65-76

[60] Gardiner S. Why geoengineering is not a 'global public good,' and why it is ethically misleading to frame it as one. Climatic Change. 2013b;**121**:513-525

[61] Banerjee B. The limitations of geoengineering: Governance in a world of uncertainty. Stanford Journal of Law, Science and Policy. 2011;**4**:15-32

[62] Bates J. Climate geoengineering and IWU's ethics bowl. The Intellectual Standard. 2013;**2**(2):15-18

[63] Baum S, Maher T, Haqq-Misra J. Double catastrophe: Intermittent stratospheric geoengineering induced by societal collapse. Environment, Systems and Decisions. 2013;**33**(1):168-180

[64] Parker A, Horton J, Keith D. Stopping solar geoengineering through technical means: A preliminary assessment of counter-geoengineering. Earth's Future. 2018;**6**:1058-1065

[65] Schelling T. The economic diplomacy of geoengineering. Climatic Change. 1996;**33**:303-307

[66] Morrow D, Svoboda T. Geoengineering and non-ideal theory. Public Affairs Quarterly. 2016;*30*(1):83-102

[67] Rayner, S. (2011). Climate Change and Geoengineering Governance. http://www3.ntu.edu.sg/rsis/nts/HTML-Newsletter/Insight/NTS-Insight-jun-1102.html. Accessed 13 March 2019.

[68] Rayner S, Heyward C, Kruger T, Pidgeon N, Redgwell C, Savulescu J. The Oxford principles. Climatic Change. 2013;**121**:499-512

[69] McLaren D, Parkhill K, Corner A, Vaughan N, Pidgeon N. Public conceptions of justice in climate engineering: Evidence from secondary analysis of public deliberation. Global Environmental Change. 2016;*41*:64-73

[70] Parthasarathy S, Avery C, Hedberg N, Mannisto J, Maguire M. *A Public Good? Geoengineering and Intellectual Property*. Work. Pap. 10-1, Sci. Technol. Public Policy Program, Univ. Ann Arbor: Mich; 2010 http://www.homepages.ed.ac.uk/shs/Climatechange/Geo-politics/Chris%20Avery%20patents.pdf. Accessed 23 April 2020

[71] Corner A, Pidgeon N. Geoengineering the climate: The social and ethical implications. Environment. 2010;**52**:24-37

[72] Gramstad, Kjetil., Sigve Tjøtta. 2010. "Climate Engineering: Cost Benefit and Beyond," *Working Papers in Economics*, Department of Economics, University of Bergen, Norway, 23 September. Available at http://mpra.ub.uni-muenchen.de/27302/ MPRA Paper No. 27302, p. 4. Accessed March 9, 2014.

[73] Elliot KC. Geoengineering and the precautionary principle. International Journal of Applied Philosophy. 2010;**24**:237-253

[74] Robock A. 20 reasons why geoengineering may be a bad idea. Bulletin of the Atomic Scientists. 2008;**64**:14-18

Chapter 18

Ethical Considerations for Global Pediatric Cardiac Surgical Assistance Programs

William M. Novick

Abstract

Global health initiatives have expanded over the last 25 years and are no longer based solely on improving public health issues like clean water and childhood vaccination programs. Global healthcare assistance has grown into programs that provide specialty services and education today. Cardiovascular diseases are causing more deaths today in low and middle-income countries today as infectious diseases cease to be the number one cause of mortality in many of these countries. Growth in cardiovascular assistance has been substantial during the last 25 years and especially in the area of pediatric cardiac care. We discuss the ethical issues that can be found when visitors assist countries with different cultural values. The success of program development depends on navigating the ethical issues such that all stakeholders are satisfied with the project and the end result. The foundation of program development should be based upon medical ethics that are sensitive to cultural differences so that a capable sustainable program is developed upon completion.

Keywords: Ethics, short term global surgery programs, cardiac surgery, humanitarian assistance

1. Introduction

Global health programs have substantially increased in number and sophistication over the last 25 years. No longer do we see only primary care programs from upper-income countries (UIC) visiting lower and middle-income countries (LMIC). Today we have specialty global health care initiatives that include pediatrics, pediatric general and cardiac surgery, orthopedic and neurological surgery. Internal medicine programs specifically for endocrine and oncology, as well as singular issues with women's health. The increase in global health initiatives has sparked a discussion in the overall purpose and ethics of these efforts. The altruistic nature of these efforts cannot be overstated, but with the health care deficits in LMIC it is vitally important to consider building capacity at the local level and not fostering a dependency on the visiting team for a higher level of healthcare that is intermittent at best. Cardiovascular diseases are the number one cause of mortality world-wide [1]. However, there are deficiencies in cardiac care in almost 90% of the world, almost all in LMIC [2]. The result is that patients in these areas are poorly served and a significant healthcare in-equality continues to exist. The need to expand cardiac care to these parts of the world must be balanced with a beneficent approach to this problem since there are several ethical issues in providing medical education, training, experience and optimal patient care which can be violated without adequate thought, preparation, and execution. Moreover, legal issues with the importation of medicines, supplies, equipment, and licensure must be considered on a country-by-country basis. Critics of surgical development assistance in general and cardiac specifically abound, some with relevant points. It is incumbent of those assisting to do so within the scope of ethical standards both from the visiting team's country and the country receiving assistance if success is to be assured.

2. Intentions of cardiac care development assistance

The overarching goal of cardiac assistance should be the development of an independent program of cardiac care capable of sustainability. Depending on the local situation such an endeavor may take only 2–3 years for a pre-existing program to as many as 7–10 years if you are building a program *de novo* [3, 4]. The timelines are not firm, as the major determinants of the speed of development are local leadership, visiting team commitment, consistent funding, and a supportive government. The key component to all the above is trust.

Good intentions without trust by all sides will result in a failure of development. Building relationships is critical to the success of advancement and attainment of the primary goal. Therefor it is important from the outset to create an environment that results in the delivery of the promise and goals. You cannot develop relationships and trust if you do not provide the opportunity for the local team to participate and grow through a program of mentored graduated responsibility. Those teams that provide services only and allowing only limited participation by the local team are not upholding the promise and as such trust will not be developed and a rapid disconnect will occur, thus dooming program development. There are few times and places where the local team cannot be developed, and the visiting team simply provides all the care without education [5]. Beneficence cannot be directed at only the children receiving operation at the time of a visit, this is only a part of truly doing good work. One must consider the local team and the children that are still waiting for surgery and those not yet born. The idea of assistance is to build a program that benefits many children for years to come.

3. Implementation of the development program

3.1 Direction and leadership

Before undertaking a program to develop pediatric cardiac services there must be an agreement between all stakeholders regarding what level of development is desired by the local team. Autonomy of choice by the local team is imperative, one should not approach this as *we know what is best for you and your country,* this is a form of neo-colonialism and is to be avoided [6]. You may be experts in pediatric cardiac care, but they are certainly the experts in the subtleties of politics and economics of their countries.

Once a decision is reached regarding the level of expertise the local team wishes to attain it is critical to identify clinical leaders in surgery, critical care, anesthesia, perfusion, and pediatrics. Once again, the autonomy of the local team is important here, they need to decide amongst themselves who will lead the various specialties involved in the care of children with heart disease. There is simply no justification for the visiting team to insist upon certain individuals being named to leadership positions. Such an attempt will lead to a fracture of trust and the development of co-dependency. However, if during the program it becomes obvious that leadership needs to change, then it is incumbent upon the visiting team to address this issue. A failure of leadership is one of the most common factors in program disruption and failure to achieve the desired goals [7].

3.2 Infrastructure issues

One does not often relate ethics and infrastructure, but when considering the creation or improvement of an existing pediatric cardiac program you must be sure that the necessary elements for providing safe pediatric cardiac care are in place or will be before program initiation. The equitable treatment of the children receiving care in LMIC must be maintained within the constraints of economic reality. The deficiency in pediatric cardiac caregivers in LMIC has resulted in large waiting lists and the needs of the many outweigh the needs of the one. Although ECMO, artificial hearts and left ventricular assist devices can provide a few children in upper-income countries (UIC) with a survivable situation, but when costs are considered many more children could be saved in LMIC with a low-risk operation. Basic infrastructure needs for pediatric cardiac care include an echocardiogram machine, an operating theater (not a hybrid OR suite) adequately provided with climate control, oxygen, air and electrical sources in addition to anesthesia machine and routine open heart surgery equipment. A basic intensive care unit with invasive monitors, adequate oxygen and electrical sources and ventilators to provide routine care. The vast majority of children requiring surgical intervention will be adequately served with these essential elements [8]. A cardiac catheterization lab is a luxurious addition to the diagnostic equipment and justification for this is difficult in a number of LMIC. However, the possibility of acquiring a refurbished catheterization laboratory rather than a state-of-the-art device is a means of providing this diagnostic capability and creating an equitable situation. A donor who provides such sophisticated equipment must be prepared to continue to support the maintenance or the equitable situation they created in diagnostic capabilities can be short-lived. There is simply no justification for providing advanced diagnostic equipment, having the local team develop capacity with it, and then having it removed because of a breakdown and an absence of support for repair [9].

3.3 Human resources

Building a team or improving an existing team requires that the visiting team provide individuals that are specialists in pediatric cardiac care and education. Frequently the visiting team will have volunteers including medical students, residents and fellows who are wanting to explore global health initiatives. The team leader of the visiting team must remember that the purpose of the visit is to increase the experience and capacity of the local team. Members of both the visiting and the receiving teams can all benefit from the educational opportunities, but it is incumbent upon the team leader to mentor the recipient team members particularly. A trip to an LMIC is not a place for medical students to learn the intricacies of pediatric cardiac surgery, or cardiology or anesthesia; you are there to teach the local team and they should always be the primary recipient of training [10]. Upper-level residents and fellows should participate in all aspects of care as the children in LMIC are not the same as the infants that they care for at their home institution. The opportunity to see the ravages of chronic congenital heart disease on a child or adolescent is a lesson in natural history for them, but a daily occurrence for the local caregivers. Global cardiac surgical initiatives are first and foremost an opportunity to exchange knowledge in both directions between visitors and the local team.

However, one should not offer to provide or perform operations beyond the scope of your practice at your home institution [11]. Moreover, one must also maintain a sense of what is actually possible not just in the operating theater but, in the intensive care unit as well. Although the surgical expertise may exist to perform a complex operation flawlessly in the operating theater there must be adequate support for the recovery of these patients.

3.4 Medications and materials

Frequently the visiting team will provide donated products to carry out the surgery and care of the children. Several medical product and pharmaceutical companies in UIC support these efforts by providing product that may have a limited expiration date remaining. Additionally, hospitals in UIC often provide products which have had the external wrapping removed but remain in a secondary sterile packet and can be re-sterilized at the local site. Working in resource limited LMIC institutions one must be sure to understand the medical importation regulations before shipping nearly expired products. There are times when a non-governmental organization (NGO) may be offered expired products and medications from hospitals and manufacturers, when is it ethical to use such products? Clearly the expiration dates for medications are arbitrarily set and publications on potency after expiration are available [12, 13]. One must ask is it ethical to use a product or medication in order to provide life-saving cardiac surgery, this is an answer that will vary by site, culture and country. Implicit in this decision is input from the local team. Consultation with the local team over this issue is both respect and autonomy for and of the local team. Moreover, the family should be involved in this decision, as patient autonomy clearly must be preserved.

Regulations for the importation of medications vary by country and similarly drugs which are registered for use vary. One must balance the good for the patient in deciding whether to adhere to local regulations but, beware of the consequences of violating such regulations [14]. A drug as beneficial as milrinone is not registered in several countries that we travel too, but we and our local colleagues know the benefits of this drug in pediatric cardiac surgery. We receive requests routinely to bring milrinone to countries where it is not registered for use and hand-carry sufficient quantities to carry-out the operative list. Clearly one must consider the ethical position of beneficence versus local regulations installed by a slowly moving bureaucracy.

3.5 Patient care

Providing safe, beneficial patient care is the first priority of the visiting team and is an excellent starting point for teaching the local team how to organize a pediatric cardiac care program [15]. We are all aware that it is not unusual for the local team to prepare a list of complex operative interventions for the visitors. Operations that the locals have never seen and certainly never performed are frequently on the list of the patient management conference. Once again it is important not to operate outside of your boundaries because you are not at home. Trust is important in developing relationships and can be eroded by an unwanted outcome as a result of operating outside normal boundaries, whether for the surgeon, perfusionist or ICU team's capabilities. Moreover, the unethical performance of an operation outside the limits of your capabilities can bring unwanted and complicating legal issues to bare as well. Teaching and reinforcing a patient-centric non-maleficence philosophy will lead to the development of a patient first approach by the locals.

Placing patient well-being above all other considerations must be balanced with your educational responsibilities to the local team. We have found that a frank conversation with the local team before the start of the surgical program is beneficial in understanding the local capabilities. A mentored program of having the local team serve as assistants first is patient-centric and provides the locals with an experience in how to perform safe surgery. The transition from assistant to primary surgeon for the local surgeon is graduated based upon the teacher's observations of progress. The principle of the placing the patient's well-being first requires that the teaching surgeon have confidence that the local can provide the operation safely. Obviously if this is not the case then the teacher must take the place of the trainee as needed. Such a switch during an operation, induction of anesthesia or care in the ICU must be done in a diplomatic fashion as not to damage the confidence of the trainee.

One issue faced by all who are involved in assistance development is the presentation of a child with either a critical defect or in heart failure from the chronicity of their defect who needs an urgent operation at the end of a visit. We must consider the issue of non-maleficence and ask ourselves if there is a team prepared to care for the child after surgery that is capable of recovering the child. Agreeing to operate on this child and then have the visiting team depart when the child still needs complex care is providing false hope to the family and ethically questionable. We have faced this problem countless times in our history and have decided that if we are operating late in the trip on critically ill children, we will leave a team of ICU caregivers behind to provide 24-hour coverage until the child is discharged from the ICU. The decision to operate on such a child must be made in concert with the local team and family, autonomy of decision for both is critical for all stakeholders.

4. Database and research needs

4.1 Database

Developing a truly equitable program in an LMIC requires the encouragement of a clinical database and research projects. The benefits of a database need to be clearly communicated to the local team and then they must be left to make the decision to proceed with the establishment of one or not. Again, this highlights the autonomy of the local team and provides them with the opportunity to display complete commitment to program development, rather than simply intermittent surgical mission trips by the visitors. Program growth cannot be judged, and corrective actions taken unless a clinical database of outcomes is implemented. Furthermore, the establishment of a database and routine review and presentation of results provides the Ministry of Health with a realistic view of program growth and development, thus justifying the continuation of Ministry level support. Ideally the site will enroll in an international database that serves LMIC, examples include the IQIC [16] and WSPCHS registries [17]. Databases which only enroll UIC programs can be discouraging when results are compared, and the site should seek to compare itself to similar programs in LMIC.

4.2 Research

Research is to be encouraged when assisting a site, whether it is pure clinical or translational, few LMIC sites will have opportunities for basic science research [18]. The research project requirements for overseas programs you are assisting as the same as those of your home institution. The project should be reviewed by the

Institutional Review Board, or Ethical Research Committee and approved before it is instituted. Moreover, it is necessary if your home institution is involved that approval from the relevant committee there is obtained. One should not be conducting research in LMIC that was not approved by the assisting team's home institution [19]. The claim of *"they are experimenting on our children"* can be a program downfall.

Research fits well with the maintenance of a database registry, providing the local team with an opportunity to publish their progress in development [20]. The presentation and publication of results builds confidence within the local team and is extremely beneficial to overall program development. Alternatively, when the visiting team publishes work that was performed at the assisted site it is beneficial to involve the local team and to invite them to contribute to the publication and therefor co-authorship [21]. Collaborative efforts like this build trust and confidence between the two teams and the benefit is program growth [22].

4.3 Parent interactions

The parents of the child with heart disease who live in LMIC and must deal with the day-by-day question of will their child live to see another day must be treated with utmost respect and clarity. Frequently local caregivers have limited understanding of the risks and complications of surgery and post-operative care and therefore should not be the sole descriptors of these risks [23]. The surgeon and anesthesiologist of the visiting team need to treat these families with the same type of informed consent as they would at home to be equitable [24]. A major difference is that a translator will frequently be needed and should be instructed to translate exactly rather than to interpret the information you are providing. Importantly the translator must be well versed in medical terminology and have an adequate understanding of the procedure and risks so they can provide a basic understanding to the parents. Similarly updates following surgery should be provided with the same level of respect and clarity by the surgeon performing the operation. Updates in the ICU should follow the same course with the Intensivist or nurse of the visiting team providing important information with a translator as needed. Once the local team becomes experienced these updates can be provided by them with the visitors available for questions from the family if needed. During all these talks it is critical that the visiting team members build confidence with the families for the local team. We have found over the years that the best approach is for the visiting and local surgeon are both present during both the pre-operative discussion as well as the post-surgical update. The format works well as the families can hear from both and it is an opportunity to build confidence in the local team in the eyes of the family.

4.4 Communications and media

The presence of a group of foreign experts in pediatric cardiac surgery in a LMIC frequently results in many media requests for access to the visitors, local caregivers, and parents [25]. The goal of the visiting team is to build confidence in the local team during these exchanges [23]. However, it is imperative that expectations be realistic so that the public clearly understands that this is a program in development and that the local team is building experience. Again, the autonomy of the local team is important, and the leaders of the local team must be included in any media interview so that they are viewed in a positive manner by the public. There is no better justification for program continuation than to have parent testimonials, but again this is a parental decision. You may request that they participate to tell their story, but you cannot coerce or demand they do so, this is not ethical, it removes parent autonomy, and the interview will not have the authenticity they need to convey.

4.5 Donors and volunteers

Donors (financial, material) and volunteers (medical, others) both contribute to the success of the program. The expectations of both groups should be known before they participate to avoid disappointment in the outcomes and program funding removed [26]. Donors should be briefed on the reality of the situation on the ground so that their expectations align with those of the visiting team. One should not make promises to donors that cannot be carried out given the local situation, this is disingenuous and not ethical. Conversely it is important to prepare volunteers for the cultural issues of the country being visited. Local medical ethics do not necessarily always coincide with those that the volunteers adhere to at home. A perfect example of this is the compassionate discontinuation of care practiced in many countries. Such medical decisions are not the norm in several countries around the world and it is important to brief the volunteers on this issue before the trip so they can decide if they can adhere to this cultural norm and refrain from inappropriate social media posts [27].

5. Summary

Pediatric cardiac global surgical initiatives have significantly increased in number and coverage over the last 25 years. The benefits to the children, their families and the local healthcare professionals are clear. However, it is possible to create discontent with the program if ethical pillars are violated. The result will be program failure. Providing autonomy for local stakeholders and parents is critical to promote confidence and trust, remembering that our primary concern is the patient and that one must practice non maleficence is important for perceptions within the local community and our first responsibility as physicians. The principle of justice runs throughout the development of pediatric cardiac programs in LMIC, it impacts all aspects from patient care to education of local healthcare professionals to allocation of resources. Following ethical principles will result in an independent locally driven program in pediatric cardiac care if all stakeholders adhere to these principles.

Author details

William M. Novick
Department of Surgery, University of Tennessee Health Science Center, Global Surgery Institute, Memphis, TN, USA

*Address all correspondence to: bill.novick@cardiac-alliance.org

References

[1] 10 Leading causes of death. WHO Factsheet. https://www.who.int/news-room/fact-sheets/detail/the-top-10-causes-of-death. Accessed 20/12/2020.

[2] Neirotti R. Paediatric cardiac surgery in less privileged parts of the world. Cardiol Young. 2004;14:341-6.

[3] Novick WM, Stidham GL, Karl TR, Arnold R, et.al. Paediatric cardiac assistance in developing and transitional countries: the impact of a fourteen-year effort. Cardiol Young 2008;18:316-323.

[4] Polivenok IV, Molloy FJ, Gilbert CL, Danton M, et.al. Results of international assistance for a paediatric heart surgery programme in a single Ukrainian centre. Cardiol Young 2019;29:363-368.

[5] Nwafor IA, Vickram A, Osenmobor KO. Surgical 'Safari' vs. Educational Program: Experience with International Cardiac Surgery Missions in Nigeria. Braz J Cardiovasc Surg. 2020 Dec 1;35:918-926.

[6] Scheiner A, Rickard JL, Nwomeh B, Jawa RS, et.al. Global Surgery Pro-Con Debate: A pathway to bilateral academic success of the bold new face of colonialism? J Surg Res 2020;252:272-280.

[7] Novick WM, Anić D, Ivancan V, Di Sessa TG. International pediatric cardiac assistance in Croatia: results of the 10 year program. Croat Med J. 2004 Aug;45:389-95.

[8] Dearani JA, Jacobs JP, Bolman RM III, Swain JD, Vricella LA, Weinstein S, et al. Humanitarian outreach in cardiothoracic surgery: from setup to sustainability. Ann Thorac Surg. 2016;102:1004-11.

[9] Pecchia L, Pallikarakis N, Magjarevic R, Iadanza E. Health Technology Assessment and Biomedical Engineering: Global trends, gaps and opportunities. Med Eng Phys. 2019;72:19-26.

[10] Bauer I. More harm than good? The questionable ethics of medical volunteering and international student placement. Trop Dis Trav Med Vaccin. 2017;3:5.

[11] Zitzman E, Berkley H, Jindal RM. Accountability in global surgery missions. BMJ Glob Health. 2018;3:e001025.

[12] Cantrell L, Suchard JR, Wu L, Gerona R. Stability of active ingredients in long expired prescription medications. Arch Intern Med. 2012;172:1685-6.

[13] Lyon RC, Taylor JS, Porter DA, Prasanna HR, Hussain AS. Stability profiles of drug products extended beyond labeled expiration dates. J Pharm Sci. 2006;95:1549-60.

[14] Sommanustweechai A, Chanvatik S, Sermsinsiri V, Sivilaikul S, et.al. Antibiotic distribution channels in Thailand: results of key-informant interviews, reviews of drug regulations and database searches. Bull World Health Organ 2018;96(2):101-109.

[15] Fenton KN, Novick WM, Entwistle JW, Moffatt-Bruce SD, et.al. Global health initiatives in cardiothoracic surgery: Ethical considerations and guidelines.

[16] Jenkins KJ, Castaneda AR, Cherian KM, et al. Reducing mortality and infection after congenital heart surgery in the developing world. Pediatrics 2014;134: e1422–e143.

[17] St Louis JD, Kurosawa H, Jonas RA, Sandoval N, et.al. The World Database for Pediatric and Congenital Heart Surgery: The Dawn of a New Era of

Global Communication and Quality Improvement in Congenital Heart Disease. World J Pediatr Congenit Heart Surg. 2017 Sep;8:597-599.

[18] Pratt B, de Vries J. Community engagement in global health research that advances health equity. Bioethics 2018 Sep;32:454-463.

[19] Rees CA, Keating EM, Dearden KA, Haq H, et.al. Improving Pediatric Academic Global Health Collaborative Research and Agenda Setting: A Mixed-Methods Study. Am J Trop Med Hyg 2020 Mar;102:649-657.

[20] Saleem K, Ahmed I, Sultan M, Haq IU, et.al. Bidirectional Glenn for residual outflow obstruction in Tetralogy of Fallot. Cardiol Young. 2019 May;29:684-688.

[21] Nwafor IA, Chinawa JM, Adiele DK, Arodiwe IO, et.al. Management of complex CHD at the National Cardiothoracic Center of Excellence, University of Nigeria Teaching Hospital, Enugu: the role of foreign cardiac missions in 3.5 years. Cardiol Young. 2017 Aug;27(6):1174-1179.

[22] Polivenok IV, Novick WM, Pyetkov AV, Cardarelli M. Perioperative complications in a paediatric cardiac surgery program with limited systemic resources. Cardiol Young 2020 Nov;30(11):1659-1665.

[23] Fenton KN, Molloy F, Novick WM. Ethics in humanitarian efforts: giving due credit to the local team. Cardiol Young. 2019 Feb;29(2):195-199.

[24] Güner MD, Ekmekci PE. A Survey Study Evaluating and Comparing the Health Literacy Knowledge and Communication Skills Used by Nurses and Physicians. Inquiry. Jan-Dec 2019;56:46958019865831.

[25] Anema A, Freifeld CC, Druyts E, Montaner JS, et.al. An assessment of global Internet-based HIV/AIDS media coverage: implications for United Nations Programme on HIV/AIDS' Global Media HIV/AIDS initiative. Int J STD AIDS. 2010 Jan;21(1):26-9.

[26] Lince-Deroche N, Leuner R, Meyer-Rath G, Pillay Y, Long L. When donor funding leaves: an interrupted time-series analysis of the impact of integrating direct HIV care and treatment into public health services in a region of Johannesburg. Cost Eff Resour Alloc 2019;17:24-37.

[27] Tabb Z, Hyle L, Haq H. Pursuit to Post: Ethical issues of social media use by international medical volunteers. Dev World Bioeth 2020 Jul 6. doi: 10.1111/dewb.12267. Online ahead of print.

Section 3

Technology

Chapter 19

Internet of Things and Distributed Denial of Service as Risk Factors in Information Security

Jairo Eduardo Márquez Díaz

Abstract

Society is increasingly dependent on technology and an example of this is the constant monitoring of large cities, which has become common and the future trend is for it to increase based on what happened with the COVID-19 pandemic. This monitoring brings with it a series of problems at the information security level at different levels or levels. Based on this fact, it addresses how the Internet of Things (IoT) can be subject to potential distributed denial of service (DDoS) attacks and the danger it poses to society. In this sense, other types of vulnerabilities are exposed, such as crypto hacking, advanced persistent threats (APT) and ransomware, which use artificial intelligence to improve their attack techniques. This poses a potential risk to society from cybersecurity regarding the use and manipulation of information, either by governments, the military and organized criminal groups, de facto violating human rights.

Keywords: advanced persistent threats, artificial intelligence, big data, cybersecurity, cloud computing, DDoS, internet of things

1. Introduction

Disruptive technologies such as artificial intelligence (AI) have been rapidly being incorporated into different scientific fields and industry, becoming a support for technologies such as: big data, data science, the internet of things (IoT), computational linguistics, intelligent computing, assisted technologies, advanced robotics, among others. Also, AI has been incorporated in fields such as medicine, an example of this are systems for the early detection of COVID-19 through the use of techniques such as deep learning and machine Learning [1, 2] for the analysis of cellular and protein images, as well as the study of molecular and cellular dynamics among other aspects.

Other fields in which AI can be found are: manufacturing and logistics, industrial processes of various kinds, finance, adaptive education, diagnostic systems, micro and nanoelectronics, precision agriculture, transport, telecommunications, defense system, etc. even in the video game and toy industry.

A peculiarity of AI is that with the current computational potential it can be applied practically in whatever is desired. For example, in the development of advanced robotic systems, both software (chatbots) [3] and hardware, which allow emulating certain traits and interaction with the human being. Similarly, AI is

IntechOpen

incorporated into information and communication technologies, in the control and monitoring devices of homes, buildings and cities (Smart Cities), which converge to the so-called Internet of Things (IoT), which involve sensory, cloud computing, data science and cybersecurity among other disciplines.

In terms of security, standard and AI-mediated IoT present a debatable level of security. This is due to the fact that the base code of the firmware or operating system of these devices [4], does not have an acceptable level of security and, as they are permanently connected to a communication network, their exposure to computer attacks is high. This type of failure is attributed in part to device design and manufacturing failures, where the safety factor was underestimated, without taking into account that the devices and sensors under the IoT scheme are supported under Internet protocols and standards, and although they do not use them in their entirety, it does not imply that they are exonerated from being exploited by some type of malware.

In this sense, hacking this type of system allows us to steal data not only from homes, but from hospitals and research centers, industries, vehicles, weapons and drones, even causing accidents or taking lives selectively. In the case of robots and cobots, cameras, toys (including sex toys), printers and household appliances, among other devices connected to the internet or through a mobile device, can be maliciously intervened if they are not configured correctly regarding their access. Cyberattacks on these devices are often attributed to botnets; since they allow attacks by distribution of denial of service (DDoS), oversaturating Internet access traffic in order to disable or take control of the network to which the devices are connected. When this is achieved, access to the privacy of the victim or target is taken for granted without their being aware of it until it is too late.

Under this type of attack, what is sought is to collect information from the victim that allows obtaining bank access codes, personal and/or corporate email codes in order to continue climbing to steal sensitive information, images or intimate videos that lead to extortion, among others. For example, in 2016 an attack on Europe and North America was used under the DDoS modality [5], using the IoT [6] to disable the DynDNS systems (Dynamic Network Services, Inc.), operated by domain name providers (DNS), this caused the denial of access to internet platforms and services. Also, this type of attack seeks to steal sensitive corporate information to be sold to the competition, destroy it if necessary when there is a contract involved, extort money from the target or destroy critical facilities for terrorist or military purposes.

The problem with botnets is that they will continue to grow as the number of vulnerabilities increases in devices connected to the Internet in the coming years, in addition to other types of vulnerabilities to which a communication network of any industry or public service is exposed or private. This statement is based on the fact that the number of IoT devices connected with other disruptive technologies are growing exponentially, where household appliances and all types of electronic devices are being permanently managed and administered via wired or wireless, making them much more vulnerable to various types of cyberattack.

2. Internet of things (IoT)

The internet of things is defined as the set of electronic devices connected to the internet, whose function is aimed at collecting various information that can be directed to the control of actuators that activate other systems (lights, blinds, thermostat, air conditioning, etc.). Also, it allows the collection of data based on the monitoring or census of physical–chemical or biological variables, communication between devices and human-devices, identification, location and monitoring, among others. The IoT is in various scenarios; from the home (Smart home), through industry and services, to the health sector (eHealth), transport systems (navigability and predictive maintenance) and infrastructures of a city (bridges, viaducts, buildings, etc.) that converge to the concept of Smart Cities (Smart energy and Smart retail). Likewise, the sensors can be controlled and/or monitored from a central or mobile device, there are even other more advanced approaches focused on the energy industry, in order to optimize communication processes and broadband efficiency, known as Internet of Things-Grid (IoT-G) [7].

A notable characteristic of the IoT is that it has diversified to such an order that there are billions of devices permanently connected to the web, and with the rise of 5G technology, even greater growth is expected in the coming years, which He envisions drastic changes in Industry 4.0, where AI is going to play a key role in this context. Under this dynamic, researchers, scientists and engineers face emerging challenges in designing IoT-based systems that can be efficiently integrated with 5G wireless communications [8]. This technology is immersed in society, which in many cases goes unnoticed. The truth is that the volume of information that is permanently recorded is colossal, where technologies such as data science, big data, advanced analytics and Artificial Intelligence, among other disciplines, contribute their own for the treatment of this information.

It is worth mentioning that in technical terms the IoT works under the TCP/IP model, in which various protocols related to data transfer operate. For example, the Internet Protocol (IP) is the one that allows interoperability between devices, where the IPv4 version is definitively replaced by IPv6 in 2020, in which the organization of the IP addresses of computers and devices is expanded and improved in various types of communication networks.

There are protocols dedicated to the IoT apart from HTTP (Hypertext Transfer Protocol) such as: OCF (Open Connectivity Foundation), MFi (Made For iPhone/iPod/iPad), AllJoyn, DDS (Data Distribution Service), Thread, HomePlug and HomeGrid, AMQP (Advanced Message Queuing Protocol), CoAP (Constrained Application Protocol), MQTT (Message Queuing Telemetry Transport), XMPP (Extensible Messaging and Presence Protocol) and OPC UA (Unified Architecture), considered as a new generation standard. The operability of these protocols is based on the TCP or UDP protocols. In the case of the UDP protocol, it presents certain limitations in terms of connectivity and functionality, specific to its architecture.

Regarding the TCP/IP model, it exhibits vulnerabilities in each of its layers (Application, Transport, Internet and Network) that can be exploited [9]. For example, at the network layer, common problems are confidentiality and access control, which can be compromised through network hardware, that is, through IoT devices. At the network layer, the attacks are carried out at the level of modifying or canceling a datagram associated with the IP of a device, using techniques such as sniffing and spoofing in the ARP protocol or disabling the MAC filter, among others.

At the network infrastructure level, the transport layer fulfills the function of transmitting data via TCP or UDP protocols over IP datagrams. At this point, security problems are presented at the level of authentication, integrity and confidentiality of the information. Consequently, denial of service attacks can be performed by obstructing the flow of data by disabling communication between client and server. Other attacks that may occur are: pseudo-random subdomain attack (PRSD), IP Flooding, distributed attack, snork, ping of death, smurf, Spoofing for SYN flood DoS attacks TCP/SYN, flooding and teardrop, NTP amplification, attacks ICMP (ping), UDP Flood, HTTP Flood, SSL (Secure Sockets Layer)/TLS (Transport Layer Security) renegotiation, among others, where each one takes advantage of the design vulnerabilities of the layer itself. In the case of the internet layer, the attacks are conceived at the level of the fragmentation of IP datagrams, masking them by others that compromise the data that circulates through different points of a network.

As can be seen, the TCP/IP model since its creation has inherent weaknesses in its own design that can be exploited to carry out various types of attacks [10, 11]. In the particular case of IoT devices, they become perfect targets for cybercrime and industrial, military and government espionage, which, as can be seen, the attack vectors come from various sources, which are not necessarily organized crime.

There are other security factors to take into account about the IoT, which is related to the use of different technologies such as Wireless Sensor Networks (WSN), Near Field Communication (NFC) and Radio Frequency Identification (RFID) that are implemented in standard mobile devices, where each of them presents its own vulnerabilities [12]. Each technology requires specific protocols [13], to which is added 5G technology, whose emerging applications open up a myriad of applications, such as new attacks on advanced networks, for example, HealthTech and BioTech-type applications.

Regarding the standard communication protocols such as Ethernet, Wi-Fi and Bluetooth, others related to the application layer are presented specifically designed for a company's own products, so they are not considered as standardized, for example: Nest, MFI, Open Interconnect Consortium (OIC) and The AllSeen Alliance. Under this scenario, each industry that works with IoT develops its own protocols without universal unification, which guarantees connectivity compatible with devices from other manufacturers; This creates a security breach that can be exploited by cybercrime. An example in this regard was an attack that occurred in 2020 in the United States, using the Drovorub malware [14], the objective of which was to massively hack IoT devices in order to access wider communication networks.

3. Distributed denial of service and IoT

The Internet of Things is found in various devices as indicated by [13], in household appliances, smartphones, smart clothes, wearables (bracelets, virtual reality glasses, etc.), smart TVs, game consoles, transportation systems, buildings (security cameras, air conditioning, access controls, etc.), public infrastructures (bridges, highways, parks, etc.), public services, industrial components (e.g. SCADA systems) [15], systems transportation, etc.

A particularity of the IoT as mentioned above, is the connection between devices and the exchange of information between them under the TCP/IP model and their own custom-designed protocols. This poses great challenges in terms of information security, which as [16] state, there are attacks on devices connected to the Internet, in which there is fear of surveillance and concern for privacy. The reason for this, underlies as [17] points out, is that the IoT is presented as a source of data collection that grows exponentially and, consequently, every object becomes a source of information.

A critical point of the IoT in terms of information security is distributed denial of service (DDoS) attacks, whose objective is focused on disabling the continuity of communication of devices connected to a network, affecting the switched flow tables, data on a network, bandwidth and latency, taking advantage of the weaknesses of the OSI (Open System Interconnection) model (see **Figure 1**), in which attacks can be carried out at the transport, network and application layers, as well as DNS, SMURF and ACK amplification type attacks, among others.

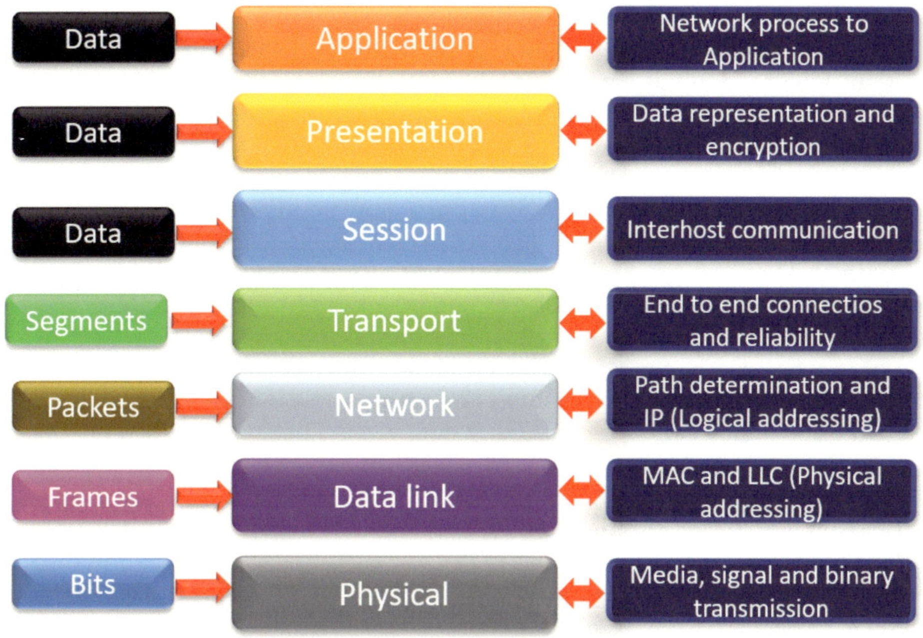

Figure 1.
Layers or levels of the OSI model, elements that it manages and functions.

Some effects of these attacks consist of making multiple requests to one or more servers (web, proxy, email, database, etc.), with the aim of saturating the network until it collapses. Also, brute force attacks can be carried out through specialized malware that is in charge of scanning the target network in search of IoT devices in order to obtain passwords, hijack them and link them to a botnet [18], which is basically a malware that takes advantage of browser vulnerabilities by installing itself on computers and/or servers.

The main characteristic of a botnet is that it infects the greatest number of systems forming the so-called "zombie" networks; which are controlled by Command & control type servers, which increase the capacity for DDoS and Spam attacks, among others, to specific objectives, which are normally companies and/or corporations, critical infrastructures such as transport, essential public services, health sector, food, etc., although attacks directed at a particular individual are not ruled out.

Regarding the defense mechanisms available to counter a DDoS attack, these present certain limitations such as the lack of resources at the software or hardware level in a network, or due to the technical and technological flexibility that a network has to deal with this. Type of attack. In this sense, the manifestation of potential risks attributed to technologies such as IoT with respect to DDoS, are expressed through security flaws that grow day by day, not only due to the number of devices, but also due to their diversification of these in multiple fields of industry, transportation, health and entertainment among others, becoming a global security problem.

The motivations for carrying out this type of attack are diverse and varied, ranging from personal or corporate resentments, through espionage, blackmail and extortion, to unfair competition or political and military ideologies. The growing reason for these attacks lies in the various vulnerabilities that can be exploited in IoT devices, whose manufacture questionable puts their security among them, as well as the poor configuration of the devices or portals by the personnel in charge.

Another aspect to be mentioned as a reference to the vulnerability of the IoT is related to the pandemic caused by COVID-19, whose attacks in the first half of 2020 increased alarmingly worldwide [19], in particular on websites of medical organizations, educational and administrative platforms, online gaming platforms and delivery services of various kinds. With this type of attack, it was shown that cybercriminals were not very interested in the social and humanitarian factor.

It is worth mentioning that DDoS attacks require poorly configured computer networks and servers, which once hijacked are connected to a Zombie network (**Figure 2**). This strategy applied to IoT devices acts as a connection bridge to be used as digital weapons of attack and espionage, expanding the coverage of the zombie network, boosting thousands or millions of times the level of request to the servers targeted by the attack. The problem with an attack on this scale is that the IoT is in continuous growth, that as [20] affirms only by 2020 there will be more than 50 billion connected devices (omnipresent) in cities, that is more than the estimated world population for this date (7.5 billion). Now, with the problem of the pandemic, there are hundreds of projects that promote the IoT for the permanent monitoring of cities, homes, hospitals and transportation systems among other critical systems of cities in the coming years, all aimed at minimizing future pandemics, for causing the number of devices to skyrocket to significantly larger numbers.

Another issue to take into account is related to metropolitan security, in which technologies such as cameras, sensors and drones are increasingly being incorporated, connected via IoT devices and mobile telephony. In the worst case scenario, when hacking this type of infrastructure, a city would be at the mercy of an attacker having access to infinite data. Now, this type of attack would not only be orchestrated by organized crime and terrorists, but by the governments themselves and the military, as noted above, with the exclusive purpose of monitoring each

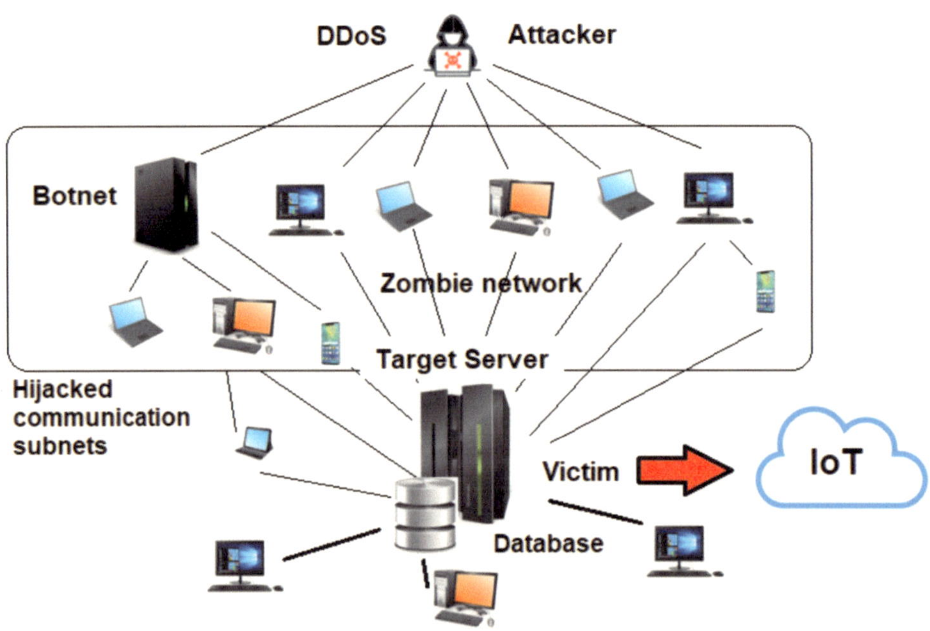

Figure 2.
Graphical representation of a distributed denial of service (DDoS) attack on an IoT system. As can be seen, a set of botnet is used to attack the victim, which in this case is a server that manages information from devices related to the IoT. The result of this attack is to have access to the database hosted on the server, to the control of the network connected to it and to the IoT devices.

individual and society permanently and with impunity, violating human rights. For example, China under his regime [21], is one of the countries that has the most information on its population using various technologies such as biometric registration and facial recognition systems, integrated with databases (includes DNA databases) managed and administered through artificial intelligence. Another example is the National Security Agency (NSA) and the CIA of the United States, which repeatedly violate human rights spying not only on their own community but on the entire world [22], as well as other agencies from other countries [23].

Returning to the topic of DDoS, there are various mitigation techniques for an attack of this type, whose large-scale feasibility is debatable. This is due in part to the efficiency and complexity of being able to implement these techniques. For example, a recent proposal is based on the use of blockchain technologies and Smart Contracts [24, 25] that have the necessary infrastructure to preserve the design and stability in terms of the development of a protocol that supports DDoS-type attacks. The proposal takes as support cloud computing, whose degree of security is high, due to the way data packets are filtered, where the system consists of a set of devices or programs (Firewalls and Proxy) configured in such a way that limits the passage or access of information in a network under certain rules and protocols.

4. Security factors in IoT

In terms of security, the IoT presents various weaknesses depending on the type of technology and application it is given, where DDoS takes advantage of, as do other variants such as low-speed DDoS (LDDoS) [26], which hides its traffic equivalent to normal traffic. Its origin is based on LDoS attack methods, which include variants such as reduction of quality (RoQ) and application servers (LoRDAS attacks). Another type of weakness attributed to the protection of information is focused on the service provider (DPS), which apart from implying additional costs, can lead to a decrease in the performance of the service and security problems, so you must be careful with whom you contract x and y services.

There are security proposals for the IoT, such as: collaborative defense using VNF (Virtual Network Functions), the use of DOTS protocols (DDoS Open Threat Signaling) [27], the exchange of events based on FLow (FLEX) and obfuscation techniques [28], among others. Although they are very good proposals, the problem is still open in establishing ideal protocols that allow confronting large-scale DDoS attacks, in which a greater degree of sophistication, duration and frequency is increasingly observed. In this sense, the use of Artificial Intelligence (AI) initially allows detection using techniques such as advanced neural networks [29] and machine learning [30], among others [31, 32].

One aspect that relates the IoT to AI and cybersecurity, are the failures at the hardware level. For example, design errors in Intel, AMD and ARM processors detected in the kernels, which were exploited by the Meltdown and Specter malware [33]. These errors allowed these malwares to access key parts of the processors by stealing security keys [34]. These failures have opened controversy, whether they were really design problems or were left on purpose for industrial or government espionage, hence policies have been implemented where countries such as the United States, China and Russia, among others, develop their own processor technology to minimize the risk of spying or hijacking in the event of a cyberattack. The implications of this type of attack show the fragility that exists in technology, where the common user has no idea what may be happening with their personal information stored on any electronic device. Seen in this way, society's ever-increasing dependence on technology poses new challenges in terms of security, which must

this perspective, the immunity of cloud computing against attacks from all types of malware was affirmed a few years ago, however, this changed, demonstrating that no system is infallible and even less against DDoS. In fact, there is evidence of DDoS-type attacks and their taxonomy on cloud computing, as indicated by [43], in which they expose the types and various counter-attack measures (detection, prevention and tolerance techniques) for mitigate DDoS attacks.

Based on the foregoing, it is worth noting that when a cloud computing system is perpetrated, it is because the attacker has managed to gain access as administrator to one of the system nodes, so he can do whatever he pleases with the data by putting in serious trouble to its objective, in which it literally has in its hands the most important asset of an organization, which is information. These types of failures are usually attributed to human failures, either due to ignorance, negligence or complicity of the administrator or a worker.

The synergy of disruptive technologies such as IoT + Big Data + Cloub computing + IA allows the creation of an unparalleled technological infrastructure for the recording, analysis, processing and storage of massive data, where the intervention of the human being will be increasingly scarce. Taking into account that, in the following years the number of IoT devices will grow exponentially, the noted synergy will be increasingly robust and autonomous with a level of security that guarantees that the information is well protected. However, it is clear that DDoS attack techniques are also evolving, giving way to what can be called intelligent distributed denial of service (IDDoS), in which advanced algorithmic techniques of artificial intelligence are integrated to attack AI-based infrastructures.

7. Crytohacking

This type of attack is constantly growing, employing malware that has the ability to hijack cloud computing systems. It is aimed at large corporations and cryptocurrency exchange houses, using the computational power of mobile devices as an attack center, mining it with cryptocurrencies, making the user believe that they are rewarded under the assumption that they are carrying out large transactions under the blockchain model. In this context, crypto hacking resembles a DDoS attack with the difference that it not only hijacks computers, servers and web pages, but also smart mobile devices, which by mining them with cryptocurrencies can make fraudulent transactions at the cost of the victim, winning money secretly, since it is not possible to make a traceability with respect to the transactions that have been carried out. An example of malware with these characteristics is coinhive and cryptoMiner [44], discovered in multinational companies such as Tesla and Avira.

One problem that continues to grow is communication with anonymous networks and the Darknet (which involves the Deep web and the Dark web). This type of network, in principle, is intended to facilitate the access and flow of information in countries whose restriction of free expression does not allow open communication. However, this network is also used for criminal purposes which, as [45] points out, is used to commit computer crimes, share compromised files (personal, pornographic, confidential, illegal software, etc.) or for the sale of goods and services prohibited. The anonymity provided by the Darknet guarantees user navigation without any restriction compared to the conventional internet, so special browsers and protocols are required [46]. For this, it is common to use ".onion" extension that guarantees an anonymous IP to access the TOR network, or networks such as ZeroNet, FreeNet or I2P.

A peculiarity of the Darknet is that, although attacks are carried out from within, it shows itself to be highly flexible, dynamic and robust enough to adapt, thus minimizing collateral damage, which is a notable differential characteristic with respect to the standard Internet. Based on this fact, when the Darknet is used for the purpose of hacking with cryptocurrencies, the probability of success is high because it operates under the blockchain model and distributed ledger [47]. This type of attack is in continuous growth parallel to ransomware, due to the ease of anonymously hijacking a device connected to the internet, added to the incessant increase in legal and illegal operations using the Blockchain as a cryptocurrency monetary system. For example, due to the particular technical and technological characteristics of the Darknet, it facilitates the exchange of sensitive information [48] between organized crime and terrorist groups, making it impossible for the authorities to intercept such as laundering. of money, planning and coordination of attacks, drug trafficking, tax evasion, hit men, kidnapping, extortion, child pornography, sale of weapons, etc.

Therefore, the combination of the blockchain with the Deep Web creates the ideal environment for the flow of legal or illegal information, in which it literally becomes almost impossible to trace [49], considering this cyberspace as a no man's land, where DDoS-type operations, among others, are carried out without any legal or police problem. Now, it should be noted that not only crime makes use of this type of network, but also government and military entities [50], institutions of higher education and research, among others, in which it seeks to guarantee anonymity and minimize risk of theft of critical information.

8. Ransomware

Ransomware is a type of cyberattack that is characterized by encrypting the files stored on a computer or web page by encoding them, where the victim must pay with cryptocurrencies for their ransom, which is why it is difficult to trace their origin or destination. This type of attack is constantly evolving in the way of encrypting information, using more sophisticated and robust algorithms that seek to hide the trail of the attacker, the form of payment and attacks on systems such as the cloud. As things are going, this type of attack will be more destructive and lethal, since it is combined with DDoS to enhance its level of hijacking, where the targets have been shifting from small companies to financial systems and industry, government and military structures and Critical infrastructures, which compromise their information and the operation of all their systems, paralyzing them, with the possibility of deleting or subtracting records and modifying them according to what the attacker or his contractor wants.

The ransomware only requires to hijack a few computers that are not updated in terms of security or to install itself by tricking its victims. Also, this malware (for example, Ekans) can be installed in SCADA-type systems [51] that are connected to the internet or to a local network whose security measures are deficient. What is critical about this type of attack is that it can be scalable, as long as the communication network infrastructure allows it, that is, when there is vulnerable software and hardware such as routers and other network devices. Also, other types of malware can be used to make way for ransomware, letting them carry out the tasks for which they were created and programmed, and then having the information as best suited. In this sense, the IoT with its various variants is not exempt from a ransomware-type attack, especially if the devices are being managed and/or administered by servers or mobile devices with an ephemeral degree of security.

The use of ransomware for targeted attacks (individuals or companies) is a great resource for organized crime, although at present it has diversified as it is a multi-platform malware, which allows it to affect Linux, Windows and MacOS operating systems alike. For example, the Tycoon ransomware. That said, the attack can vary, encrypting personal or corporate files (web server, for example WastedLocker), locking the PC screen (lock screen), locking the hard drive and backups, blocking access to a mobile device, etc. The problem does not end here, since, at the time of the seizure of information, the attacker has unlimited access to the information, which allows the tracking of other potential victims, their computers and networks. An example that occurred in mid-2020 was through the Netwaklker malware, in which critical information was seized from the migration computer systems in the United States, which contained data from the Federal Intelligence Agency, some embassies and consulates, in which was asked for a ransom in the amount of 4 million dollars. It was not paid for it, but it exposed the vulnerabilities to which any system considered safe is found.

The ransomware attack feature consists of hiding it within files, which when executed by the victim installs a Trojan in the operating system, which internally begins to make changes to some registries, such as the keyboard, disables the antivirus and any other program security, among other critical protection functions. The next step is to connect to the victim's network that is supposed to be vulnerable and enter via remote connection from the computer's desktop, which uses various protocols such as RDP (Remote Desktop Protocol) - Also, previously through social engineering, having guessed the password, but rather employ a brute force attack to find it. In general, there are a large number of tools to violate the system. Then, the process of encrypting files on the computer begins, including critical databases such as backups that correspond to servers -physical or in the cloud-, although more recently ransomware has been found that also encrypts data stored on network drives. Once the encryption is completed, the victim receives a message indicating immediate payment for the information seized through cryptocurrencies, otherwise it will be destroyed. Being able to decrypt ransomware is complex, especially since some of them already use symmetric encryption algorithms such as Galois/Counter (GCM) mode3 with a length of 16 bytes.

9. Artificial intelligence and advanced persistent threats

Although publications about cyberattacks using software based on artificial intelligence (AI) are scarce, it does not imply that they do not exist, since what is least wanted is publicity about it. AI can be used to find vulnerabilities in software such as hardware connected to a network, where appropriate equipment and resources are required for this purpose. For example, data can be searched on the darknet on activities related to clients or organizations that may be compromised and involve a security threat that is exploited by cybercriminals; this includes documentation and private information that has been infiltrated (personal and financial information, intellectual property, access credentials, etc.).

AI has already started to play a critical role when it comes to cybersecurity. Currently security companies use predictive models based on machine learning combined with neural networks and other disruptive technologies, in order to anticipate attacks on computer systems and critical infrastructures, as well as detect what is happening in a particular network. From this perspective, reverse engineering it to carry out attacks based on found vulnerabilities is viable, where robust datasets used as libraries can be used for brute force attacks. In fact, the creation of AI algorithms with programming that attacks certain systems already exists, the control of which is carried out by "intelligent" malware.

Although it is based on an assumption, with AI applied to carry out cyberattacks it compromises all the security of a system, including the lives of people and society in general. An attack of this type would be planned to be executed on several fronts, using various resources such as advanced persistent threats (APT), DDoS, ransomware and other intelligent malware, hijacking certain critical systems, temporarily disabling them or destroying them, in such a way that any functionality or functionality collapsed. Operation of these in cyberspace, in this particular case of the different governmental and industrial organizations of a nation.

AI can not only threaten the security of an organization but that of any country, which can be orchestrated by organized criminal groups or by groups funded by governments and militia. An example of this are APTs, which are a highly specialized type of malware that is custom designed to infect and disable systems at the software and hardware level. The objective of this type of malware is the theft, modification, destruction, espionage and sabotage of industrial and corporate information. APTs possess stealth type attack traits, combining advanced encryption techniques with close polymorphic algorithms with AI. [52] points out that APTs can persist inside a computer system for a long time without being detected, taking advantage of the vulnerabilities of the infrastructure or the architecture of the communication protocols in the packaging of data in a network.

Based on the above, an APT is a cyber weapon designed for specific attacks on targets, particularly critical infrastructure. From this perspective, the IoT is no exception to an attack of this type, since communication between devices can be intercepted and disabled or modified. This is because APTs can leak through software or hardware that is not properly protected and from there scale the systems, so blocking or hijacking using a DDoS-type attack is feasible.

APTs are exclusive, so they are not abundant on the internet, this is because their managers are not just any organized criminal group, but governments, rival corporations and large criminal syndicates that have unlimited financial, technical and technological resources, which allows them undertake this type of development and carry out targeted cyberattacks. Under this model [53] point out that a variant of the APT called S-APT is used, whose action is focused on creating attack vectors based on disinformation strategies within the framework of the military.

The IoT within the framework of industry 4.0 increasingly incorporates AI in its developments, where connectivity to the internet and mobile devices is constantly increasing. Under this scenario, the introduction of an APT or malware similar to these technologies taking advantage of their vulnerabilities is feasible, either when they are already on the market or from their own manufacture, as demonstrated by [54]. Consequently, countless plausible scenarios are opening up to carry out cyberattacks, to and from drones, autonomous vehicles, advanced robots (military, industrial, leisure, etc.), smart electrical grids, even the IoT infrastructure that a smart city has. Consequently, the concern arises of programming errors in AI-based systems, which are exploited and taken advantage of to violate other systems, as demonstrated by the DeepXplore intelligent system [55].

10. Discussion from the bioethical plane

In the IoT industry, the term Edge Computing has recently been coined, which is the next step in Cloud Computing technology, in which all the information from intelligent IoT devices connected to a network is collected, to be stored and processed in large database repositories arranged for this purpose. The implications of this new proposal are broad and complex, because the data collected from sensors and various devices, combined with advanced AI algorithms, make inferences that lead to decision-making both human and automated devices. The density of data and its variety under this scheme will increase exponentially for the next few years, exceeding zettabytes (10^{21} bytes), so technologies such as 5G, next-generation communication networks including the quantum internet, will accelerate and optimizing information traffic without saturating networks by reducing latency, incorporating other tools such as Edge/Fog Computing. It is worth mentioning that these technologies are characterized by the fact that the data is managed in the form of a chain of blocks or blockchain to guarantee a high level of security, which may possibly be migrated to specific applications such as the health field, minimizing the risk of compromising clinical information from the patients.

In the case of edge computing, it does not work alone apart from IoT, but is linked to other technologies such as Mobile Cloud Computing [56] and Collaborative Mobile Edge Computing [57], an example in this regard, are the Google Cloud IoT technologies, which are active in today's market. As they are considered as emerging technologies, the level of security is still in question, so the risk of compromising sensitive information of users and services through a crypto-hacking attack is high. Let us remember that the security infrastructure in the cloud is high, but not that of the IoT, added to the bad practices that inevitably lead to unauthorized access to a network.

Normally, unnecessary or insecure network services are activated, being exposed to attacks where unauthorized control of any service can be assumed, violating the confidentiality, integrity, authentication or availability of the information. Along the same lines, there are often interfaces that are managed by proprietary or third-party devices, such as mobile applications, data repositories in the cloud, the corporate website itself and the backend APIs. These flaws lead to vulnerabilities such as weak encryption (or lack thereof) on the data circulating on the network, as well as the absence of input/output filters.

Other common failures found in IoT devices are: failure to update firmware or manage related processes such as encrypting in transit and validating updates without appropriate mechanisms for doing so; use of insecure or outdated software components and libraries; inappropriate use of personal information stored on a device whose degree of security is questionable, in addition to the absence of a formal permission or informed consent; absence of data encryption and access control.

There are variants of the IoT, such as the industrial and services field, known as the Internet of Robotized Things (IoRT), which is gaining strength due to the continuous industrialization that demands the attention of robots, particularly in industrialized countries. There is also the Internet of Things on the Battlefield (IoTotBF) [58]; which combines various advanced communication network technologies (including quantum ones) with massively interconnected systems, thus taking warfare to a new level of technicality. In this context, the technicality of the military is increasing and AI together with robotics are frequently used in the development of new intelligent weapons, of which there is no guarantee that something can go wrong in the field of cybersecurity. Viewed in this way as [59], oversight at the cybersecurity level by human operators is going to be increasingly difficult, if not impossible. This opens a strong discussion about the role of the human being in military operations, since the responsibility of decision-making is transferred to a machine about destroying a target in which it implies the death of innocents.

From the above, a number of questions are presented related to how to minimize the risk of a cyber-attack on a military infrastructure with technologies such as IoTotBF or similar, by foreign militias, terrorist groups, organized crime or by advanced automatic systems based on AI. We must not forget that the militias of various nations of the world are constantly developing new robotic and cybernetic technologies, aimed at improving their attack and defense systems while minimizing the number of casualties.

The IoT presents great benefits for society, as well as great challenges in terms of security, due to its integration with various standard and advanced communication technologies, which manage multiple devices in the home, industry, health and transportation, among others. This trend must be taken into consideration not only by manufacturers and governments, but by society itself, since the risk of information collection by third parties is high and the uncertainty of its handling remains between said. In fact, the tradeoffs of transparency in the management of information by governments and large corporations are critical, and this problem will be further accentuated with the advent of next generation technologies.

As for cyberattacks such as DDoS combined with other techniques mentioned throughout the chapter, the spectrum of damage to private and public computer networks is broadened, including devices connected to it such as the IoT, mobile devices and other emerging technologies. In this sense, the authorities and governments in general must take the potential cyberattacks that can be carried out on critical infrastructures such as health very seriously, since not only information is compromised, but people's lives are compromised. For example, zero-day or volume-based DDoS attacks, which are difficult to avoid due to the speed with which they run. In fact, it only takes one flaw for a botnet to saturate its target's network and fully control it. Along the same lines, there are other types of more sophisticated, highly destructive and selective attacks that take control of a system, such as protocol attacks, in this case TCP directed at networks that communicate with servers, firewalls (physical and logical), gateways and load balancers, where damage to an infrastructure can be severe.

To recap, although the attacks mentioned in this document are attributed to organized groups, a person with minimal knowledge could put an institution, industry and even a nation in serious trouble, since some of the information to create malware does not It is only found on the Darknet, but on the conventional internet, where with a minimal payment you can find programs to create ransomware and other types of computer viruses. Likewise, you can hire the service of any type of malware, the packages are sold on the dark web for reasonable prices, even malware kits. Most of the public is unaware of this type of thing and in this way is exposed to their personal or corporate information being stolen by cybercriminals.

Based on what is stated in this document, it is evident that special attention must be paid to privacy, ethical, bioethical and legal aspects, security and rights, among other elements that threaten human dignity. Under this fact, there is a constant concern about the unauthorized access and manipulation of personal and massive data concentrated in technologies such as big data, IoT, cloud computing, among others, which contain sensitive information at a clinical, ethnic, sociocultural, economic, financial and industrial, etc., which require a thorough examination from the bioethical and biopolitical point of view that guarantee respect for the protection of information. At this point, a number of elements arise to evaluate, because not only is reference being made to the seizure of information and sale of it to third parties, but also to irreparable damage to the individual in terms of inequity or damage generated by the interference to the private life of the victim or victims.

Under the exposed characteristics of a cyberattack, the violation in terms of property, rights, use, exploitation, maintenance and licenses for the administration of massive data, means little or nothing for the attackers, but if a great legal, ethical weight, bioethics and security for the organization and/or personnel in charge of managing and administering this data. From this point of view, there are gray areas regarding the formulation of public policies that guarantee an adequate safeguard on the property of the data, protection and prohibition of use for other purposes, so it is expected that in the coming years letters will be taken on this matter will require the collaboration of various groups of experts and disciplines that seek to minimize risks, both in the handling of massive data, and in cyberattacks by various means.

11. Conclusions

Society is increasingly dependent on technology, examples of which are: the internet, mobile technology, artificial intelligence, big data, cloud computing and blockchain among others, which facilitate the management and administration of massive information. In the case of the IoT, it has been becoming widespread in various environments such as health, industrial, transport and services, among others, progressively incorporating the aforementioned technologies. In this sense, there is growing concern about the fragility of this technology, which proves to be notoriously vulnerable to cyberattacks. The reason for this is the continuous proliferation of IoT devices that do not meet the minimum-security standards; thus, they expose an individual and society in general to being spied on and possibly attacked. Added to this panorama are the vulnerabilities inherent to communication architectures, which have yet to be resolved, and the lack of management and administration of devices by the personnel in charge, which increase the risk of unauthorized access to an information system. What is critical about this matter is that the health sector has been incorporating the IoT into its services and although it takes their security very seriously, the spectrum of vulnerabilities to which this technology is subjected is alarming.

The IoT is expanding its range of action by integrating with 5G communication networks, and the hospital environment is no stranger to this. With this in mind, the diversification of services and connectivity will be reflected in Smart City, Green Systems and Transport Systems, which will facilitate the analysis and visualization of large volumes of data that the IoT generates permanently. This implies that secure communication architectures are required, capable of withstanding attacks of various kinds, particularly those of the DDoS type that have been expanding their modalities by integrating other advanced malware technologies. Consequently, the development of networks of sensors, actuators and remote diagnostic systems will require a unification of standards and protocols that guarantee that the IoT devices that are or are released on the market present a minimum risk that compromises critical or sensitive information, well of a person, institution, industry or government.

Based on the current global instability attributed to social, political, economic, health and environmental factors, cyberattacks have not diminished. In fact, with the problem of the COVID-19 pandemic, remote work skyrocketed and with it the objectives of cybercriminals were diversified, where resources such as corporate VPN gateways and non-public web resources such as emails have been compromised by the high risk of being hijacked by malware, such as APTs, ransomware and botnet, to name a few, giving way to the growth of the DDoS market.

Finally, in the coming years an increase in IoT devices is predicted in large cities in their critical infrastructures, expanding their services and promoting permanent monitoring in search of anomalies of various types: climate, environmental pollution, security (citizen, computing, biosafety, etc.), mobility and health, among others, which is why the use of other technologies for the analysis and treatment of massive data is expected to explode, and with it the risk and vulnerabilities that need to be addressed from now on. The task in this sense is not easy but it is not impossible either, technologies such as quantum encryption, quantum internet and AI processors that reduce the risk of attacks on hardware such as system software,

are some advances that promise to reduce the gap to information security. However, there is a problem regarding the role that governments play under what is stated in this document, since in the end their transparency in the handling of information is debatable, especially when it has an incalculable value.

Conflict of interest

The author declares no conflict of interest.

Author details

Jairo Eduardo Márquez Díaz
Universidad de Cundinamarca, Chía, Colombia

*Address all correspondence to: jemarquez@ucundinamarca.edu.cor

References

[1] Ozturk, T., Talo, M., Yildirim, E. A., Baloglu, U. B., Yildirim, O., & Rajendra, A. U. Automated detection of COVID-19 cases using deep neural networks with X-ray images. Computers in Biology and Medicine, 2020, 103792. Advance online publication. https://doi.org/10.1016/j.compbiomed.2020.103792

[2] Jiang, X., Coffe, M., Bari, A., Wang, J., Jiang, X., Huang, J., et al. Towards an Artificial Intelligence Framework for Data-Driven Prediction of Coronavirus Clinical Severity. Computers, Materials & Continua (CMC). 2020, 63(1). 537-551. http://doi.org/ 10.32604/cmc.2020.010691

[3] Adam, M., Wessel, M., & Benlian, A. AI-based chatbots in customer service and their effects on user compliance. Electronic Markets. 2020. http://doi.org/10.1007/s12525-020-00414-7

[4] Mäki, P., Rauti, S., Hosseinzadeh, S., Koivunen, L., & Leppänen, V. Interface diversification in IoT operating systems. Proceedings of the 9th International Conference on Utility and Cloud Computing - UCC '16. 2016. doi:10.1145/2996890.3007877

[5] Mahjabin, T., & Xiao, Y. Mitigation Process for DNS Flood Attacks. 2019 16th IEEE Annual Consumer Communications & Networking Conference (CCNC), 2019. http://doi.org/10.1109/ccnc.2019.8651715

[6] Abdmeziem, M., Tandjaoui, D. & Romdhani, I. Lightweighted andenergy-aware mikey-ticket for e-health applications in the context ofinternet of things. International Journal of Sensor Networks, 2018, 26(4), 227-242.

[7] Hao, H., Wang, Y., Shi, Y., Li, Z., Wu, T & Li, C. IoT-G: una arquitectura de comunicación inalámbrica de energía privada de baja latencia y alta confiabilidad para redes inteligentes. IEEE International de 2019 Conferencia sobre tecnologías de comunicaciones, control y computación para redes inteligentes (SmartGridComm), Beijing, China, 2019, 1-6, doi: 10.1109/SmartGridComm.2019.8909773.

[8] Ejaz, W., Anpalagan, A., Imran, M. A., Jo, M., Naeem, M., Qaisar, S. B., & Wang, W. Internet of Things (IoT) in 5G Wireless Communications. IEEE Access, 2016, 4, 10310-10314. doi:10.1109/access.2016.2646120

[9] Acharya, S. & Tiwari, N. Survey of DDoS Attacks Based On TCP/IP Protocol Vulnerabilities. OSR Journal of Computer Engineering (IOSR-JCE), 2016, 18(3), 68-76. http://doi.org/0.9790/0661-1803046876

[10] Alotaibi, A. M., Alrashidi, F. B., Naz, S. & Parveen, Z. Security issues in protocols of TCP/IP Model at Layers Level. International Journal of Computer Networks and CommunicationsSecurity, 2017, 5 (5), 96-104

[11] Elejla, O. E., Anbar, M., & Belaton, B. ICMPv6-Based DoS and DDoS Attacks and Defense Mechanisms: Review. IETE Technical Review, 2016, 34(4), 390-407. doi:10.1080/02564602.2016.1192964

[12] Santiago, A. et al. Modelo de Seguridad para Garantizar la Integridad de Pagos Móviles sobre Near Field Communication (NFC). Rev. Espacios, 2018, 39(19), 6-30.

[13] Márquez, D. J. Riesgos y vulnerabilidades de la denegación de servicio distribuidos en internet de las cosas. Revista Biòtica i Dret. Rev Bio y Der. 2019; 46: 85-100. https://doi.org/10.1344/rbd2019.0.27068

[14] NSA and FBI. National Security Agency Federal Bureau of Investigation

Cybersecurity Advisory. Russian GRU 85th GTsSS. Deploys Previously. Undisclosed Drovorub Malware. Rev 1.0, U/OO/160679-20, PP-20-0714.

[15] Coffey, K., Smith, R., Maglaras, L., & Janicke, H. Vulnerability Analysis of Network Scanning on SCADA Systems. Security and Communication Networks, 2018, 1-21. doi:10.1155/2018/3794603

[16] Rose, K., Eldridge, S. & Chapin, L. La internet de las cosas - Una breve reseña para entender mejor los problemas y desafíos de un mundo más conectado, Internet Society (ISOC). Available June 10, 2020: https://www.internetsociety.org/wp-content/uploads/2017/09/report-InternetOfThings-20160817-es-1.pdf

[17] Barrio, A. M. Internet de las cosas. Madrid, España, 2018, Editorial REUS.

[18] Antonakakis, M., April, T., Bailey, M., Bernhard, M., Bursztein, E., Cochran, J., Durumeric, Z., Halderman, J. A., Invernizzi, L., Kallitsis, M., Kumar, D., Lever, C., Ma, Z., Mason, J., Menscher, D., Seaman, C., Sullivan, N., Thomas, K. and Zhou, Y. Understanding the mirai botnet. 26th USENIX Security Symposium. Vancouver, BC: USENIX Association, 2017, 1093-1110.

[19] Kupreev, O., Badovskaya, E. & Gutnikov, A. Ataques DDoS en el primer trimestre de 2020. Available June 13, 2020: https://securelist.com/ddos-attacks-in-q1-2020/96837/

[20] Márquez, D. J. Seguridad metropolitana mediante el uso coordinado de Drones. Ingenierías USBMed, 2018, 9(1), 39-48. http://doi.org/10.21500/20275846.3299

[21] Nardi, D. Country update: China. Religious Freedom in China's High-Tech Surveillance State. Washington, DC., United States of America, United States Commission on International Religious Freedom (USCIRF), 2019.

[22] Bignami, F. Human Rights Extraterritoriality: The Right to Privacy and National Security Surveillance. GWU Law School Public Law Research, 2018, 1-20.

[23] Lemieux, F. Intelligence and state surveillance in modern societies: an international perspective. Howard House, United Kingdom, Emerald Publishing Limited, 2019.

[24] Feliu, R. J. Smart Contract: Concepto, ecosistema y principales cuestiones del Derecho privado. In: La Ley mercantil: Contratación mercantil, 47, ed. Wolters Kluwer, 2018, 7-10.

[25] Alharby, M. & Moorsel, A. Blockchain-based smart contracts: a systematic mapping study. Computer Science & Information Technology, 2017, 125-140. http://doi.org/ 10.5121/csit.2017.71011

[26] Savchenko, V. Detection of Slow DDoS Attacks based on User's Behavior Forecasting. International Journal of Emerging Trends in Engineering Research.2020, 8. 2019-2025. 10.30534/ijeter/2020/90852020.

[27] Rashidi, B., Fung, C. CoFence: a collaborative DDOS defence using network function virtualization. In: 12th International Conference on Network and Service Management (CNSM 16), October 2016.

[28] Hosseinzadeh, S., Hyrynsalmi, S., & Leppänen, V. Obfuscation and diversification for securing the internet of things (IoT). Internet of Things, 2016, 259-274. doi:10.1016/b978-0-12-805395-9.00014-9

[29] Wang, M., Lu, Y. & Qin, J. A Dynamic MLP-Based DDoS Attack Detection Method Using Feature Selection and Feedback. Computers & Security, 2019, 1-15. https://doi.org/10.1016/j.cose.2019.101645

[30] Zekri, M., Kafhali, S. E., Aboutabit, N., & Saadi, Y. DDoS attack detection using machine learning techniques in cloud computing environments. 2017 3rd International Conference of Cloud Computing Technologies and Applications (CloudTech), 2017. http://doi.org/10.1109/cloudtech.2017.8284731

[31] Aljumah, A. Detection of Distributed Denial of Service attacks using artificial neural networks. (IJACSA) International Journal of Advanced Computer Science and Applications, 2017, 8(8), 306-318.

[32] Zhang, B., Zhang, T., & Yu, Z. DDoS detection and prevention based on artificial intelligence techniques. 2017 3rd IEEE International Conference on Computer and Communications (ICCC), 2017. doi:10.1109/compcomm.2017.8322748

[33] Innus, M. D., Simakov, N. A., Jones, M. D., White, J. P., Gallo, S. M., DeLeon, R. L., & Furlani, T. R. Effect of Meltdown and Spectre Patches on the Performance of HPC Applications. 2018. arXiv preprint arXiv:1801.04329

[34] Giles, M. Estos han sido los peores ciberataques en lo que llevamos de 2018. [Internet]. 2020. MIT Technology Review, Available from: https://www.technologyreview.es/s/10339/estos-han-sido-los-peores-ciberataques-en-lo-que-llevamos-de-2018 [Accessed: 2020-june-14]

[35] Metcalf, D., Milliard, S. T. J., Gomez, M., & Schwartz, M. (2016). Wearables and the Internet of Things for Health: Wearable, Interconnected Devices Promise More Efficient and Comprehensive Health Care. IEEE Pulse, 7(5), 35-39. doi:10.1109/mpul.2016.2592260

[36] Spender, A., Bullen, C., Altmann-Richer, L., Cripps, J., Duffy, R., Falkous, C., ... Yeap, W. (2019). Wearables and the internet of things: considerations for the life and health insurance industry. British Actuarial Journal, 24. doi:10.1017/s1357321719000072

[37] Qureshi F. & Krishnan, S. (2018). Wearable Hardware Design for the Internet of Medical Things (IoMT). Sens Basel. 18(11):3812

[38] Rodríguez, G. R. (2019). Internet de las cosas: Futuro y desafío para la epidemiología y la salud pública. Universidad Y Salud, 21(3), 253-260. https://doi.org/10.22267/rus.192103.162

[39] Tao, H., Bhuiyan, M. Z. A., Abdalla, A. N., Hassan, M. M., Zain, J. M. and Hayajneh, T. (2019). Secured Data Collection with Hardware-Based Ciphers for IoT-Based Healthcare. IEEE Internet of Things Journal, vol. 6, no. 1, pp. 410-420, Feb. 2019, doi: 10.1109/JIOT.2018.2854714.

[40] Elkhodr, M., Alsinglawi, B. & Alshehri M. (2019) A Privacy Risk Assessment for the Internet of Things in Healthcare. In: Khan F., Jan M., Alam M. (eds) Applications of Intelligent Technologies in Healthcare. EAI/Springer Innovations in Communication and Computing. Springer, Cham. https://doi.org/10.1007/978-3-319-96139-2_5

[41] Alraja, M. N., Farooque, M. M. J. & Khashab, B. (2019). The Effect of Security, Privacy, Familiarity, and Trust on Users' Attitudes Toward the Use of the IoT-Based Healthcare: The Mediation Role of Risk Perception. IEEE Access, 7,111341-111354, doi: 10.1109/ACCESS.2019.2904006.

[42] Thibaud, M., Chi, H., Zhou, W., & Piramuthu, S. (2018). Internet of Things (IoT) in high-risk Environment, Health and Safety (EHS) industries: A comprehensive review. Decision Support Systems, 108, 79-95. doi: 10.1016/j.dss.2018.02.005

[43] Qureshi F. & Krishnan, S. (2018). Wearable Hardware Design for the

Internet of Medical Things (IoMT). Sens Basel. 18(11):3812

[44] Deshmukh, R. V., & Devadkar, K. K. Understanding DDoS Attack & its Effect in Cloud Environment. Procedia Computer Science, 2015, 49, 202-210. http://doi.org/10.1016/j.procs.2015.04.245

[45] Eskandari, S., Leoutsarakos, A., Mursch, T., & Clark, J. A First Look at Browser-Based Cryptojacking. 2018 IEEE European Symposium on Security and Privacy Workshops (EuroS&PW). 2018, 58-66. http://doi.org/10.1109/eurospw.2018.00014

[46] Arenas, A. La resiliencia de la red oscura. Un estudio logra describir las propiedades estructurales de la «Internet invisible» y explica por qué esta se muestra tan inmune a los ataques informáticos. Rev. Investigación y Ciencia, 2018, 498, 12-14.

[47] Beshiri, A. S. and Susuri, A. Dark Web and Its Impact in Online Anonymity and Privacy: A Critical Analysis and Review. Journal of Computer and Communications, 2019, 7, 30-43. https://doi.org/10.4236/jcc.2019.73004

[48] Lipton, A. & Pentland, S. Hacer saltar la banca. Investigación y Ciencia, 2018, 498, 16-23.

[49] Mirea, M., Wang, V., & Jung, J. The not so dark side of the darknet: a qualitative study. Security Journal, 2018. doi:10.1057/s41284-018-0150-5

[50] Bautista, D. L. Deep web: aproximaciones a la ciber irresponsabilidad. Revista Latinoamericana de Bioética, 2015, 15(1), 26-37.

[51] Chertoff, M. A public policy perspective of the Dark Web. Journal of Cyber Policy, 2017, 2(1), 26-38. doi:10.1080/23738871.2017.1298643

[52] Ibarra, J., Javed Butt, U., Do, A., Jahankhani, H., & Jamal, A. Ransomware Impact to SCADA Systems and its Scope to Critical Infrastructure. 2019 IEEE 12th International Conference on Global Security, Safety and Sustainability (ICGS3). doi:10.1109/icgs3.2019.8688299

[53] Márquez, D. J. Armas cibernéticas. Malware Inteligente para ataques dirigidos. Ingenierías USBMed, 2017, 8(2), 48-57. http://doi.org/10.21500/20275846.2955

[54] Ahmad, A., Webb, J., Desouza, K. C., & Boorman, J. Strategically-Motivated Advanced Persistent Threat: Definition, Process, Tactics and a Disinformation Model of Counterattack. Computers & Security, 2019. doi:10.1016/j.cose.2019.07.001

[55] Choi, C. Q. Lo siento, Dave. Rev. Investigación y Ciencia, 2018, 499, abril, p. 8.

[56] Pei, K., Cao, Y., Yang, J. & Jana, S. DeepXplore: automated whitebox testing of deep learning systems. Communications of the ACM. 2019, 62. 137-145. http://doi.org/10.1145/3361566.

[57] Tran, T., Hajisami, A., Pandey, P. & Pompili, D. Collaborative mobile edge computing in 5g networks: new paradigms, scenarios, and challenges. IEEE Commun Mag, 2017, 55(4), 54-61.

[58] Márquez, D. J. Nanotecnología. Internet de las cosas. [Internet]. 2019. Available from: http://doi.org/10.13140/RG.2.2.33697.66402 [Accessed: 2020-oct-04]

[59] Theron, T. P. et al. Towards an Active, Autonomous and Intelligent Cyber Defense of Military Systems: the NATO AICA Reference Architecture. International Conference on Military Communications and Information Systems, Warsaw, Poland, 22nd - 23rd May 2018.

Section 4

Society

Chapter 20

Evolution of Catholic Marriage Morality in the Twentieth Century from a Baby-Making Contract to a Love-Making Covenant - Part I: Era of John Paul II

Edward Collins Vacek S.J.

Abstract

For nearly a century Catholic Church teaching on Sexuality evolved greatly. Changes in science and the teaching of other Christian churches begged for a fresh start. John Paul II, elected pope in 1978, attempted to update that teaching by providing new background arguments, without changing any of the strictures in the foreground. John Paul II insisted on necessity of total love, allowing for no exceptions. He claimed that divorce was impossible because the spouses retained no control over their marriage promises. Homosexual activity was judged to be morally deficient. Likewise the recent arrival of reproductive technology was largely condemned for breaking the sexual unity of the spouses. But fertile sexual activity was newly appreciated as the important activity of spouses cooperating with the creative activity of God. In the twenty-first century the Church's official teachings continues to be reformed, but their relevance is widely questioned as social norms continue to drastically change.

Keywords: John Paul II, total love, contraception, natural family planning, divorce, homosexuality, reproductive technology, canon law

1. Introduction

As the twentieth century wore on, the Catholic Church's teaching on sexual ethics became ever more embattled. After the 1968 teaching on the wrongness of birth control, the teaching became more and more contested. The new pope, John Paul II, who inherited the enthusiasm of the second Vatican Council, tried to give a new defense of the teaching. This effort was somewhat unexpected, since he had previously skipped the commission that was working on the topic. John Paul, confident in his unique abilities, decided to rethink the teaching from a personalist perspective.

2. John Paul II

In his major document on marriage, *Familiaris consortio*, John Paul II's central criterion for marital morality is "total love." His thesis is considered by some of his admirers to be a brilliant flowering of personalism, and it is considered a romantic flight of fantasy by some of his critics. He writes

> *The total physical self-giving would be a lie if it were not the sign and fruit of a total personal self-giving, in which the whole person, including the temporal dimension, is present. If the person were to withhold something or reserve the possibility of deciding otherwise in the future, by this very fact he or she would not be giving totally... This totality which is required by conjugal love also corresponds to the demands of responsible fertility.*

Total love leads to the "greatest possible gift, the gift by which spouses become cooperators with God for giving life to a new human person." Total love proclaims "the central word of Revelation." More broadly, John Paul II writes, all human beings have a "dignity" that "demands that they should be always and solely the term of a self-giving love without limitations of time or of any other circumstance". It seems clear that John Paul II used inflated language ([1], #11–14, 19–20, 37, 80). One might recall that when Jesus spoke of "total love," including mind, heart, and body, he was speaking about love for God. Further, whatever his divine nature, the human Jesus always acted under limitations of time and circumstance. When Jesus spoke of the greatest gift, he referred to dying for a friend. These reservations notwithstanding, John Paul II's introduction of the concept of "total love" served for the next decades as a new foundation for a marital ethic.

Using this foundation, John Paul II attempted to shore up the prohibition against contraception. He no longer defended the Church's prohibition by repeating arguments about the nature of sperm or the nature of the biological organs or the nature of sex or the ends of marriage. Instead, he offers a novel argument that is far from the tradition. He argues that contraception fails because it prevents the self from "total self-giving." Further, he adopts the metaphor of sex as expressive language. Thus contraceptors are said to tell a "lie" by pretending to offer their total self while withholding some aspect of their self. They practice a "falsification of the inner truth of conjugal love." They use "objectively contradictory language, namely, that of not giving oneself totally to the other" while professing total self-gift to the other. By contrast, those who practice natural family planning are said to respect the "inseparable connection between the unitive and procreative meanings of human sexuality" ([2], #11, 32).

The term "objectively" is designed to forestall the obvious possibility that the spouses might honestly tell one another that they do not intend particular sexual acts to express a desire to become parents at the present moment. John Paul II asserts that bodily activity necessarily speaks a word (that is, has a meaning) quite apart from the intentions of those who speak that word. And it cannot have any alternative meaning. This emphasis on objective language of the body itself, furthermore, also enables John Paul to skip over the possibility that the recipient of this sexual word might be harmed by that word. For example, a pregnancy might kill a wife or overburden a family. One speaks the truth, even if it kills.

John Paul continues his personalist argument by asserting that those who practice contraception degrade one another because they split the personal unity of body and soul, nature and person. A distinction between nature and person had become central in the theological debate over contraception. John Paul characterized the distinction as leading to "two irreconcilable concepts of the human person and of human sexuality." In John Paul II's view, those who say they prioritize the needs and life of the "person" over the natural patterns of their bodily cycles in effect create a harmful separation of the person from his or her nature ([3], #32). Instead, only those who accept the natural cycle accept the person. As the Vatican incautiously wrote two years later, "The biological nature of every human is untouchable, in the sense that it is constituent of the personal identity of the individual" ([4], #6). On that criterion, kidney surgery would be unacceptable. As has been noted above, the "sanctity" of the body refers primarily to the sexual organs.

Do those who practice natural family planning fail in total love since they do not totally give themselves during fertile periods? On two occasions, John Paul II proposed that using natural family planning was wrong, if used to not bear children: "it is not possible to practice natural methods as a 'licit' variation on the decision to be closed to life, which would be substantially the same as that which inspires the decision to use contraceptives" ([5], #6). Later, John Paul II said that "natural methods of fertility regulation" should not be considered merely in their functional aspect. Otherwise people would properly "speak of them as if they were another form of contraception" ([6], #2). This position is contrary to that of Paul VI. Presumably, the fact that it was put forward in minor addresses and has rarely been repeated means that it is not a position the Vatican wants to publicize. Nevertheless, this prohibition against choosing natural family planning methods as a way of avoiding procreation is quite consistent with John Paul II's emphasis on total love.

In his lengthy "Letter to Families," John Paul II tries to ground the indissolubility of marriage also in the total love that spouses have for one another. The very nature of love is that it "must be lasting and irrevocable. The indissolubility of marriage flows in the first place from the very essence of that gift: the gift of one person to another." This claim strongly grounds the moral obligation to keep a marriage thriving. Still, John Paul II does not make clear how it grounds the ontological impossibility of divorce. That is, divorce typically is a moral failure in marital love on the part of one or both spouses; but a love union that has become loveless has in practice ceased to exist. It has dissolved. John Paul II responds by explaining that by "the intimate truth" of the mutual communion of spouses, the Church does not mean something subjective; rather it is the objective truth of the spouses ([7], #11–12). This line of argument follows the pattern that a practice means something objective (here, marriage is essentially a love union), even when subjectively there is no love present. This position has much to commend it. Moral norms often critique an absence of what should be. But it also leads to a rather severe disjunction between an "objective nature" and reality as people experience it. For instance, if those who (attempt to) divorce seem happily married in a new loving marriage, they are objectively not married and not really loving. Sexual intercourse with their new partner is adultery, while having an "affair" with their (previous) spouse would be virtuous. The disjunction, which goes back to Jesus [Matt 19:9], has widened in recent times. It is a disjunction between an ahistorical, non-narrative view of marriage and the actual lives of people.

John Paul II closed out the twentieth century when he addressed this tension between the canonical tradition of indissolubility and the reality that marriages were breaking up in ever growing numbers. He began by asserting to jurists that "the central core and foundation" of the canon law view of marriage "is the authentic concept of conjugal love." But then he interpreted this love in a way that makes it loveless. By conjugal love, he meant "essentially a commitment to the other person ... made through a precise act of the will." In other words, conjugal love is simply the exchange of marriage vows at a wedding. John Paul rightly noted that affections of love or mutual attraction cannot provide the necessary stability. Hence he resorted to traditional language: "marriage consists essentially, necessarily and solely in the mutual consent expressed by those to be married. This consent is nothing other than the conscious, responsible assumption of a commitment through a juridical act by which, in reciprocal self-giving, the spouses promise total and definitive love to each other." He added that only a "reciprocal commitment of self-giving ... can guarantee its permanence" ([8], #3–4). The problem, as is evident, is that even a strong commitment of the will at the time of marriage does not guarantee permanence except on the parish registry of sacraments.

3. Vatican documents during John Paul II's papacy

During the long pontificate of John Paul II, various Vatican offices published several documents in response both to the changing culture in the West and to the lack of reception of official teachings by many Catholics. Six documents show the Church adapting to its time, even as it tried not to change its norms.

3.1 Code of canon law 1983

The authors of the 1983 revised *Code* had to incorporate the new understandings of marriage in Vatican II while still preserving the kind of minimalism and precision that is appropriate for law. They made several changes. Where the first *Code* of 1917 said marriage was a contract, the second begins by saying it is a covenant. But, since the biblical notion of covenant, while doubtless richer than contract, is also less clear in what it entails, the 1983 *Code* returns in the very next sentence (and generally throughout) to describing marriage as a "contract." The new *Code* thereby embodies the vexing tension between a more personalistic and more legalistic understanding of marriage. Where the old *Code* spoke of mutual help and allaying of concupiscence, the new *Code* speaks of "the good of the spouses.". In the earlier *Code,* the good of the spouses was not envisioned. Here it is placed first, ahead of procreation; and the updated *Code* highlights a "partnership of the whole of life." Where the old *Code* claimed that husbands and wives exchange "rights over the body," in the new *Code* the exchange is that of personal selves. The personalism of the new *Code* adds that a marriage is not ratified until the spouses have performed the "conjugal act" "in a human manner." Finally, while the old *Code* spoke of "each party," the 1983 *Code* speaks of "a man and a woman," thereby excluding not only polygamy but also–a looming problem–same-sex marriages ([9], #1055.1, 1057.2, 1061.1).

3.2 Congregation for Catholic education

The influence of the psychologies of the twentieth century is manifest in the Congregation for Catholic Education's (CCE) assertion that "sexuality is a fundamental component of personality, one of its modes of being, of manifestation, of communicating with others, of feeling, of expressing and of living human love. Therefore it is an integral part of the development of the personality." The animal model for understanding human sexuality is no longer appropriate. Complementing the usual view that the human spirit must shape and control human bodies, the CCE observes that sexuality shapes the psychological and spiritual levels of human existence. In this new framework, the CCE describes chastity as "the capacity of guiding the sexual instinct to the service of love and of integrating it in the development of the person" ([10], #4–6, 18).

Sexual activities are still not evaluated in terms of stages of growth. For example, the CCE sees the immaturity of adolescent masturbation not as a developmental phase, but as a symptom of profound problems. Similarly, youthful sexual intercourse outside the context of marriage is described as "not personal, but instinctive." That is, sexual activity must be fully personal, or else it is animalistic, without meaning, and simply selfish ([11], #5, 95, 97, 99).

The claims that sexuality is integral to a person's identity raised new questions about celibacy. The CCE boldly admits that virginity is, in one sense, a vocation not to love; it requires one to renounce the love that typifies marriage. This means that it renounces "the maximum expression on the physical level of the married persons' communion of love." Nevertheless, the CCE argues that the real dynamism in sexuality is that of self-giving openness to others. It rightly insists that the dynamism of love itself can be expressed both in marriage and celibacy. Still, it claims--counter-intuitively and without explanation--that those who are virgins can exercise this virtuous love more profoundly than married people ([12], #5, 31).

4. Congregation for doctrine of the faith

The issue of homosexuality became ever more public in the Western world, and the CDF had to address it again. In its earlier document, it had described homosexuality as a pathology, but admitted that it might not be possible to alter this condition. In the meantime, many advocates asserted that homosexuality, like heterosexuality, was not alterable because it was a healthy, constitutive part of a person's identity. As such, they concluded, those who were homosexually oriented should be allowed or even encouraged to fulfill their own distinctive nature. The Congregation reverses this analysis. It does not go from the homosexual orientation to the moral legitimacy of acts, but from the moral wrongness of the acts to the "disorder" of the orientation to those acts. Homosexual acts are said to "annul the rich symbolism and meaning, not to mention the goals, of the creator's sexual design." Homosexual activity, since it is not part of a union that is able to transmit life, "thwarts the call to life of that form of self-giving which ... is the essence of Christian living" ([13], #3, 7). In spite of any experience to the contrary, it cannot be an act of love.

Throughout the twentieth century, there grew a great divergence between Church teaching and the changing experience of people. Whenever there is a divergence between experience and a normative pattern, it can be that the experience is partial or illusory and so should change. But it can also be that what is taken to be the normative pattern itself needs to change. Thus, Church teaching insisted on the procreative meaning of sexuality and marriage. But the Church accommodated for sterility, most commonly in the case of post-menopausal women. The Church seems to have done so out of recognition that marriage and sexual activity can greatly help human beings, even when procreation is impossible.

When, however, it comes to sexual activity that in principle and not just in fact is sterile, the Church will not allow that there can be any good whatsoever involved. Heterosexual acts, even when they are sterile, are the "type" of act that is procreative, while homosexual acts are not. The Vatican then concludes that if a sexual act is not the type that can be procreative, it is by that fact also incapable of expressing love. The inseparability principle is read to mean not merely that these two meanings ought not to be separated, but also that they cannot be separated. Absent one meaning, the other too is gone. Neither of these claims matched the experience of many persons, whether heterosexual or homosexual. The CDF concedes that homosexually active persons may be otherwise generous and self-giving. But it holds that their sexual activity cannot be anything other than self-indulgent. The Vatican admits that homosexual abstinence itself is a denial of self and thus leads to a lack of human fulfillment. Rather than conclude from this deficiency that some accommodation should be made, the Vatican points to this loss as a way of embracing the cross of Christ ([14], #7, 12).

In 1987, the CDF issued an exploratory document, *Donum vitae*, on reproductive technology. Paradoxically, St. Augustine, the source of so much of the Church's sexual ethics, likely would have gladly welcomed these modern technologies since they make it possible to fulfill the reproductive task of marriage without the moral dangers involved in sexual acts. The CDF reports that it is not opposed to reproductive technology merely because its interventions are artificial. Such technological interventions in other parts of human life and in all sub-human animals are permissible. Rather, the immutability and inviolability of the laws of nature given by God refer only to the sexual transmission of life among humans. "The gift of human life must be actualized in marriage through the specific and exclusive" sexual acts of the husband and wife ([15], Intro. #2–4).

The inseparability principle became a double-edge sword. Where previously the Church forbade love-making without openness to baby-making, now it forbids baby-making without a spousal-act of love-making. These new means to achieve the end of marriage are morally prohibited. That is, it is not legitimate to judge the morality of, say, in vitro fertilization "from the totality of conjugal life" to which it contributes nor to view it as part of "the conjugal acts which may precede or follow it." Therefore, surrogacy, heterologous fertilization, and homologous artificial fertilization, even when they might enhance the marriage, are excluded. It is better that marriage be imperfect than that conception take place in an imperfect way ([16], #II.B.4–5).

5. Pontifical council for family

The shift to a sexual ethic based on love is seen unambiguously in the Pontifical Council for Family's (PCF) document, "Truth and Meaning of Human Sexuality." For the PCF, "the body also expresses spiritual love." Instead of being described as a remedy for concupiscence, sexual activity in marriage has the lofty goal of enabling spouses to grow, "and at the same time it contributes to building up the civilization of love." This love includes and surpasses friendship and "is achieved when they give themselves totally" ([17], #3, 11, 14).

The problem with the shift to a sexual ethic based on love appears in the challenge faced by the PCF's document, "Family, Marriage, and 'De Facto' Unions." More and more people were cohabiting with no commitment to marriage. For many, marriage was not the first step into adulthood but the last step after adulthood had basically been attained. The PCF wisely judged that many de facto unions, regrettably, spring from and contribute to an individualism and a privatism that neglects the common good. Somewhat surprisingly, compared to other Church documents, the PCF sensitively recognizes that, in these de facto unions, there often is "reciprocal affection." What they lack is the "marriage bond, with its original public dimension." Established by love and free consent, these cohabiting unions still do not have the public and formal commitments and responsibilities of marriage, enforceable by law. Therefore, the PCF strikingly concludes, it is inadequate to speak of love "as the basis of marriage." While de facto unions can describe themselves as "a community of life and love," marriage is distinguished by being an institution of conjugal love ([18], #2, 9, 11, 12, 20, 34). This reference to "conjugal love," as distinct from other kinds of love, makes a significant advance. Unfortunately, its distinctiveness is rarely clarified.

6. Catechism of the Catholic Church

The *Catechism of the Catholic Church* (CCC) announces that it intends not to say anything new, but only to present "an organic synthesis of the essential and fundamental contents of Catholic doctrine." It asserts that God is the author of marriage, though it also admits that the institution has undergone many variations in different cultures. With a nice turn of phrase, it interprets the biblical role of women as "helpmate" to mean that she "thus represents God from whom comes our help." In a balanced assessment, the *Catechism* explains, "After the fall, marriage helps to overcome self-absorption, egoism, pursuit of one's own pleasure, and to open oneself to the other, to mutual aid and to self-giving." (It assumes that pursuing one's own pleasure is wrong.) The *Catechism* recognizes that some people are perplexed by the Church's "unequivocal insistence" on such things as the indissolubility of the marriage bond. Nevertheless, it claims the "Church does not have the power to contravene" the "disposition of divine wisdom" ([19], #11, 1602, 1605, 1609, 1615, 1640).

The *Catechism* urges spouses to understand that they are "*cooperating with* the love of *God the Creator* and are, in a certain sense, its interpreters" ([20], #2367–70). This continues one of John Paul'II most significant theological changes. The spouses' interpretation, however, would not be responsible if they thought it might be permissible to take an action that would make conception impossible. Unfortunately, the ready identification of God with biological nature but not with human ingenuity was becoming less plausible as contemporary medicine made ever greater alterations in human life.

The *Catechism* exhorts people to "acknowledge and accept" their sexual identity, which includes "physical, moral, and spiritual *difference* and *complementarity*." The *Catechism* updates the traditional description of chastity. The first aspect of chastity is what traditionally was called temperance. The *Catechism* calls it the "integrity of the person" and describes it as "self-mastery." It acknowledges that there are "stages of life," and draws from this that one should never consider self-mastery to be fully acquired. The second aspect is described in a neologism, "the integrality of the gift," which is a new name for love of another ([21], #2332–2333. 2338, 2342, 2346). Its importance is that chastity, unlike its presentation in much of the tradition, now includes more than self-regulation. It also includes how one sexually relates to others.

The *Catechism*, as one might expect, proscribes a list of sexual sins. Masturbation, fornication, and pornography are wrong since they seek pleasure outside the marital relationship. Prostitution violates both the dignity of the prostitute and the virtue of the one who pays for sex. The *Catechism* recognizes that a homosexual orientation is not chosen. Still homosexual acts are contrary to the natural law since they do not proceed from sexual complementarity and since they "close the sexual act to the gift of life" ([22], #2351–2355, 2357). The first reason given is simply a description of homosexuality, but it implicitly presupposes the normativity of dimorphic complementarity. The second charge is less clear, since homosexuals directly neither physically nor psychologically close sex to fertility; rather they practice sexual activity that, de facto, is not fertile and so their acts are not completely different from what infertile heterosexuals do.

The *Catechism* bans the use of reproductive technology because such interventions "infringe the child's right to be born of a father and mother known to him and bound to each other by marriage. These medical practices betray the spouses' 'right to become a father and a mother only through each other'" ([23], #2376). Apart from the appearance that these two rights seem to have been invented after the fact for the purpose of proscribing the new technologies, the difficulty remains that the "child" who would exercise this right would never come to existence. Nor would the spouses who exercised this right ever become parents. Indeed, by these criteria, adoption would be proscribed.

Lastly, the *Catechism* says that attempting divorce is a grave offense: "It claims to break the contract, to which the spouses freely consented, to live with each other till death." That point notwithstanding, the text then says that separation, in which the spouses also do not "live with each other till death," can be legitimate. Similarly, the *Catechism* requires a polygamist to "repudiate one or more wives with whom he has shared years of conjugal life." But then it adds, the polygamist "has a grave duty in justice to honor the obligations contracted in regard to his former wives" ([24], #1649, 2383–84, 2387). Those very obligations, however, previously included living sexually with such wives until death. These two contradictory points are minor in the large scheme of marital theology discussed in this essay; raising them, however, calls attention to the reality that complications abound in marital lives. In allowing for "separation" and in insisting on justice for former wives, the Church rightly accommodated reality.

Lest there be further disjunction between the Church and the experience of the people of God, the Church may have to accommodate further revisions to meet new realities in the twenty-first century. Some theologians, for example, have argued that it is morally wrong to break the marital promise of life-long fidelity. But this sin, like other sins, can be forgiven. They hold that divorce is a moral wrong, to be repented and reversed where humanly possible, but not an ontological impossibility.

7. Conclusion

In his "Letter to Families," John Paul II acknowledged that the Church's teaching alienates people. He wrote:

> *The Church's Magisterium is often chided for being behind the times and closed to the promptings of the spirit of modern times, and for promoting a course of action which is harmful to humanity, and indeed to the Church herself. By obstinately holding to her own positions, it is said, the Church will end up losing popularity, and more and more believers will turn away from her ([25], #12).*

These are serious charges. The Church does not adequately respond by repeating ideas that no longer compel assent. At some point, Haidt's observation that the tendency to find or create reasons to justify prior intuitions breaks down. Cognitive dissonance sets in. Then new intuitions about the meaning of sexuality may supplant the older intuitions.

Speaking in a very different context, Pope Benedict XVI wrote in *Spes salvi* that "every generation has the task of engaging anew in the arduous search for the right way to order human affairs; this task is never simply completed. Yet every generation must also make its own contribution to establishing convincing structures of freedom and of good" ([26], #25). The popes of the twentieth century have engaged in this arduous search. But many of the people of God have not been convinced by the results. To them the absolutisms the Church claims for marriage too often do not promote freedom and the good of persons. Unfortunately, many have drifted away or even deliberately stomped out of Catholicism. Unfortunate too is that many more no longer pay heed to the Magisterium's statements on sexuality. Perhaps some will take heart in the recent letter by Pope Francis, ([27], #301), in which he writes, concerning second marriages, "it can no longer be said..."

Our current culture should hardly be fully embraced. The contemporary loss of a concern for authority, loyalty, and the sacred has opened the door to severe problems. Current cultural patterns of sexual activity underscore the danger of forgetting the older, negative concerns about the hazards of sexuality and the need for a communal ethic. In the current "hook-up" era, among 18–23 year old, non-married Americans, 71% have had oral sex and 73% have had sexual intercourse. In the twenty-first century, "the average age for both first oral sex and first sexual intercourse is 16 years old" ([28], p. 149, 153). Unwed mothers account for over 40% of births ([29], p. 1). Further, in the past 20 years, the percentage of people in the United States who think children are very important to marriage has declined by over 35%; and approximately 20% of women who reach menopause remain childless ([30], p. 3). The pendulum is swinging nearer to the other end. As is almost always true, older understandings of the dangers and problems involved in sexuality, based also on human experience, had considerable purchase on the truth.

Haidt's conservative triad of sacredness, authority, and loyalty, in addition to promoting needless taboos, denigrating women, and fostering haphazard procreation, also imparted a sense of humanity to sexual activity. There is something about the "sacredness" of sexuality that is being lost. The loss of any cultural, legal, or ecclesial authority in sexual matters—authority at least in the sense of credible teaching and teachers—leaves people without adequate guidelines for sexual activity. The loss of an appropriate loyalty to one's sexual partner appears in uncommitted sexual activity, adultery, divorce, and single-parent children. The pendulum should oscillate nearer to the middle of its arc.

Vatican II said that the Church "requires special help, particularly in our day, when things are changing very rapidly and the ways of thinking are exceedingly various. She must rely on those who live in the world" ([31], #44). In the twentieth century, the Church did rely on those married people who experienced that sexuality is good and that love is part of both sexuality and marriage. Doing so had the unexpected consequence of opening the understanding that there is a spousal love, expressed sexually, that was worthy in itself whether or not it produced children. Once the Church accepted that infertile spouses could engage in love-making without the possibility of baby-making, the next question was whether one could deliberately make procreation unlikely or impossible. This became more and more plausible at least in those cases where strictly adhering to biological "nature" seemed unloving because it threatened harm to spouses, to other children, to the marriage union itself, or to the common good. Further, after the Church had reworked the foundations for its sexual morality, it appeared to many of the people of God that there should be a similar reformation of several sexual norms.

Many of the "people of God," in concert with their non-Catholic counterparts with whom they shared in various other communities, began that reformation. They adapted and changed various norms in light of their actual personal, interpersonal, and social lives. They seem to have turned to what the Congregation for the Doctrine of the Faith recently said is the genius of women, namely, "A sense and a respect for what is concrete develop in her, opposed to abstractions which are so often fatal for the existence of individuals and society" ([32], #13). In the twenty-first century, with the rise of the Church's appreciation for women's dignity, this respect for the concrete continues to motivate the search for a better marital morality.

The present century has led to even more radical changes in the sexual mores of the West, but that is a tale for another occasion.

Author details

Edward Collins Vacek S.J.
Loyola University New Orleans, Louisiana, United States of America

*Address all correspondence to: evacek@loyno.edu

References

[1] Familiaris Consortio, 1981, http://www.vatican.va/holy_father/john_paul_ii/apost_exhortations/documents/hf_jp-ii_exh_19811122_familiaris-consortio_en.html.

[2] Familiaris Consortio, 1981, http://www.vatican.va/holy_father/john_paul_ii/apost_exhortations/documents/hf_jp-ii_exh_19811122_familiaris-consortio_en.html.

[3] Familiaris Consortio, 1981, http://www.vatican.va/holy_father/john_paul_ii/apost_exhortations/documents/hf_jp-ii_exh_19811122_familiaris-consortio_en.html.

[4] Educational Guidance in Human Love, 1983, http://www.vatican.va/roman_curia/congregations/ccatheduc/documents/rc_con_ccatheduc_doc_19831101_sexual-education_en.html.

[5] "Natural Family Planning," Origins: CNS Documentary Service 20 no. 31, January 10, 1991, 508, no. 6. Speech given on 14 December 1991, http://www.vatican.va/holy_father/john_paul_ii/speeches/1990/december/documents/hf_jp-ii_spe_19901214_fertilita_it.html.

[6] John Paul II, "Message of his Holiness Pope John Paul II to the Director of the Centre for Research and Study on the Natural Regulation of Fertility," February 27, 1998, http://www.vatican.va/holy_father/john_paul_ii/speeches/1998/february/documents/hf_jp-ii_spe_19980227_fertilidade_en.html.

[7] "Letter to Families," http://www.vatican.va/content/john-paul-ii/en/letters/1994/documents/hf_jp-ii_let_02021994_families.html.

[8] http://www.vatican.va/content/john-paul-ii/en/speeches/1999/january/documents/hf_jp-ii_spe_19990121_rota-romana.html.

[9] The Code of Canon Law, 1983, http://www.vatican.va/archive/ENG1104/_INDEX.HTM.

[10] http://www.vatican.va/roman_curia/congregations/ccatheduc/documents/rc_con_ccatheduc_20051996_profile_en.html.

[11] http://www.vatican.va/roman_curia/congregations/ccatheduc/documents/rc_con_ccatheduc_20051996_profile_en.html.

[12] http://www.vatican.va/roman_curia/congregations/ccatheduc/documents/rc_con_ccatheduc_20051996_profile_en.html.

[13] http://www.vatican.va/roman_curia//congregations/cfaith/documents/rc_con_cfaith_doc_19861001_homosexual-persons_en.html.

[14] http://www.vatican.va/roman_curia//congregations/cfaith/documents/rc_con_cfaith_doc_19861001_homosexual-persons_en.html.

[15] Donum Vitae [1987] http://www.vatican.va/roman_curia/congregations/cfaith/documents/rc_con_cfaith_doc_19870222_respect-for-human-life_en.html.

[16] Donum Vitae [1987] http://www.vatican.va/roman_curia/congregations/cfaith/documents/rc_con_cfaith_doc_19870222_respect-for-human-life_en.html.

[17] http://www.vatican.va/roman_curia/pontifical_councils/family/documents/rc_pc_family_doc_08121995_human-sexuality_en.html.

[18] http://www.vatican.va/roman_curia/pontifical_councils/family/documents/rc_pc_family_doc_20001109_de-facto-unions_en.html.

[19] https://www.vatican.va/archive/ENG0015/_INDEX.HTM.

[20] https://www.vatican.va/archive/ENG0015/_INDEX.HTM.

[21] https://www.vatican.va/archive/ENG0015/_INDEX.HTM.

[22] https://www.vatican.va/archive/ENG0015/_INDEX.HTM.

[23] https://www.vatican.va/archive/ENG0015/_INDEX.HTM.

[24] https://www.vatican.va/archive/ENG0015/_INDEX.HTM.

[25] http://www.vatican.va/content/john-paul-ii/en/letters/1994/documents/hf_jp-ii_let_02021994_families.html.

[26] Spe Salvi, 2007, http://www.vatican.va/holy_father/benedict_xvi/encyclicals/documents/hf_ben-xvi_enc_20071130_spe-salvi_en.html.

[27] Amoris Laetitiae, 2016, http://www.vatican.va/content/francesco/en/apost_exhortations/documents/papa-francesco_esortazione-ap_20160319_amoris-laetitia.html.

[28] Christian Smith et al., Lost in Transition: The Dark Side of Emerging Adulthood [New York: Oxford University Press, 2011], 284 p. ISBN-13: 978-0199828029.

[29] http://www.cdc.gov/nchs/fastats/unmarry.htm.

[30] Joel Kotkin, Research Report for the Seminar on the Rise of Post-Familialism: Humanity's Future? Singapore: Civil Service College, 2012, 3.

[31] Gaudium et Spes, 1965 http://www.vatican.va/archive/hist_councils/ii_vatican_council/documents/vat-ii_cons_19651207_gaudium-et-spes_en.html.

[32] Letter to the Bishops of the Catholic Church on the Collaboration of Men and Women in the Church and in the World, 2004. http://www.vatican.va/roman_curia/congregations/cfaith/documents/rc_con_cfaith_doc_20040731_collaboration_en.html.

Chapter 21

Medical Ethics

Dabota Yvonne Buowari and Kehinde Kazeem Kanmodi

> *"Without ethics, everything happens as if we were all five billion passengers on big machinery that nobody handles, and its going faster and faster but we don't know where".*
>
> *'Jacques Cousteau'*

Abstract

Medical ethics is very important at this time of medicine due to the covid-19 pandemic which has caused a lot of mortality and morbidity world wide. Medical is important in guiding clinicians and other healthcare workers not to cause harm while caring out their duties. There are various aspects of medical ethics such as negligence, do not cause harm, beneficent and others. Not practicing medical ethics has caused some physicians to face litigation by their patients and clients and also face medical disciplinary boards. Some have even been suspended from medical practice for a number of months or years or even have their names struck off from the register of the licensing board of their country.

Keywords: medicine, ethics, litigation

1. Introduction

Every profession has guidelines, rules and regulations guiding its practice, this is what ethics is all about. It has to do with the professional conduct of the practice of medicine including conducting research for the advancement of medicine and the making of new discoveries. One of the necessary aspects of human life is health [1] and it is the duty clinicians to care for the health needs of the populace [2]. Medical ethics has been from time immemorial. In medical education, medical ethics is taught at both the undergraduate and postgraduate levels to equip a new graduate in the dental and medical profession. Clinicians' face various challenges in the course of their work some of which may have ethical issues, worst are when there is poor clinical state and death is anticipated [3]. Decision making on patients care is a critical responsibility of every physician practicing medicine [4]. Physicians who err in their practice may be tried by the medical council. Each national medical council has guidelines for erring doctors who derail in their practice of medicine. Medical councils have a panel of enquiry and disciplinary boards for doctors who do not practice medicine without the application of the medical ethics. Various disciplinary measures include apology to the patient, suspension from practice for a period of time, or the extreme of which is the permanent seizure of the medical practicing license and striking off the doctors' name from the medical practicing register. Due to the importance of ethics the World Health Organization in 2002 established an ethics team known as the global health ethics unit, a unit dedicated to ethics [5].

IntechOpen

Through this ethics team, the World Health Organization works in close collaboration with the United Nations interagency committee, non-governmental organizations and other international organizations [5].

2. Overview of ethics

Ethics is a concept in the field of philosophy. The word ethics is derived from the Greek word 'Ethnos' which means character, customs and habits. It is closely linked to the word 'moral' which is derived from the Latin word 'Mos' (mores) but they are different. Both ethics and morals generally mean the customs and socially accepted norms. Ethics is a branch of philosophy and includes the values, guidelines, rules and regulations and the justification for these values, and guidelines [6, 7]. Hence ethics is required in our daily lives in making choices from various available options and alternatives [8]. Ethics guides various professions in carrying out their duties while morals have to do with the way of life. Ethics is concerned in the concept of right and wrong. This was first conceptualized and structured by many Greek scholars and established by Aristotle by the third century BC. Various factors have affected the concept hence it has evolved over many centuries. It is the systematical philosophical study of morality. Ethics guides human beings where they are as it is the set of principles and values governing a group or even an individual. It also extends to the consequences of wrong misdoings. The application of values and moral to human activities also constitute ethics [9]. The conduct of right and wrong are both related to ethics and morals. The principles developed by an individual regarding what is perceived as right or wrong are morals. Morality can sometimes transform to cultural and regional norms while ethics is not affected by societal, cultural or religious norms. The standard of conduct which guides and governs an occupation or profession is knowns as professional ethics. Ethics is connected to the code of conduct, it guides how employees and employers to conduct themselves at the workplace. The understanding of moral values is related to ethics hence the relationship between ethics and morals [10]. Ethics is widespread in all areas of life as decision making is a part of life [11].

3. Overview of medical ethics

Medical ethics is a type of professional ethics. In modern medicine, ethics is believed to have started in the 18th century; a physician by name Thomas Percival authored a book on medical ethics which is believed to be the beginning and development of medical ethics code of conduct [7, 12, 13]. It is Dr. Thomas Percival who coined the term medical ethics after publishing his book in 1803. Medical ethics is the conduct required from any medical practitioner, it is necessary for the physician as it acts as a guide in making clinical decisions [14]. Medical ethics is the ethical, morals and values aspect that guides the medical profession and its allies and it consists of interdisciplinary knowledge [15, 16]. It guides decision making, medical practice, medical education and research in medicine [10, 12, 17]. There have been ethical guidelines which must be followed by healthcare workers in the history of the medical profession [6]. In 1949, the international code of medical ethics has been adopted by the World Medical Association in London [17]. Every national medical board has a code of conduct to guides the practice of medicine in that country. For instance, in Nigeria, medical ethics is governed by the Medical and Dental Council of Nigeria (MDCN) which published a booklet titled 'A Code of Conduct' which is handed over to every new medical and dental graduate at the time of induction into the medical and dental profession. Medical ethics is a type of applied ethics. The basis of the medical ethics is centred on the Hippocratic Oath [7] which is an oath is taken by every dental and medical graduate at the time of been inducted into the medical and dental profession. They are set of rules, regulations and guidelines that guides and governs physicians in carrying out their duties [18]. Medical ethics focuses on the relationship that exists between the doctor and their patient, which includes the legal and ethical implications. Hospitals also ensure that their employees practice medical ethics to prevent litigations which can cause loss of resources to the health facilities. Medical ethics guides decision making of medical practitioners as patient centred care is based on medical ethics. It is expected that medical practitioners are well equipped with medical skills and knowledge and also they are familiar with the medical ethics and its legal implications [9]. Medical ethics is a branch of applied ethics and bioethics. Medical ethics promotes the respect of patient and confidentiality. Medical ethics is important in the practice of medicine [10] therefore it is taught in most medical and dental schools both at the undergraduate and postgraduate levels. It is applied in all clinical settings as well as the medical workplace and in research; it encompasses other disciplines such as history, philosophy, theology, anthropology and sociology [18]. The different categories of professionals involved in providing healthcare practice medical ethics [19]. Therefore for a better understanding of the concept training in medical ethics should be incorporated into undergraduate and postgraduate education curriculum [10, 15, 20]. It is the duty of every physician to practice medical ethics in every consultation with the patient.

4. Importance of medical ethics

Medical ethics is important both in medical practice which involves the patient doctor relationship and in medical research.

Some of the roles of medical ethics are:

1. It provides standards in the professional relationship between the physician and their clients or patient hence provides guidelines in the prevention of litigation [17, 21, 22].

2. The social capital in the professional relationship is established with members of the community [21].

3. Medical ethics is implemented in decision making by both the physician and the patient [23].

4. Medical ethics provides moral values necessary in providing solutions to ethical dilemma [23].

5. It provides privacy, confidentiality and truthfulness in the doctor-patient relationship [19].

6. Medical ethics promotes health, wellbeing, respect decision making, dignity, justice and accountability in the medical profession [17, 19].

7. Medical ethics helps in promoting good and quality medical care by identifying, analyzing and attempting to resolve the ethical problems that arise in medical practice [10, 17, 24].

8. Medical ethics promotes diligence and proper training skills among healthcare professionals [7].

9. Medical ethics helps in the prevention of unethical practices such as negligence and malpractice.

5. Ethical dilemma

Controversies and conflict sometimes occur in the practice of medicine especially in the decision making process. Sometimes, these conflict occurs in the shared decision making as conflicts arise when the doctors and patient decisions contravenes medical ethical principles. Generally everyday there are ethical dilemma occurs daily in the practice of medicine [14, 25, 26]. Ethical dilemma has been described by the World Health Organization as a dilemma between different values which are seen as necessary particularly in cases and circumstances in conflict with each other [10]. It occurs when all the possible remedy for a clinical care will lead to moral violation [8, 23]. Sometimes, there are no answers to ethical dilemmas [27]. These ethical dilemmas arise when there are options for the decisions which may be compelling reasons and actions [19]. Ethical dilemma consists of a type of ethical problem. From several researchers, there is a relationship between ethical dilemma and ethical principles [28]. No ethical principle can explain adequately ethical dilemma [2]. There is a connection between facts, values and morals conflict, some of the problems associated with ethical dilemma have been there for centuries with medical ethics [29]. Ethical dilemma is a product of conflict arising from ethical principles and options [30, 31]. When ethical dilemma occurs, it can be resolved using individualistic approach as there are no general principles in tackling it [17, 23]. Ethical dilemma is not limited to medical practice only as it also occurs in medical research [14]. The establishment of a comprehensive ethical framework and legal framework will guide medical health care workers in resolving ethical dilemma.

Ethical dilemma can be found in telemedicine, artificial intelligence, COVID-19 testing, management of near end of life care, medical error, priority testing, biotechnology, medical ethics education, e-health and bioethics [10]. Some of the factors and barriers to ethical dilemmas are connected to medical facts, individual characteristics and unclassified factors [31].

Medical Factors: Some of the medical factors associated with ethical dilemmas are the patient's history, diagnostic results, risks, complications and previous intervention associated with the illness the patient is suffering from.

6. Individual characteristics

There are some characteristics that are peculiar to everyone that can lead to ethical dilemma. These factors affect the decision process of the patient. These include values, culture, religion, relationships and previous experiences [31]. The belief system affects medical care.

7. Unclassified factors

There are some factors not associated with individual characteristics and medical affects that affect decision making process. These include the logistics facility, competing interests and Interprofessional perspectives [31]. In some countries, there have been Interprofessional rivalry between the different categories of health care workers and the clients and patients are the ones who suffer when such occur.

8. Some ethical dilemmas

Healthcare professionals regularly encounter ethical dilemmas while carrying out their duties [3, 17, 27]. Some situations in which ethical dilemma can occur are discussed below:

1. **Near End of Life**: End of life is a topic and phenomenon that has been debated for several decades. There is the ethical dilemma if end of life care should be provided or should such patients be abandoned? Is end of life care a waste of resources? Especially in low resource counties where health insurance is not available. Near end of life care is expensive as it may involve artificial nutrition hydration, telling the patient's care givers the truth and disagreement that may arise in the course of management. Such disagreements may include continuation of artificial ventilation or administration of oxygen, care in the intensive care unit. Moral distress sometimes is experienced by the physician, patient and their caregivers when there are unexpected clinical developments as death approaches [3].

2. **Telemedicine:** Telemedicine means "healing from a distance" and it is very beneficial [32]. Telemedicine is becoming popular due to the need for social and physical distancing especially in this time of the COVID-19 pandemic. During telemedicine proper history taking may not be gotten and a comprehensive physical examination may not be done. Physical examination is important in every medical consultant. Ethical dilemma occurs in telemedicine due to the conflict in productivity and patient confidentiality which cannot be obtained. Privacy cannot be attained in the patient-doctor relationship while using telemedicine [30]. This is worst if artificial intelligence is also used. During consultation using telemedicine, it must be done in a secure and safe way so that medical electronic information is not leaked out. Therefore passwords security should be maintained always [32].

3. **Coronavirus-2019 (COVID-19) Testing**: The ethical considerations in the management of epidemics are different [16]. This is not different for the Covid-19 pandemic as several ethical principles are affected by the pandemic. This includes autonomy, truthfulness, confidentiality and justice. In the Covid-19 pandemic, testing is done without consent of the patient. This is also to help in the control and to protect others. Also the ethical principle of autonomy is not respected during contact racing [33]. Confidentiality is not maintained in the Covid-19 testing in order to protect the public. This is why it is required before air travel especially international travel and also admission and entry into certain public places such as schools and camps. There is the ethical dilemma in the allocation of the scare resources and medical supplies, therefore question to be answered are who need what? Hence the ethical principle of justice has to be practiced. The physician decides which patient may likely die or lives, which patient should be connected to the ventilator, which is in short supply worldwide. Allocation of these scares medical resources must be done fairly and with justice [16, 34]. This has created a high ethical dilemma which intensivists have to deal with [35]. Rationing of medical equipment is a great dilemma ravaging

the world due to the covid019 pandemic [34]. Also coupled with the increase patient load in intensive care units [35]. In other to protect the public, there is the ethical dilemma when a patient with the signs and symptoms of the Severe Acute Respiratory Syndrome-2 refuses to get tested.

4. **Medical Error**: Errors can occur in any profession but it can lead to fatalities in some sectors. Some of such sectors are the aviation, architecture and medicine as any mistake can lead to disability and even death. Not all medical errors can lead to disability and death but it may increase hospital stay and loss of work days. This is a serious issue especially as there are increase cases of litigation. There is the ethical dilemma of weather the patient or client should be informed about medical errors when they occur.

Ethical dilemma is also encountered when dealing with "Do Not Resuscitate Order".

9. Ethical dilemma: case report

A 72 years old woman was brought to the emergency room in a developing country by her children with complaints of unconsciousness and difficulty in breathing of 24 hours duration. She is a known hypertensive not compliant with her medications. On examination, she was not pale, unconscious with a Glasgow Coma Scale of verbal response – 1, eye opening – 1, and motor response – 1 with a total of 3/15. She was not cyanosed, in respiratory distress evidenced by flaring alae nasi and subcostal and intercostal recession. Oxygen saturation was 76%, she was gasping for breathing, auscultation of the chest yielded vesicular breath sounds. Respiratory rate was 60 cycles per minute and pulse rate was 100 beats over minute, regular and full volume. Blood pressure was 150/100 mmhg. A diagnosis of cerebrovascular accident was made. She was placed on oxygen, 20% mannitol and normal saline. After admission, the patient was not improving and the patient's caregiver was paying out of pocket for the management of the patient. After twenty four hours of hospital admission, the patient's daughter requested that the patient was in a bad clinical state therefore she wants the oxygen to be discontinued, that she (the daughter) feels that the mother may not survive the illness. She was counseled on her mother's condition but she still insisted that the oxygen should be discontinued. After much argument, she was asked to write an undertaken that she is the one requesting for the discontinuation of oxygen therapy. At this point she refused to put it down in writing and started crying. The oxygen was never discontinued.

10. Discussion of the case report

This is a case of ethical dilemma on end of life care and care of a patient unable to make decisions. When patients are unable to make decisions for themselves such decisions are made by a legal caregiver who may be patient, child, guardian or legal representative. In some developing countries like Nigeria where the extended family system is practiced and the people live a communal life, the caregiver may be any relative or even a neighbor; the group of people who cannot make decisions for themselves are the unconscious patient, children and minors and the mentally impaired. Some of the decisions made by a caregiver may not always be in the best interest of the patient. For instance in this case, the patient is unconscious and the caregiver is her daughter. Even though the patient was in bad clinical state, oxygen therapy was necessary for her management hence the dilemma of whether or not to discontinue the intranasal oxygen. There is conflict between the ethical principles of autonomy and maleficence. This can lead to litigation against the attending physician. If the physician decides to obey the wish of the caregiver, even though she takes an undertaken, the conflict of maleficence in which a doctor is required not to do harm may arise. The physician will have no justification to discontinue the administration of oxygen if taken to a court of law as he or she has received medical training.

11. Institutional review boards

In the practice of medicine, research is an important and essential tool [36–38]. Institutional review boards are ethical committees in institutions that analyze and review proposed research protocols, it serves as a deliberation forum in which ethical issues in medical researchers are analyzed (WHO). Some institutions have more than one institutional review board [39]. Institutional review board plays an important role in medical practice. Some of these roles of the institutional review boards are to protect human subjects in the course of any medical research [39]. This is because some researches may have detrimental effects on research participants, some of these effects may lead to morbidity and even mortality. Institutional review board act as risk benefit analysis intermediary between research participants and the researcher. They also determines if the research should be commenced or not [38]. They ensure that the research is conducted as specified by the researcher in the research protocol. Institutional review boards are important in the improvement of medical practice by working with researches to apply good ethical principles in their research [40]. Each institutional review board has its guidelines. Some have documents that must be used while applying for ethical approval for any study. The institutional review boards are made up of persons with expertise in medical ethics and medical specialties. There are different types of institutional review board namely national ethics committee, research ethics committee and clinical ethics committee (WHO). Globally, there are different local and medical national institutional review boards and communities [17]. An example of a national ethics committee is the National Health Research Ethics Committee of Nigeria (NHREC). In medical research, the human subjects are a very valuable resource hence there safety has to be protected [37]. Extra scrutiny is done on research that will be carried out among vulnerable populations [38]. This is to avoid coercion in the recruitment of research participants. The vulnerable populations are children and minors, older persons, the mentally retarded, pregnant women and people in conflict and war zones.

12. Principles of medical ethics

Since the olden days, it is believed that the doctor knows it all and also knows the best [41]. These have been challenging as there is an increase in litigation against doctors, practice and implementation of ethical principles helps as the preventive measure against such medical litigation and jurisprudence. Modern medicine is faced with several ethical problems [29]. Some of these problems can be abated by ethical principles. Clinicians try to do their best for their patients by providing the best medical care available. These ethical principles guide physicians in decision making in the course of their work especially where there are ethical dilemmas and helps in the resolution of ethical conflict [27]. Hence physicians generally act in the best interest of their clients and patient though conflict may sometimes occur as they weigh the risks and benefits of all available choices but act in the best interest of the patient [2, 27]. The principles of medical ethics [3, 4, 10, 27] are:

1. Autonomy

2. Justice

3. Beneficence

4. Non maleficence

5. Confidentiality

6. Truth telling

12.1 Autonomy

The fundamental principle of medical ethics is autonomy [42] Autonomy is the freedom of patients and clients to make their decisions on their conditions without the interference, pressurization and duress. It means giving adequate information to the patient respectfully and disclosure of information about a patient after obtaining informed consent to do so [4, 43, 44]. Trust is always key in every doctor patient relationship as the patient rusts the doctor to the best for him or her therefore the patient is entitled to autonomy. It is the duty of the doctors to counsel and explain to the patient the diagnosis, proposed management and treatment options. The doctor is not expected to impose any decision on the patient. Hence, the freedom of thought, intention and decision making process especially in the new era of shared decision. For this to be complete, the patient should be counseled in simple language so that they understand the risks and benefits of the procedures. This is also the principle for not to do evil or inflict harm. Autonomy allows the freedom of choice and action by the patient. The principle of autonomy requires the physician should provide all available therapeutic options to the client [21]. This also shows that the patient even after all the counseling has the right to refuse and reject treatment [10]. Ethical principles are affected by cultural and traditional beliefs and practices. Some cultures frown against being told that the clinical condition is poor worst is if death is anticipated. Since time immemorial, doctors have been faced with the notion of to what extent and how much clinical information should be released to a patient especially when it is bad news [45]. This is in conflict with the ethical principle of autonomy for which people should be allowed to make their decisions without any influence or been coerced to do so. The other medical ethical principles of truth telling, and confidentiality including informed consent are all based on autonomy [4]. Autonomy allows the patient to choose from every available treatment options depending on their goals and values [41, 44, 46]. Although sometimes doctors are able to convince their patients to accept what they believe is the best for the patient [41]. Autonomy therefore makes patients to be responsible for their health needs and wishes [11].

12.2 Beneficence

The principle of beneficence is that everything done by the medical practitioner should be in the best interest of the patient [4, 10, 47]. Therefore this guides the decision making process of the physician. This means that all negative options which will not be in the best interest of the patient should not be offered to the patient. An example is administering a medication to a patient because of the side effect of the drug. The summary of beneficence is to do good always. Though every physician has to practice the ethical principle of autonomy, options that shall be beneficial to the patient should only be offered to be patient [47]. This ethical principle is implemented in the choice of drugs as all medications have adverse effects some of which may be mild which can be tolerated while some others are severe which can worsen the clinical state and can even cause death.

12.3 Non-maleficence

The ethical principle of non-maleficence is related to beneficence but they are different [44]. The principle of non-maleficence states that no harm should be done to the patient or other people in the community [4, 9, 10, 17, 48]. Implementing the principle of non-maleficence means that any treatment option that will be harmful to the patient should not be offered to the patient [44] as the patient will also exhibit autonomy. Hence the patient's medical condition may worsen. Violation of non-maleficence can lead to litigation and malpractice. Negligence is a consequence of the violence of non-maleficence. Medical error can result in non-maleficence. This also applies to research in which the research protocol must be reviewed by an institutional review board so that none of the research participants suffer harm. All the benefits, risks and consequences of all treatment are weighed in the course of the medical consultation [49]. It is the duty of the physician to protect their patients [48]. Though all health practitioners encounter ethical challenges in the course of their work, the principle of not to do harm is always a priority [17].

12.4 Justice

Justice requires fairness in the management of patients and distribution of resources especially in the time of scarcity and when priority needs to be maintained such as during mass casualty and pandemics. Individuals at all time should be treated fairly when they visit a health facility [10]. The distribution of health resources requires justice for it to be done fairly and equitably [1, 4, 19, 47]. Justice is also necessary in respecting the rights of patients [19]. In times of scarcity, the ethical principle of justice is used to determine areas of priority in the distribution of health resources [10].

12.5 Confidentiality

Confidentiality and privacy generally is required in any human relationship [43]. A physician is expected to maintain confidentiality of all discussions made with the client or patient [50]. The physician requires permission and consent of the patient before divulging such information to any other person even among fellow physicians and health care professionals. The ethical principle of confidentiality is related to other ethical principles of autonomy and truthfulness. Patients trust their physicians hence that can open up their privacy to the doctor and trust that the doctor will keep secret all information confided in him or her to be private. Patients and their caregivers hold different preference in the disclosure of medical information about them or their loved one to third parties [44]. The ethical principle of confidentiality is based on trust hence patients trust the clinician therefore they tell them the truth and except the doctor to keep it private to themselves only without their authorization or informed consent [4, 43, 51]. Part of a patient's right dwells on the physician respecting patient confidentiality. Trust in the patient doctor relationship is based on trust confidentiality and trustfulness. This issue of confidentiality is also maintained even after the death of the patient [32]. The doctor patient relationship is strengthened by confidentiality including communication between them [51]. It is ethical and legally binding on doctors to respect patient confidentiality always [50, 52]. Sometimes, the ethical principle of confidentiality is breached when the medical information is required by a court or law or when the illness is a threat or will be harmful to others and the public. Patients put their trust in their physician hence it is importance to maintain this trust to meet all legal requirements [32]. Confidentiality is breached when communicating with patients through an interpreter.

12.6 Truthfulness

Generally, most people say that it is hurtful to be told the truth but it has to be told whether it is palatable or not [45]. The physician is expected at all times to tell the patient the truth about their clinical condition. Truthfulness is also known as truth telling is one of the principles of medical ethics. Truth telling is guided by trust and confidentiality. Every patient expects the physician to tell them the truth about their illness always. It is very important in the advancement of patient autonomy [42]. Trust telling is an important ingredient in the physician doctor relationship as lack of it leads to distrust [4]. It is expected that all healthcare professionals including doctors should always tell their patients the truth always even if it will become bad news [45]. This is why medical ethics and breaking of bad news is incorporated in medical training. Some cultures and religions forbid been told the truth when it is bad news. Sometimes there is a contradiction between respect for the person as an individual and with the patient's right not to know due to patient autonomy [44, 45]. Therefore the issue of truth telling has been debated in biomedical ethics [2]. This is because truth telling has been a challenge in medical ethics as there is no guideline on the limit to how much information should be given to the patient which sometimes leads to ethical controversies [2, 44]. How the information is relayed to the patient is also very important [2]. Even with all these controversies, challenges and ethical issues associated with truth telling, it is always the right of the patient to be told the truth always [53].

13. Ethical considerations in conducting medical research

The Nuremberg Code is a code which was a part of the wins of the famous 'Nuremberg Trial' (1945–1946) and the "Doctors' Trial" (1946–1947) which indicted and tried major World War II (WWII) criminals and lesser WWII criminals, respectively, on war crimes, crimes against peace and other crimes committed against humanity. This trial began shortly after the end of WWII [54, 55]. If not for the Nuremberg Code, the whole world might not have been safer due to abuses against humanity. The Nuremberg Code is a set of 10 ethical principles that guides research involving human experimentation (**Table 1**) [56–58]. The Code emphasized four basic ethical principles of research which are 'informed consent', 'beneficence', 'non-maleficence', and 'non-coercion' [56, 58]; it was drafted in 1947 during the trial of some German physicians who were indicted of conducting heinous, unethical and invasive experiments on people incarcerated in concentration camps during the WWII [56, 57].

Through the creation of the Nuremberg Code, the whole world became awakened to the urgent need for the creation of policies and laws that guide the ethical conduct of research involving human subjects [57, 59]. However, over the years, many scientists had criticized the Nuremberg Code on issues pertaining to plagiarism and

Article	Code
1	The voluntary consent of the human subject is absolutely essential.
2	The experiment should be such as to yield fruitful results for the good of society, unprocurable by other methods or means of study, and not random and unnecessary in nature.
3	The experiment should be so designed and based on the results of animal experimentation and knowledge of the natural history of the disease or other problem under study that the anticipated results will justify the performance of the experiment.
4	The experiment should be so conducted as to avoid all unnecessary physical and mental suffering and injury.
5	No experiment should be conducted where there is an *a priori* reason to believe that death or disabling injury will occur; except, perhaps, in those experiments where the experimental physicians also serve as subjects.
6	The degree of risk to be taken should never exceed that determined by the humanitarian importance of the problem to be solved by the experiment.
7	Proper preparations should be made and adequate facilities provided to protect the experimental subject against even remote possibilities of injury, disability, or death.
8	The experiment should be conducted only by scientifically qualified persons. The highest degree of skill and care should be required through all stages of the experiment of those who conduct or engage in the experiment.
9	During the course of the experiment the human subject should be at liberty to bring the experiment to an end if he has reached the physical or mental state where continuation of the experiment seems to him to be impossible.
10	During the course of the experiment the scientist in charge must be prepared to terminate the experiment at any stage, if he has probable cause to believe, in the exercise of the good faith, superior skill and careful judgment required of him that a continuation of the experiment is likely to result in injury, disability, or death to the experimental subject.

Table 1.
Articles in the Nuremberg code [56, 57].

embedded loopholes in the Articles 4, 5, 9 and 10 of the Code (**Table 1**) [56–60]. Nonetheless, the Code is still widely regarded as the most important document in the history of medical research ethics [57–59].

Over the years, many countries had formulated and enacted policies and laws guiding the ethical conduct of medical research within their borders [61–63]. In fact, multilateral and globally recognized health organizations like 'World Medical Association', 'World Health Organization', and 'Council for International Organizations on Medical Research' also have their own guidelines on the ethical conduct of biomedical research involving human subjects [57, 64, 65]. This shows the huge gravity of global concerns regarding the ethical conduct of research involving human subjects.

Ethical considerations are issues that are widely considered as a keystone part of medical research [66, 67]. In medical research, all stakeholders involved must consider the ethical implications of their conduct. The thoughts on these implications are, in order words, known as ethical considerations. There are so many issues of ethical consideration in the field of Medicine. Ethical considerations regarding the conduct of medical research centre on the basic ethical principles discussed in one of the preceding paragraphs. Pertinently, these considerations apply to all forms of medical research irrespective of where, when, how, and why it was conducted.

It is noteworthy to discuss extensively on some peculiar issues of ethical concerns that needs to be considered in medical research carried out at/on some peculiar (or special) periods or population groups. These discussions are below:

a. **Ethical considerations in conducting research in public health emergencies (PHEs):** Disease pandemics, flooding, earthquake, tsunamis, etc. are PHEs [6]. The periods of PHEs are tragic periods. They are characterized by chaos, panics, untold economic hardship, movement restrictions, stigmatizations, and other unforeseen irregularities [68]. Public health emergencies have been a public health problem that had been bedeviling humanity for thousands of years; unfortunately, the rate of occurrence of such emergencies has increased in its frequency due to many factors [68]. During PHEs, medical research are being conducted either to solve the problems that are associated with the PHE or other problems that are not related to the PHE. However, irrespective of the aims of the medical research endeavors conducted during PHEs, the point still boils down to the fact that such medical research endeavor is being carried out when a PHE is on.

In 2020, the whole world is battling with the COVD-19 pandemic [69]. In a situation like this, many medical researchers are facing enormous difficulties when it comes to the conduct of medical research [70]. Despite these enormous problems, yet a medical researcher still needs to be very mindful about the ethical implications of his/her conduct in all research endeavors carried out during the period. This issue will take us to have a look at a publication of the World Health Organization (WHO) on the ethical standards to be followed during PHEs [71]. Research is part of the key response to public health emergencies; hence, the conduct of medical research during disease pandemic is an ethical imperative as some research hypotheses can only be best answered during the period of a pandemic [71]. This implies that it becomes necessary, from an ethical point for view, for medical researchers to conduct research that will manage a pandemic.

However, it was further mentioned that:

i. Medical research should be conducted only if it does not impede emergency response efforts.

ii. When doing collaborative medical research, be it an international or a local collaboration, medical researchers should work as a team to jointly prioritize the challenges faced in the outbreak, determine the best fit research project that will provide answers to those challenges, conduct the research, and ensure that the research ultimately benefits the affected communities.

iii. The affected communities must be meaningfully and fairly engaged in the research process, and they should be involved in inclusive decision making efforts.

iv. Independent research ethics review committee should be strengthened and the ethical review process of research protocols should be expedited.

v. Appropriate research methodologies should be adopted for all medical research conducted during public health emergencies.

vi. Research participants' selection and treatment must be done safely, fairly, justifiably, and with respect.

vii. The research participants must be fully aware of the implications of their participation in such research, their participation must be voluntary, and their identity must be treated with strict confidentiality.

viii. Individuals and communities that participated in a research must have access to the benefits accrued from such research.

b. **Ethical Considerations in Conducting Research on Clinical Emergencies:** Medical and surgical emergencies are clinical emergencies that often occur in the hospital setting. Some of these conditions include asthma, trauma, cerebrovascular accidents, sepsis, and myocardial infarction [72]. These emergency conditions constitute a large proportion of global disease burden [72]. Through medical research, many clinical emergency problems had been solved. A very good example is the 1922 discovery of insulin by Frederick Banting which has helped in improving the clinical outcomes of diabetic emergencies [73]. There are also many other examples of great discoveries in medical research.

The ethical conduct of medical research more especially in the area of clinical emergencies is a very crucial one. There are many factors that pose as problems to the proper conduct of medical research in clinical emergencies. These factors are diverse and they range from governmental regulations to time of decision making/intervention [72]. To start with, many nations do not have a specific regulation regarding clinical emergency research [72]; this is a fundamental problem that needs to be solved. Also, the issue of informed consent taking is an issue central to the ethical conduct of research in clinical emergencies, more especially clinical trials. For instance, in unconscious patients, the decision on what clinical intervention (a trial intervention or an approved intervention) to do on such patient must be made in few minutes; unfortunately, such patients are unable to give their consent [74]. Also, the legally authorized representative (LAR) of such patients are often unavailable or in a state of emotional imbalance [74, 75]. In such kind of situation, some authorities recommended the use of 'deferred consent' taking from the LAR (e.g. next of kin) of such patient while some recommended that the medical researchers involved in such kind of clinical research (such as trials) can proceed with the study must be able to prove that:

i. Such patient is in a life-threatening situation.

ii. Available treatment options are unsatisfactory or untested.

iii. Such patient is unable to give his/her informed consent due to his/her current clinical condition.

iv. Such patient might have direct benefits from participating in the clinical study.

v. The time to seek an informed consent from the LAR prior to the clinical intervention is not available [74–76]. All the recommendations in this paragraph operate at the clinical level.

At the community level, it is recommended that medical researchers should engage the communities in inclusive decision making on how best to handle informed consent taking in clinical emergency research, make relevant public disclosures about the scope of their study, work closely with the Institutional Review Board (IRB) to guide the study protocol [74, 77]. This approach at the community level is plausible, since one of the ultimate goals of medical research is to improve the health and wellbeing of our communities.

c. **Ethical Considerations In Conducting Research On The Vulnerable Population Group:** The vulnerable population groups are the disadvantaged sub-segment of the community; they include minors (people below 18 years), the mentally impaired, children, prisoners, pregnant women, fetuses, older persons, displaced persons, and other categories of disadvantaged people [65, 78–80]. Due to the inherent characteristics of this population group, they are not capable to protect their own interests [80, 81]. Among all medical research types, trials are one of the most crucial types due to its complexity and the inclusion of an intervention in its scope. A study by Welch et al. extensively described the peculiarities and limitations of each particular sub-group in the vulnerable population group, when it comes to the ethical issues surrounding their participation in a clinical trial [80]. Based on these peculiarities and limitations, it was concluded that:

i. Vulnerable population group should not be unnecessarily excluded from a trial because their participation in a trial will also provide informative outcomes that might benefit their group.

ii. When designing a trial, the issues concerning the inclusion of vulnerable people and how the ethical and regulatory requirements of such people will be evaluated and managed must be addressed in the protocol of such study

iii. Activities that could result into the stigmatization of vulnerable groups, such as unnecessary exclusion, should be avoided as it could result into the violation of confidentiality and even loss of vital research data about vulnerable groups.

iv. The risks associated with the participation of vulnerable people in a trial should be properly evaluated and the protections of such group of people should be properly addressed.

v. There should be regular revisits of the laid down regulations of the appropriate authorities governing the ethical conduct of research on vulnerable groups during the course of the trial so as to ensure that the conduct of the study does not violate ethical standards.

vi. There should be regular review of the incremental risk of the study design with respect to the participating vulnerable persons so as to determine if there is a need for the provision of further protections to this peculiar group [80].

14. Informed consent taking in medical research

In medical research involving human subjects, taking consent from the subjects is very necessary. In fact, it is not just about consent taking; rather, it is about taking a consent that is informed. Informed consent refers to the voluntary agreement of a human subject regarding his/her participation in a medical research as a subject. Informed consent taking constitutes a major aspect in the ethical conduct of medical research. As non-coercion is one of the standing pillars in the principles governing medical ethics, all persons participating in a medical research must not be coerced to participate in any way; rather, their participation should be completely voluntary. In the course of recruiting humans into a medical research, they should be given prior information that is of relevance to the protocol and safety profile of the study. However, informed consent taking applies only to adults, i.e. people aged 18 years or above. It is believed that only people in this category are mentally and psychologically capable to make decisions regarding consent. As for children, due to their age and level of psychological development, they can only give assent.

Informed consent can be taken in two ways: written or verbal. In written informed consent, the human subject gives a written documentation of him/her agreeing to participate in a study while in verbal informed consent, the human subject gave his/her agreement verbally. Between the two forms of informed consent, the written type is more reliable.

15. Conclusion

Medical ethics is a branch of ethics is a branch of philosophy that guides all human endeavors. Medical ethics are sets of regulations that guide physicians in their work and protect them against litigation. Ethical principles that guide the medical profession are autonomy, beneficence, non-maleficence, justice, confidentiality and truth telling. Sometimes these ethical principles have to be breached to protect the public and when required by a court of law. These ethical principles also help in there resolution of ethical dilemmas and conflict.

Author details

Dabota Yvonne Buowari[1]* and Kehinde Kazeem Kanmodi[2]

1 Department of Accident and Emergency Services, University of Port Harcourt Teaching Hospital, Port Harcourt, Nigeria

2 Cephas Health Research Initiative Inc, Ibadan, Nigeria

*Address all correspondence to: dabotabuowari@yahoo.com

References

[1] Widjaja G, Sijabat HH. The importance of bioethics subject in healthcare education in Indonesia. International Journal of Multidisciplinary and Current Educational Research (IJMCER). 2020, 2 (5), 21-26

[2] Amer AB. Understanding the ethical theories in medical practice. Open Journal of Nursing. 2019, 9, 188-193

[3] McKillip KM, Lott AD, Swetz KM. Respecting autonomy and promoting the patients good in the setting of serious terminal and concurrent mental illness. Yale Journal of Biology and Medicine. 2019, 92, 597-602

[4] Varkey B. Principles of clinical ethics and their application to practice. Medical Principles and Practice. 2020, doi.10.1159/000509119

[5] World Health Organization. Global health ethics, key issues. Global network of the World Health Organization collaborating centres for bioethics. 2015, www.who.int

[6] Jain S. Ethics in the genetics of epilepsy: what are the pros and cons? Neurology Asia. 2010 (1)(Supplement 1): 33-35

[7] Sheikh SA. The importance of ethics in health care system. JDUHS. 2007, 1 (1), 46-48

[8] Figar N, Dordevic B. Managing an ethical delimma. Economuc Themes. 2016, 54(2), 345-362

[9] Monsudi K, Oladele TO, Nasir AA, Ayaniyi AA. Medical ethics in sub-saharan africa: closing the gaps. African Health Science. 2015, 15 (2), 673-677

[10] Talukder MHK, Nazneen R, Hossain MZ, Ishrat JC. Basic ideas on medical ethics. Bangladesh Journal of Medical Science. 2010, 9 (3), 131-135

[11] Cowley C. The dangers of medical ethics. Journal of Medical Ethics. 2005, 31, 739-742

[12] Ramana KV, Kandi S, Boinpally PR. Ethics in medical education, practice and research: an insight. Annals of Tropical Medicine and Public Health. 2013, 6 (6), 599-602

[13] Patuzzo S, Goracci G, Ciliberti R. Thomas Percival discussing the foundation of medical ethics. ACTA Biomedical . 2018, 89(3), 343-348

[14] Gomez-Gomez C, Gonzalez-Melendez R, Riquelme-Heras H. Journal of Human Health Research. Ethical Delimma and Research. 2018, 1: 101

[15] Sharifi K, Hoghooghi M, Sadoni H. Importance of medical ethics philosophy and necessity of medical ethics teaching. International Journal of Bioassays. 2015, 4 (03), 3753-3756

[16] Jegede A, Ajayi I, Akintola S, Falade C, Dipedu IO et al. Ethical issues in the COVID-19 pandemic control preparaedness in a developing country. Pan African Medical Journal. 2020, 35(2),95. doi.10.11604/pamj.SUPPL.2020.35.23121

[17] Elsayed DE, Ahmed REM. Medical ethics : what is it? Why is it important? Sudanese Journal of Public Health. 2009, 4(2), 283-286

[18] Munyaradzim M. Critical reflections on the principle of beneficence in biomedicine. Panafrican Medical Journal . 2012. http://www.panafrican-med.journal.com/content/article/

[19] Canadian Nurses Association. Code of ethics for registered nurses. Can.aiic.ca/ethics. www.cna.aiic.ca

[20] Humayun A, Fatima N, Naqqash S, Hussain S, Rasheed A,

Imtiaz H, Iman SZ. Patients perception and actual practice of informed consent, privacy and confidentiality in general medical out patient departments of two trtiary care hospitals of labore. BMC Medical Ethics. 2008, 9.14. doi.10.1186/1472-6939-9-14; www.biomedcentral.com/1472-6939/9/14

[21] Moosavi SR, Mousavi SM, Heydari A, Meddikhah Z. Role of medical ethics in health economics. International Journal of Medicine and Medical Sciences. 2010, 2(12), 387-390

[22] Sheather J, Shah T. Ethical dellimma in medical humanitarian practice: cases for reflection from medecins sans frontiers. Journal of Medical Ethics. 2011, 37, 162-165

[23] Josipovic-Jelic Z, Soljan I. Clinical ethics delimmas in the theory and practice. ACTA Clinical Croatia. 2007, 46, 325-330

[24] Pimentel DE, Oliveira CB, Vieira MJ. Teaching of medical ethics: students perception in different period of the course. Review of Medicne in Chile. 2011, 139: 36-44

[25] Zubovic S. Ethical delimmas of nurses and physicians in the primary healthcare settings. Hospice and Palliative Medicine International Journal. 2018, 5 (2), 280-284

[26] Miljeteig I, Defaye F, Desalegn D, Danis M. Clinical ethics delimmas in a low income setting - a national survey among physicians in Ethiopia. BMC Medical Ethics. 2019, 20:63

[27] Chargani SMI. Telling the truth - a tussle between four principles of ethics. Journal of Clinical Research and Bioethics. 2014, 5: 21. Oi.10.4172/2155-9627.1000172

[28] Jie L. The patient suicide attempt: an ethical delimma case study. International Journal of Nursing Sciences. 2015, 408-413

[29] Maedia Y. A method to solve ethical delimma as medicine. Journal of Philosophy and Ethics in Health Care and Medicine. 2012, 6, 9-28

[30] Yeo CJJ. Ethical delimma of the practice of medicne in the information technology age. Singapore Medical Journal. 2003, 44(3), 141-144

[31] Ong WY, Yee CM, Lee A. Ethical delimmas in the care of cancer patients near the care of life. Singapore Medical Journal. 2012, 53 (1), 11-15

[32] Ateriya N, Saraf A, Meshram VP, Setia P. Telemedicne and virtual consultation, the Indian perpective . The National Medical Journal of India. 2018, 31(4), 215-218

[33] Russ MJ, Sisti D, Wilner PJ. When patients refuse COVID-19 testing, quarantine andsocial distancing in inpatint psychiatry: clinical and ethical challenges. Journal of Medical Ethics. 2020, 46, 579-580

[34] McGuire AL, Aulisin MP, Davis DF, Erwin C, Harter TD et al. Ethical challenges arising in the COVID-19 pandemic: an overview from the Association of Bioethics Program Directors (ABPD) taskforce. The American Journal of Bioethics. 2020, 20(7), 15-27.

[35] Robert R, Kentish-Barnes N, Bayer A, Laurent A, Azoulay E, Rieignier J. Ethical delimmas due to the COVID-19 pandemic. Annals of Intensive Care . 2020. 10:84. doi.org/10.1186/s/3613-020-00702-7

[36] Nesom G, Moore TM, Petrof I. Operational characteristics of institutional review boards (IRB) in the United States. AJOB Empir Bioethics. 2019, 10 (4), 276-28

[37] Qiao H. A brief introduction to institutional review boards in the United States. Pediatric Investigation. 2018. doi.10.1002/ped4-1/2023

[38] Nurunnabi ASM. Institutional review board (IRB) its role and responsibility in making research ethical. Bangladesh Journal of Bioethics. 2014, 5 (1), 5-10

[39] Kotsis SV, Chung KC. Institutional review boards whats old, whats new, whats needs to change. Plastic Reconbstructive Survey. 2015, 133 (2), 439-445

[40] Enfield KB, Truwit JD. The purpose, composition and function of an institional review board: balancing priorities. Respiratory Care. 2008, 53 (10), 1330-1336

[41] Dzeng E. Habermasion communication pathologies in Do-Not-Resuscitate discussions at the end of life: manipulation as an unintended consequnce of an ideaology ofpatient autonomy. Social Health Illn. 2019, 41(2), 325-342

[42] Murgic L, Hebert PC, Sovic S, Pavlekovic G. Paternalism and autonomy: views of pateints and providers in a tranbsitional (post- communist) country. BMC Medical Ethics. 2015, 16: 65. doi.10.1186/s/2910.015.0059.2

[43] Gray DP. Confidentiality: a core feature of general practice. British Journal of General Practice. 2010, 780-782

[44] Zhang Z, Muri X. The ethical delimma of truth telling in healthcare. Bioethical Inquiry. 2020, 17, 337-344

[45] Zolkefli Y. Patient autonomy in healthcare ethics - a concept analysis. Brunei Darussalam Journal of Health. 2017, 7(1), 43-52

[46] Entwistle VA, Carter SM, Cribb A, McCaffey K. Supporting patient autonomy: the importance of clinican - patient relationships. JGIM. 741-745

[47] Zapata JA, Moriates C. The high - value care considerations of inpoatient versus out patient testing. American Medical Association Journal of Ethics. 2015, 17 (11), 1022-1027

[48] American Dental Association. Principles of ethics and code of professional conduct with offical advosory opinions revised. 2018

[49] Girdler SJ, Girdler JE, Tarpada SP, Morris MT. Nonmalefecence in medical training . Balancing patient care and effocoent education. Indian Journal of Medical Ethics. 2018. doi.10.20529/IJME.2018.100

[50] Knapp VBD, Ogunbanjo DA. Ethics in health care: confidentilaity and information technologies. South African Family Practice. 2014, 56(1)(Supplemtal 1): 53-55

[51] Cipi B. Ethical juridical and historical aspects of medical confidentiality. JAHR. 2012, 3 (5), 139-146

[52] O'Brien J, Chantler C. Confidentilaity and the duties of care. Journal of Medical Ethics. 2003. www.jmedethics.com. doi.10.1136/jme.29.1

[53] Amer AB. Understanding the ethical theories in medical practice. Open Journal of Nursing. 2019, 9, 188-193

[54] Oddi LF, Cassidy VR. Nursing research in the United States: the protection of human subjects. Int J Nurs Stud. 1990;27(1):21-33. doi: 10.1016/0020-7489(90)90020-j.

[55] Burns N, Grove SK. The practice of nursing research: Conduct, critique, and utilization, 5th ed. St Louis, MO: Elsevier/Saunders, 2005.

[56] Ghooi RB. The Nuremberg Code-A critique. Perspect Clin Res. 2011;2(2):72-6. doi: 10.4103/2229-3485.80371.

[57] Shuster E. Fifty years later: the significance of the Nuremberg Code. N Engl J Med. 1997;337(20):1436-40. doi: 10.1056/NEJM199711133372006.

[58] Vollmann J, Winau R. Informed consent in human experimentation before the Nuremberg code. BMJ. 1996;313(7070):1445-9. doi: 10.1136/bmj.313.7070.1445.

[59] Grodin MA, Annas GJ. Legacies of Nuremberg: medical ethics and human rights. JAMA 1996;276:1682-1683.

[60] Moreno JD. Reassessing the influence of the Nuremberg Code on American medical ethics. J Contemp Health Law Policy 1997;13:347-360.

[61] Molyneux S, Geissler PW. Ethics and the ethnography of medical research in Africa. Soc Sci Med. 2008;67(5):685-95. doi: 10.1016/j.socscimed.2008.02.023. Epub 2008 May 2.

[62] Zielinski C, Kebede D, Mbondji PE, Sanou I, Kouvividila W, Lusamba-Dikassa PS. Research ethics policies and practices in health research institutions in sub-Saharan African countries: results of a questionnaire-based survey. J R Soc Med. 2014;107(1 suppl):70-76. doi: 10.1177/0141076813517679. Epub 2014 Mar 18.

[63] Hedgecoe A, Carvalho F, Lobmayer P, Raka F. Research ethics committees in Europe: implementing the directive, respecting diversity. J Med Ethics. 2006;32(8):483-6. doi: 10.1136/jme.2005.013888.

[64] Council for International Organizations of Medical Sciences (CIOMS). International ethical guidelines for health-related research involving humans, fourth edition. Geneva. 2016. Available from: https://cioms.ch/wp-content/uploads/2017/01/WEB-CIOMS-EthicalGuidelines.pdf. Accessed: November 12, 2020.

[65] World Medical Association. WMA declaration of HELSINKI – Ethical principles for medical research involving human subjects. Available from: https://www.wma.net/policies-post/wma-declaration-of-helsinki-ethical-principles-for-medical-research-involving-human-subjects/. Published: July 9, 2018. Accessed: November 12, 2020.

[66] World Health Organization. Ethical standards for research during public health emergencies: Distilling existing guidance to support COVID-19 R & D. Available from: https://www.who.int/blueprint/priority-diseases/key-action/liverecovery-save-of-ethical-standards-for-research-during-public-health-emergencies.pdf. Accessed: November 12, 2020.

[67] Akaranga SI, Makau BK. Ethical considerations and their applications to research: a case of the University of Nairobi. J Educ Policy Entrepr Res 2016;3(12):1-9.

[68] Madhav N, Oppenheim B, Gallivan M, et al. Pandemics: Risks, Impacts, and Mitigation. In: Jamison DT, Gelband H, Horton S, et al., editors. Disease Control Priorities: Improving Health and Reducing Poverty. 3rd edition. Washington (DC): The International Bank for Reconstruction and Development / The World Bank; 2017 Nov 27. Chapter 17. Available from: https://www.ncbi.nlm.nih.gov/books/NBK525302/. doi: 10.1596/978-1-4648-0527-1_ch17. Accessed: November 12, 2020.

[69] Morens DM, Breman JG, Calisher CH, Doherty PC, Hahn BH, Keusch GT, Kramer LD, LeDuc JW, Monath TP, Taubenberger JK. The

Origin of COVID-19 and Why It Matters. Am J Trop Med Hyg. 2020;103(3):955-959. doi: 10.4269/ajtmh.20-0849.

[70] Tuttle KR. Impact of the COVID-19 pandemic on clinical research. Nat Rev Nephrol 2020;16:562-564. doi: 10.1038/s41581-020-00336-9.

[71] World Health Organization. Ethical standards for research during public health emergencies: Distilling existing guidance to support COVID-19 R & D. Available from: https://www.who.int/blueprint/priority-diseases/key-action/liverecovery-save-of-ethical-standards-for-research-during-public-health-emergencies.pdf. Accessed: November 12, 2020.

[72] Millum J, Beecroft B, Hardcastle TC, Hirshon JM, Hyder AA, Newberry JA, Saenz C. Emergency care research ethics in low-income and middle-income countries. BMJ Glob Health. 2019 Jul 29;4(Suppl 6):e001260. doi: 10.1136/bmjgh-2018-001260.

[73] Frederick Grant Banting (1891-1941). Codiscoverer of insulin. J Am Med Assoc. 1996;198(6):660-61. doi: 10.1001/jama.1966.03110190142041.

[74] Tisherman SA. Defining “community” and “consultation” for emergency research that requires an exception from informed consent. AMA J Ethics. 2018;20(5):467-474. doi: 10.1001/journalofethics.2018.20.5.stas1-1805.

[75] Dutton RP, Stansbury LG, Hemlock B, Hess JR, Scalea TM. Impediments to obtaining informed consent for clinical research in trauma patients. J Trauma. 2008;64(4):1106-12. doi: 10.1097/TA.0b013e318165c15c.

[76] Abramson NS, Safar P. Deferred consent: use in clinical resuscitation research. Brain Resuscitation Clinical Trial II Study Group. Ann Emerg Med. 1990;19(7):781-4. doi: 10.1016/s0196-0644(05)81703-2.

[77] Dickert N, Sugarman J. Ethical goals of community consultation in research. Am J Public Health. 2005;95(7):1123-7. doi: 10.2105/AJPH.2004.058933. Epub 2005 Jun 16.

[78] Shivayogi P. Vulnerable population and methods for their safeguard. Perspect Clin Res. 2013;4(1):53-7. doi: 10.4103/2229-3485.106389.

[79] Waisel DB. Vulnerable populations in healthcare. Curr Opin Anaesthesiol. 2013;26(2):186-92. doi: 10.1097/ACO.0b013e32835e8c17.

[80] Welch MJ, Lally R, Miller JE, Pittman S, Brodsky L, Caplan AL, Uhlenbrauck G, Louzao DM, Fischer JH, Wilfond B. The ethics and regulatory landscape of including vulnerable populations in pragmatic clinical trials. Clin Trials. 2015;12(5):503-10. doi: 10.1177/1740774515597701. Epub 2015 Sep 15.

[81] Levine RJ. Ethics and regulation of clinical research. 2nd ed. Yale University Press; New Haven, CT: 1988. Available from: https://yalebooks.yale.edu/book/9780300042887/ethics-and-regulation-clinical-research

Chapter 22

The New Challenges for Medical Ethics

Liliana Lorettu, Jocelyn Aubut and Rosagemma Ciliberti

Abstract

The evolution of medicine confronts healthcare professionals with new ethical challenges. Elements such as professional secrecy, patient benefit, justice in the distribution of resources are put in crisis by the evolution of medical procedures. Today, doctors must make life-and-death decisions about many patients. As the resources are not enough for all patients, the 'first-come, first-served' criterion crumbles under the weight of the overwhelming demand for treatment. Consequently, they can no longer make treatment decisions based only on proportionality and clinical appropriateness criteria. They must take into account the availability of resources and prioritise patients with 'the longer life expectancy'. This amounts to saying 'the weakest will die' ... with the doctors' consent. While the guidelines issued by scientific societies may well protect doctors from lawsuits, the choice of who to treat and who to let die is left to the conscience of the individual doctor; and it is a choice sharply clashing with the Hippocratic oath and with professional and personal ethics. This and others are a real ethical problem.

Keywords: medical ethics, professional responsibility, availability of resources

1. Introduction

Biomedical ethics has made giant strides over the past decades and has come to be recognised as integral to medical education. This has encouraged the growing inclusion of the teaching of medical ethics, together with that of the human sciences, in the syllabi of medical and nursing schools. In the 1980s, increased awareness of ethical issues shone a light on some excesses of medical research and medical paternalism which conflicted with ethical principles. The 1990s saw the establishment of the first medical ethics committees in hospitals, overseeing both research and clinical practice. Since the 2000s, the various bodies regulating the doctors' right to practice have issued regulations, guidelines and recommendations laying down formal ethical rules for medical practice, together with a system of penalties for infringement of these rules.

Many social and cultural factors have contributed to the increase in ethical concerns. The increase in individual civil liberties, codified in various Charters of Citizens' Rights, has fuelled a growing drive to claim new rights in previously unexplored areas. The development of biomedical technologies has created new frontiers, such as the attempt to shape one's own medical fate, as in the case of the actress Angelina Jolie, who chose to undergo preventive double mastectomy and subsequent ovariectomy because she carried a gene that greatly increased (over 80%) her risk of developing an aggressive and often fatal type of breast cancer, or the decision of a

British manager to have his prostate removed for the same reason. In the meantime, the constant budget cuts have increased the need to make very complex choices.

Recently, the Covid-19 pandemic has confronted us with specific ethical dilemmas, in particular the choice about who to treat or not to treat in a health emergency with scarce resources.

The growing ethical concerns have highlighted the fact that doctors only receive very basic training in medical ethics during their studies and practical training. Some studies even show that the awareness of ethical issues of students and trainees decreases as they advance in their studies [1, 2].

Most doctors trust their ethical judgement and believe that their decisions are morally sound. Yet most doctors lack adequate training and theoretical knowledge of ethical issues to support their beliefs and choices in a manner that stands up to scrutiny. The ethical judgement of most doctors is based on their professional life experience, personal opinions, beliefs and values, but few know the theoretical foundations of biomedical ethics and moral decision-making.

The first part of this paper outlines the key theoretical concepts framing ethical decision-making by physicians. Next, the principles governing the ethical decision-making process are presented. This is important because ethics is not only about the medical decision, but also about the process for reaching that decision. Certain issues in the application of ethical principles and the challenges brought by current events to medical ethics are also discussed.

2. Historical overview and remarks on the relationship between medical ethics and bioethics

The birth of bioethics as understood today is closely linked to the giant strides made by the biomedical sciences and technologies (most notably molecular biology and genetic engineering) around the 1970s.

The gradual unlocking of the mechanisms of life, coupled with the possibility of manipulating and modifying living beings, enabled a number of procedures that gave rise to widespread ethical concerns: medically assisted reproduction, tissue and organ transplantation, genetic intervention, the possibility of artificial life independent of 'natural' life, euthanasia, cloning, etc.

The word *Bio-Ethik* was coined by German Protestant pastor and ethicist Fritz Jahr, who used the term to propose a new bioethical imperative that extended to all living beings Kant's categorical imperative of respect for all persons [3, 4].

However, the current meaning of bioethics can be ascribed to American oncologist Van Rensselaer Potter, who used this term in a paper entitled *Bioethics: the science of survival* [5] and later in his best-known work *Bioethics: a bridge to the future* [6].

According to Potter, building an ethic based on scientific knowledge is necessary to ensure the very survival of *Homo sapiens*, which could be threatened if research were allowed to proceed unchecked and unfettered. Potter rejected merely speculative knowledge and stressed the need to connect ethical values, traditionally confined to the realm of the humanities, with biological facts and thus build a 'bridge to the future'.

Potter himself defined bioethics as the 'knowledge of how to use knowledge', to highlight the distinctive nature of this discipline as a dialogical meeting point between the natural sciences, the social sciences and philosophy.

In his subsequent book, entitled *Global Bioethics*, Potter made the by now well-established subdivision of bioethics into three branches: medical ethics, environmental ethics and animal ethics [7].

It is interesting to note that originally, the scope of bioethics was not restricted to medical practice, even though in subsequent years this came to be considered its main, if not exclusive, area of concern. Indeed, differently from Potter's definition of bioethics (later followed by Jonas in his work *The Responsibility Principle* [8]) the term has mostly been applied in the narrower sense given to it by Dutch obstetrician Andre E. Hellegers, co-founder of the Kennedy Institute, who considered bioethics as ethics applied to the biomedical sciences [9]. This narrowing of the scope of bioethics from its original reflection on the ethical problems relating to life, '*bios*' in all its complexity, is partly due to the fact that the two centres where bioethics research and teaching were first developed (the Kennedy Institute in Washington and the Hastings Center in New York) focused on medical issues, specifically, on medically assisted reproduction. This meant that issues such as the treatment of animals or environmental risks were not considered to fall within the scope of bioethics proper.

The close links between the different facets of bioethics and the high complexity of the problems addressed require constant cross-disciplinary dialogue among scientists and scholars from a range of disciplines such as philosophy, law, economics, sociology, ethology, psychology and anthropology [10].

The interdisciplinary nature of bioethics is also in evidence in the current definition of this discipline, contained in the 2nd edition of the Encyclopedia of bioethics: '*Bioethics is the systematic study of the moral dimensions - including moral vision, decisions, conduct, and policies - of the life sciences and health care, employing a variety of ethical methodologies in an interdisciplinary setting*' [11].

The relationship between ethics and science is certainly at the heart of philosophical reflection and may be summed up in one question: should we do everything we can do?

In the United States, the debate on ethical issues had already started long before the breakthroughs in genetics: it was prompted by news of gross abuses committed in several clinical trials, namely at the Jewish Chronic Disease Hospital in Brooklyn, the Willowbrook State Hospital in New York and in the famous 'Tuskegee Study of Untreated Syphilis in the Negro Male' which began in 1932 and continued until 1972 [12].

However, the historical roots of bioethics and, in particular, of medical ethics, can be traced further back in time by a deeper examination of the relationship between science and ethics.

The atrocities committed in the experiments on concentration camp prisoners in Nazi Germany dramatically revealed, well before the later events that prompted the appearance of the term 'bioethics' in the literature, the need to investigate the relationship between ethics and science.

The Nuremberg Code was the first document to enshrine in specific rules the ethical principles that govern research on human subject. The Code, which although it never attained legal value has a universal moral value, established for the first time the following standards for human experiments:

- The voluntary consent of the human subject is absolutely essential: this means that the person involved must be given detailed prior information about the nature, purpose, duration, means and risks of the experiment;

- the experiment must be justified in terms of necessity, anticipated results and avoidance of injury;

- the risks of the experiment must be carefully weighed against the expected benefits;

- the personnel conducting the experiment must be appropriately trained and qualified;
- appropriate equipment and facilities must be used;
- it must be possible to bring the experiment to an end at any time on the initiative of either the human subject or the scientist.

Thus, the Nuremberg Code is a landmark document in the development of medical ethics, paving the way for a gradual and profound revision of the doctor-patient relationship in order to shed the traditional paternalistic approach in favour of the principles of consent, shared decision-making and therapeutic alliance.

Following the Nuremberg trial and the consequent drafting of the Nuremberg Code (1946), several international instruments on human rights were drafted, starting from the Universal Declaration of Human Rights (1948), which laid down the first legal principles of bioethics. The Declaration contains strong statements on the right to life and physical integrity, together with other fundamental civil and political freedoms. In so doing, it opened up a new legal and regulatory path for bioethics and inspired and influenced the subsequent development of international legislation.

The global and regional documents, charters, declarations and conventions that followed explicitly refer to the Universal Declaration of Human Rights as the foundation of their statutes and precepts, including the WMA Declaration of Geneva and the International Code of Medical Ethics of 1948 and the WMA Declaration of Helsinki on Ethical Principles for Medical Research Involving Human Subjects of 1964 (with its several subsequent amendments.

However, with regard to the aims of bioethics, it would be reductive and historically incorrect to limit its statutory, founding aims to the need to fix the ethical boundaries for the technical progress of science. As Mori [13] pointed out, what the Nuremberg trial itself so dramatically exposed is the need to set limits not to the technological advances of science but to the abuse of those advances. Thus, Mori reminds us that the core problem of bioethics is not to trace the boundaries of technological advancement, pitting science against ethics, but to identify the reasons that justify a specific moral judgement. Thus, as remarked by Schiavone [14], a crucial premise for any ethical approach to be legitimate and justified is that any critical reflection on scientific areas and disciplines should originate and develop within science itself and the scientific and technical advances achieved by it, instead of referring to a source of regulation outside science.

Far from being a system distinct from science and which attempts to stem its progress, bioethics aims to pursue critical and coherent reflection on human dignity, as an instrument of moral control (in the secular sense) over science in terms of its impact on human beings and the environment.

The subject matter of bioethics (which concerns itself with the sphere of '*bios*', i.e. living beings) is associated with the theme of the destiny of human beings, and thus is an emotionally charged topic, inevitably subject to strong pressures. Bioethics is constantly at risk of sliding from the role of neutral and unbiased observatory – to the extent that such a role can effectively be achieved and maintained – onto the dangerous terrain of ideology and its associated dogmatic views.

Returning to the question of the origins of bioethics, it should be noted that ethical reflection in medicine dates back to long before Potter's text. The Hippocratic oath is significant evidence of this. The oath, which evidently reflects the philosophy and culture of a time when the medical profession had a hieratic character, contains the seed bioethics in its principles of *non nocere* (i.e. 'do no harm' to the patient) and 'beneficence' as cornerstones of the doctor's activity.

Hippocrates (460-377 BC)	Hippocratic Oath
Sefer Asaph ha-Rofe (6th century AD)	Oath of Asaph (Book of Asaph the Physician)
Moses ben Maimon (Maimonides, 1135-1204)	The Physician's Daily Prayer (Maimonides)
Mohamad Hosin of Shiraz (1770)	A Physician's Ethical Duties
Military Tribunal of Nuremberg, 1946	Nuremberg Code
World Medical Association, 1948, 1968, 1984, 1994, 2005, 2006. 1949. 1964, 1975, 1983, 1989, 1996, 2000, 2002, 2004, 2008, 2013. 1968. 1970, 1983, 2006, 2018.	Declaration of Geneva International Code of Medical Ethics Declaration of Helsinki (on Medical Research) Declaration of Sidney on the determination of the time of death Declaration of Oslo on therapeutic abortion Declaration on terminal illness
European Court of Human Rights Council of Europe, 1950.	European Convention on Human Rights Protocol N. 14 (CETS No. 194)
Supreme Soviet of the USSR, 16 March 1971	The Soviet Union Medical Oath
United Nations, 1971 1975, 1984 1975	Declaration on the Rights of Mentally Retarded Persons Declaration on the Protection of All Persons from Being Subjected to Torture and Other Cruel, Inhuman or Degrading Treatment or Punishment Declaration on the Rights of Disabled Persons
American Hospital Association, 1973	Patient's Bill of Rights
Council of Europe, 1976 1979	Recommendation on the rights of the sick and dying Draft recommendation on artificial insemination of human beings
State of California, 1976	Natural Death Act
Hospital Committee of the European Economic Community, 1979	Charter of the hospital patient
Tribunal for Patients' Rights, Italy, 1980	*Carta dei 33 diritti del cittadino* (Charter of the 33 rights of citizens)
International Conference of Medical Professional Associations and Bodies with similar remits, EEC, 1982	European Guide to Ethics and the Professional Conduct of Doctors
World Medical Assembly (WMA), Marbella, Spain, 1992, 2005, 2015. 70th WMA General Assembly, Tbilisi, 2019	Statement on Physician Assisted Suicide WMA Declaration on Euthanasia and Physician-Assisted Suicide
Council of Europe, (Oviedo), 1997	Convention for the Protection of Human Rights and Dignity of the Human Being with regard to the Application of Biology and Medicine: Convention on Human Rights and Biomedicine (ETS No 164)

Figure 1.
Main documents on medical ethics.

The Western world has adopted this approach and has formulated codes of medical ethics and laws inspired by ethical principles to regulate the exercise of the medical profession. These sets of rules are regularly updated in response to cultural and ethical developments and to the growing demand for professional standards to safeguard not only the interests of medical professionals, but also, and most importantly, those of their patients.

In this regard, medical ethics and standards of professional conduct play a major role in the physician-patient relationship. This is the setting where protecting the patient's fundamental rights is crucial and where the risk that medical practice may infringe the individual's rights protected by the Constitution is highest. Indeed, since ancient times, the power imbalance inherent in the patient-provider relationship has required a framework of principles and rules specifying the physician's duties, in order to protect the patient (**Figure 1**).

3. General principles

Ethical theories can be grouped for simplicity into two main currents.

One is **teleological ethics**. Teleological theories focus on the purpose of the decisions taken and on their positive and negative impacts, and assess the consequences of the action [15]. These theories are deductive and pragmatic. Among the best known are John Stuart Mill's utilitarianism [16] and American principlism [17]. The latter is undoubtedly one of the most widespread currents in medical ethics, at least in the United States, and will be discussed below. These theories focus on doing good for each individual, but also for the community.

The second current is **deontological ethics**; this differs fundamentally from the teleological approach in that its focus is not on achieving a good outcome but on doing what is morally right. The deontological approach is based on a series of '*prima facie*' principles; it is an inductive principle focused on processes rather than on the final decision and it refers to the theories of Kant and Habermas [18]. Deontological ethics recognises absolute prohibitions, which admit no exceptions for any reason, override other duties, are fixed '*a priori*' and are unchangeable. However, since conflict may arise between different duties, priorities must be identified in the hierarchy. Thus, a shift occurs from a hierarchy with absolute '*a priori*' duties to an ethics with '*prima facie*' duties, which also requires examination of the circumstances.

Teleological ethics and deontological ethics are two alternative ethical theories that determine the moral good or evil of an action.

The key difference between the two theories is that teleological ethics weighs the good or evil of an action according to its consequences. By contrast, deontological ethics determines the good or evil of an action on the basis of an examination of the action itself. Its vision is based on rules that determine the action.

Application of these two theories to end-of-life care can help to clarify the difference between them. Under the teleological framework, doctors who practice assisted dying focus on the purpose of decisions. They respect the patient's choice to end her suffering when there is no hope of improvement. By contrast, under the deontological approach, doctors may refuse to provide assisted dying care on the basis of the a priori principle that doctors are trained to treat and not to take life. These are two diametrically opposed positions, which require different ethical frameworks.

General principles that state universal values of common morality also contribute to the basic reasoning on medical ethics . Beauchamp and Childress [17] have identified a model consisting of four moral principles that constitute the most common framework for achieving what is 'good' and what is 'right' in healthcare. 'Principlism' is a basic framework because it identifies four fundamental principles that come into play in most medical decisions, across the different medical specialities, countries and continents. These principles do not constitute a moral system or theory, but offer a framework for reflection on the moral problems encountered, and provide a starting point for making a moral judgement and assessing the procedure to be followed. The main principles are:

- respect for autonomy/the individual
- beneficence

- non maleficence
- justice.

The principle of *autonomy* refers to liberal thought, which has always emphasised individual rights and freedom of choice as an expression of the individual's free will. The patient is recognised as possessing critical thinking and decision-making skills that must be respected. The model that emphasises the autonomy principle aims to oppose and overcome the paternalistic approach that has long dominated the doctor-patient relationship. The paternalistic model was based on an asymmetric relationship between the doctor (acting as a good parent) and the patient, who was treated as a 'child', unable to make decisions because of his lack of scientific and, especially, medical knowledge. This model has been discarded by reversing the patient's role, from a passive one, to that of an autonomous person, capable of self-determination according to the principle of individual autonomy. The principle of autonomy ensures that the patient is involved in the medical decision-making process and protects his right to choose, accept, refuse or stop treatment. This is an absolute right of the individual, even where the refusal or interruption of treatment might cause adverse health consequences or even death. Autonomy implies respect for an individual's physical and mental integrity. A person cannot be forced to receive treatment against her will. The patient cannot be subjected to any physical or mental coercion. The principle of autonomy also underpins the patient's right to accurate and exhaustive information on the proposed treatment. Recognition of this right has led to development of the informed consent procedure. However, for certain specifically identified medical conditions that pose a public health threat, the government has the coercive power to impose treatment; this can occur, for instance, in the case of acute psychiatric patients or highly infectious diseases. However, even in these cases, the dignity of the person must always be respected. To apply these rules, doctors must know the legislation in force in the country in which they work; in any case, they must take all proper actions to minimise the need for coercion and maximise the patient's consent.

The principle of *beneficence* states that the patient's well-being is the ultimate goal of care. This principle lies at the heart of medicine, whose mission is precisely to prevent, diagnose and treat illness in order to promote the patient's health. It is a question of proposing a treatment that is proportionate to the patient's needs and whose benefits for the patient outweigh its possible harms. This principle means that doctors may act in the patient's best interest also by refraining from acting and/or by acting prudently, always from the viewpoint of the benefit for the patient. Traditionally, this principle has been focused on 'objective' good, i.e. the outcome considered to be good by the doctor. However, cultural and ethical developments have gradually led to add to this principle that of autonomy, supporting a more subjective interpretation of the patient's 'best interest'.

The principle of *non-maleficence* has been well known to doctors since the time of the Hippocratic precept of *primum non nocere*. Non-maleficence encompasses two key concepts. The first is that of not causing harm to patients, even before doing them good. The second is the need to properly assess the risks and the benefit/risk balance of a treatment, and hence to refrain from prescribing a treatment that, although effective, could be harmful to the patient.

The non-maleficence principle is reflected in a number of legal provisions regarding wilful medical malpractice, where the patient was intentionally injured, or negligent malpractice, where the harm was caused by negligence, inexperience, recklessness or failure to comply with laws, regulations, orders or standards.

The principle of **justice** requires that all people be treated fairly. It is difficult to provide a single definition of justice, as various theories have produced different versions. Egalitarian theories stress the importance of universal access to basic necessities [19]. Libertarian theories affirm the right to social and economic freedom [19]. Utilitarian doctrines require the balancing of the two principles in order to maximise public and private utility [17]. Moreover, the principle of justice includes the concept of distributive justice, which states that resources should be allocated so as to ensure that access to care is not affected by socio-economic, ethnic or other factors which could favour certain sectors of the population to the detriment of others. The problem of resource allocation arises at different levels. For example, a national government decides which share of funding to allocate to finance social and healthcare relative to other sectors such as education, labour, transport. Moreover, the healthcare budget is in turn distributed differently among the different specialties. Thus, in practice, implementing the distributive principle raises complex issues; for instance, to what extent can expensive experimental treatments be justified in patients who have not responded to conventional approaches? Some of these treatments can cost more than €100,000 per year and clearly erode the sums available to treat other patients.

These four principles are not independent of each other. Rather, they interact in all medical situations of varying complexity, engaging in a dialectical relationship which requires their careful balancing. The clinician's art is to fully understand how to best weigh these factors on a case-by-case basis, to reach the most appropriate decision for the individual patient.

4. The decision-making process

In modern biomedical ethics, the process by which a decision is reached is as important as the decision itself. This is why it is necessary to have a clear approach that takes into account the problems to be addressed and all the persons concerned.

Figure 2 shows a decision making process according to Jonsen's four box model for decision making which evaluates four fundamental variables: medical indications, patient and family preferences, quality of life and contextual features [20].

The approach proposed here is one example, among the many available, of a framework to guide the decision-making process. The approach is based on a series of questions, which are set out and explained below.

1. What are the facts, the circumstances? This question prompts a description of the clinical problem, concurring factors and psychosocial and environmental aspects. The starting point is awareness that the interaction is not with an illness, but with a sick person with a life history, family, affections, job and deep personal, existential and ideological values. Each participant will, in their own way, experience the impact of the decision. Clearly, at the centre of the decision is the patient, being the person that will ultimately make the decision and bear the consequences. The available options should be assessed from a clinical standpoint, considering the likelihood of success of the option chosen. For example, what are the chances that a patient with aggressive cancer will survive mutilating surgery which may have major adverse effects? Besides the purely clinical assessment, the human and emotional costs involved must also be considered.

2. What is the 'spontaneous' option? What do the patient, their family members, the treating physician, the nursing staff and the medical team want? What is

Medical indications	Patient and family preferences	Quality of life	Contextual features
The principles of beneficence and non-maleficence	The principle of respect for autonomy	The principle of beneficence, non-maleficence. And respect for autonomy	The principle of justice (loyalty and fairness)
What is the prognosis including survival, function, and quality of life? How certain or uncertain is the prognosis? What are the treatment options? What are the potential benefits and burdens of disease-focused treatment? What are the potential benefits and burdens of palliative-focused treatment?	How much informations does the patient want about prognosis? How involved does the patient want others (friends, family) to be in decison-making? Does the patient have decision-making capacity? If the patient does not have decison-making capacity, who isthe appropriate surrogate decision maker? Has the patient previously expressed preferences in an Advance Directive or physician order for life-sustainig treatment (POLST)?	What are the most important things for the patient at this time (eg. survival, pain control, family wellbwing, etc)? What is the patient hoping for? What is the patient fearful of? What is the patient expecting? If tradeoffs between duration and quality of life need to be made, wich side should be more heavily weighted? Are there quality-of-life-states (eg. permanent coma) where it would not make sense to continue life-sustainig treatment or disease oriented care?	Is significant moral distress among staff members present, and how is this affecting care? Are there important family dynamics that must be considered? Are there financial or economic factors that must be considered, such as access to hospice care? Are there religious or cultural factors? Are there problems related to allocation of resources? How does the law affect treatment decisons? Is there any conflict of interest present?

Figure 2.
Jonsen's four box for medical decision-making.

the impact of pressure from fellow doctors or hospital managers, for instance in the event of a shortage of inpatient beds. What is the possible impact of pressure from the media?

3. What are the values at stake for each of the parties concerned? To answer this question it is necessary to draw up a personalised list of the hierarchy of values at stake, in the specific clinical situation, for the main parties concerned, mainly the patient, but also her family members (clearly where they have a say) and the medical team. For example, in the case of surgery entailing the risk of serious adverse effects and disability, the patient might refuse the surgery if she feels that the degree of beneficence, as perceived by her, is not adequate; the patient might instead wish to retain her current physical status, refusing a procedure that she considers to be invasive and destructive; this because the patient fears that after surgery, she might not recognise herself as the person she was before. On her part, the doctor may feel that the surgery will enable the patient to survive with what the doctor considers an acceptable quality of life (beneficence/maleficence). In other cases, the reverse may happen: the patient and his family members may want the surgery to be performed

no matter what, even if its positive impact may be minimal or zero (patient autonomy vs. doctor autonomy vs. fair allocation of resources).

4. What is the moral dilemma? The matter here is not to choose the best course of action, but to identify clearly the moral dilemma faced by the doctor and the whole team, spelling it out in the most explicit and detailed way. What must be decided is not whether to operate on a patient who demands a treatment that will yield little or no benefits, but whether to prioritise the patient's autonomy, and what he considers to be beneficial, or to prioritise the professional autonomy of the doctor, expressed through his clinical judgement on a procedure that he considers to be maleficent (a useless operation that will cause suffering to the patient) and to entail an unfair allocation of resources.

5. What are the alternatives? All too often, emotionally charged situations lead to a polarisation of views between just two possibilities. In the example in point 3, the only two options considered are surgery versus non-surgery. Instead, all options should be considered and presented to both the patient and his family members: chemotherapy, palliative care, home care, etc.

6. Which was the initial spontaneous choice? It is always advisable to return to the first spontaneous choice and assess whether the position of the main parties has evolved, and whether they have moved closer or farther apart from each other or have otherwise changed their views. If a change of position did happen, it should be considered whether this could help to reduce the conflict.

7. Making the decision. The decision must be made after consultation with the main parties involved, first and foremost the patient, but also his family members (where their involvement is authorised by the patient), the medical team, etc. It is important to have an open attitude and to truly listen. The patient must be seen not only from a medical point of view, but as an all-round individual with a life story, beliefs and concerns. As J. F. Malherbe [21] said, the patient remains the protagonist of his illness and not just the object of treatment. One should not hesitate to consult a colleague to get a second opinion, or even the hospital's ethics committee. After exhausting all these steps, a decision must be made. The decision must be justified by taking into account the medical evidence for each situation, but also the ethical issues specific to the situation. It is essential to specify which elements justify the principles that were given priority in the decision-making process.

5. Issues in implementing ethical principles

In theory, the description of ethical principles seems to give a clear overview of medical ethics and the procedures to be followed when making treatment decisions.

However, in clinical practice, the application of ethical principles is increasingly complex and is often affected by issues that complicate the decision-making process and come into conflict with ethical principles. Some issues arise when different principles clash with each other; others are linked to patient-specific situations, while yet others are linked to the organisation of services.

With regard to the conflict between principles, a common opposition may arise between the principles of autonomy and beneficence, for example in terminal cancer patients. According to the principle of autonomy, the patient should be told that her condition is now terminal, to allow her to freely choose among treatment options and decide what to do with the time she still has to live. However, under the principle of beneficence, one might argue that providing such accurate information might cause deep pain, and hence be harmful to the patient, affecting negatively her will to live and her quality of life in the time left to her. Moreover, the conflict between the two principles is not an abstract one; on the contrary, it is experienced by the parties to the decision-making process, with real consequences. The principle of autonomy can be interpreted in very different ways by doctors. For example, some doctors might resort to the legacy of medical paternalism and feel authorised to deliver all the bad news to the patient; other doctors could rely on the principle of autonomy to avoid making difficult decisions by shifting the responsibility onto the patient and/or her family members, placing a heavy emotional burden on the patient; still other doctors may not provide the full set of options to their patient to prevent her from making decisions that the doctor does not consider beneficial to her, resorting to a sort of 'palliative paternalism' [22] and thereby arbitrarily reducing the patient's free choice.

Conflict may also occur between the principles of beneficence and non-maleficence. An example is found in pain management for terminal patients, where the use of opioids relieves pain and meets the beneficence principle, but may shorten life, thereby violating the non-maleficence principle. Both principles are not absolute and are often combined, as in the above example, giving rise to the 'double effect' phenomenon, a term that in bioethics refers to an action that can have more than one result and contrasts two principles [17].

1. Other issues in the application of ethical principles arise when healthcare systems have to contend with limited resources. In these cases, the first ethical problem is patient selection for access to and discharge from care, which clashes with the principles of beneficence, non-maleficence and justice [23, 24]. The American Medical Association [25] has provided guidance on the ethical implications of the allocation of organs for transplant, which may be helpful in the task of determining priority of access to scarce and costly medical resources. The AMA paper has identified five criteria related to the patient's Medical Needs, which should be considered when making resource allocation decisions: likelihood of benefit

2. the improvement in quality of life

3. the duration of the benefit to the patient

4. the urgency of the patient's condition

5. only in some cases, the amount of resources required for successful treatment

These criteria help to maximise three primary goals of medical treatment: number of lives saved, number of years saved and improvement in quality of life. A hierarchy of objectives prioritises the goal of saving the greatest number of lives. [25] While the AMA document makes an important contribution to ethical decision-making, many questions about distributive justice and discrimination against older people remain open.

Furthermore, major social changes have affected the organisation of health systems and have further complicated the application of ethical principles. The globalisation of modern society, with its marked contradictions, inequalities and injustices has also inevitably affected healthcare systems. The undoubtedly successful McDonaldization phenomenon, [26], characterised by efficiency, productivity, cost reduction, procedural standardisation and control, has also influenced the organisation of healthcare services. The double pressure to cut costs and make a profit has impoverished the healthcare system, hitting hardest the most vulnerable and deprived citizens and generating major inequalities in the access to healthcare services: this has deeply affected the ethical principle of justice and beneficence and has altered the doctor-patient relationship [27].

6. Current issues

In 2020, the whole world was struck by the Covid-19 pandemic. The pandemic disrupted life for every person with an unexpected, novel situation and caused an unprecedented humanitarian emergency. Its sudden outbreak has put the health systems under massive strain, causing a number of ethical problems for healthcare staff and managers, and giving rise to real challenges to basic ethical principles.

Compounding the existing problems in applying ethical principles, the pandemic has brought about new complex scenarios and issues, which have not always been addressed appropriately and in line with ethical principles.

The first moral dilemma posed by the pandemic relates to the strain on healthcare quality caused by the surge in demand. The pandemic has spread quickly, catching the health structures unprepared to handle the rapid increase in workload. At the height of the crisis, the number of patients rose dramatically and the hospitals soon ran out of beds. The number of healthcare workers (doctors and nurses) was also insufficient to deal with the surge in cases. Many health workers faced the additional workload with great dedication and sense of responsibility, aware that their patients' lives also depended on their willingness to put in the extra hours. They prioritised the beneficence for their patients over their personal well-being. Many healthcare workers fell ill and many died [28]. At the peak of the pandemic, medical and nursing staff worked 12–14 hours a day wearing uncomfortable face masks, visors and coveralls. It is fair to assume that fatigue and stress at work may have affected the quality of care, hence the actual beneficence for patients. It can also be presumed that the quality of the care provided at the start of a work shift was higher than that provided by the same worker after 12 hours of gruelling work. Thus, the actual working conditions undermined both the principle of beneficence and the principle of justice, according to which all patients must be treated equally.

Moreover, the spike in patient numbers was so high that it produced an imbalance between the healthcare needs of the population and the availability of intensive care resources. The situation that came about was and still is an exceptional one, to the extent that it has been classified as 'disaster medicine' [29]. With regard to intensive care, in addition to the criteria for access to and termination of care, traditionally based on the appropriateness and proportionality of care, the criteria of distributive justice and appropriate allocation of limited health resources had to be applied. The 'first-come, first-served' criterion for access could not be applied. Healthcare workers were forced to carry out an unusual triage, in which they often had to apply the criterion of 'greater life expectancy'. In Italy, SIAARTI (the Italian Society of Anaesthesiology, Analgesia, Resuscitation and Intensive Care) issued 'Clinical ethics recommendations for the allocation of intensive care treatments, in exceptional, resource-limited circumstances' [29]. The recommendations are solidly grounded in ethical principles, to relieve clinicians from the burden of making subjective decisions, and establish explicit resource allocation criteria [29]. (SIAARTI). Robert et al. highlighted the ethical issues in patient management in intensive care units during the pandemic in France [30]. Despite the guidance provided, the dramatic pressure of the situation often forced physicians to grapple alone with the final decision about who should get life-saving care. While admittedly it was necessary to make a selection among the patients, we must also note that a dramatic discrimination occurred by age group, comorbidity and patient type. Elderly patients, patients with comorbidities and frail patients were often denied access to the ICU.

The pandemic emergency also gave rise to other issues. Many patients could not even reach the hospital and died at home while waiting for an ambulance that never arrived. In those cases, the decision was not guided by any particular and specific recommendations, but was simply left to chance: the lottery of life decided for them.

For the patients' protection, during their stay in hospital, the patient-family and healthcare worker-caregiver connection was severed, counter to more than 20 years of research and care practice aimed at improving those relationships for the patient's benefit [30]. Many patients were left to face death alone, without the comfort of family members, without any spiritual or religious care. As hospitals were overwhelmed, much was attempted to provide the benefit to the body but little was done to provide psychological and emotional care; healthcare moved back from caring for the whole person to focusing on the illness alone.

Yet other decisions have impacted ethical principles and good clinical practice in the management of chronic patients. For a long time now, the healthcare system has placed emphasis on prevention and early diagnosis programmes, educating the public about the importance of health screening and monitoring. The emergency has deeply disrupted this approach. Many cancer patients have been unable to attend their routine checks, and the same has happened to patients with heart conditions or diabetes. The principles of beneficence and non-maleficence have been severely compromised. An increase in deaths due to cardiovascular diseases has already been recorded, and the number of deaths secondary to cancer is also expected to rise [31].

7. Conclusions

The above overview confirms that the practical application of ethical principles in medicine is fraught with difficulties that may complicate the decision-making process. The current pandemic is confronting us with novel organisational, social and ethical challenges.

As a rule, major changes in healthcare occur at a much slower pace, giving us enough time to process them, adapt and make decisions. Today's explosive crisis calls instead for urgent emergency measures. The assessment tools we have used so far have been made obsolete by the extraordinary pace of the crisis. In the health sector, clinical guidelines have traditionally been the gold standard for good clinical practice, in addition to providing some protection from medical liability. However, many guidelines have lost their relevance in the pandemic, which has created an unprecedented health situation for which no specific guidance could be prepared. The dramatic developments have put ethical principles under strain in various circumstances and cases. Moral dilemmas have severely affected the emotional resilience of clinical staff; in the near future we will have to deal with the moral distress they experienced.

Ethics, once a discipline of interest to scholars, has nowadays taken on a prominent role in the social debate. However, moral questions must be addressed and analysed critically, in order to define not only what is right, but also why it is right [32].

Hopefully, we can draw some lessons from this tragedy.

The rationalisation of healthcare resources – through major budget cuts, the push for standardised care processes according to the McDonaldization model, the emphasis on hi-tech and highly specialised care – has not withstood the test of the pandemic. While of course it is hard to say which model would withstood the Covid crisis, it remains a fact that the current one failed, and this requires some reflection.

First, we should strengthen the human dimension of the physician-patient relationship. The focus on performance and profit has reduced the time available for listening to patients and their family members; as medical professionals, we have contributed to the achievement of the productivity targets set by the health authorities, but we have not always respected the ethical principles of an authentic doctor-patient relationship based on caring for the individual as opposed to simply treating a medical condition. Health professionals should take the brave step of fostering the relationship with their patients and prioritising quality over quantity, eschewing the industrial assembly line model: people are not machines and do not function like machines.

Social systems as a whole should revisit their resource allocation models. For a long time now, policy makers from all sides have made major cuts to health care; the pandemic has shown that 'sick countries' with difficulties in the delivery of healthcare are also countries with persistent economic problems. The share of public spending allocated to healthcare should be fairer, instead of treating the health service as the poor relation.

During the pandemic, we helped the patients with the greatest chance of survival, but we were unable to help the frailest ones. We went back to the model of Sparta, the ancient Greek city where frail male infants were tossed off a cliff, to train the others to become strong and valiant warriors. However, the Spartan model was not the one that prevailed in ancient Greece, nor the one that produced the greatest protagonists of classical culture. Healthcare systems, with the contribution of medical ethics, should develop care models that protect the frailest and shelter them from 'competition' for survival in which they would be doomed from the start.

We should also send the message that medical ethics is not just a matter for the individual health professional but is the responsibility of the whole community. The pandemic is teaching us that the responsible behaviour of each of us plays a key role in preventing the spread of the infection. The principles of medical ethics, beneficence and non-maleficence should be better known, understood and applied not only by health workers but by all persons.

Last but not least, the expectations placed on doctors today are very high, if not excessive, as concerns both clinical skills and patient relations. Although ethical issues are now on the front line, there is still very little training in biomedical ethics for health professionals. The development of science and technology require that physicians be knowledgeable of ethical issues pertinent to end-of-life care [33, 34]. It is crucial to invest more in this of training, to ensure that the new generations of doctors and other health professionals, within their respective roles, are better equipped to face the new challenges for medical ethics.

Author details

Liliana Lorettu[1]*, Jocelyn Aubut[2] and Rosagemma Ciliberti[3]

1 Psychiatric Clinic, Department of Medical, Surgical and Experimental Sciences, University of Sassari, Italy

2 Department of Psychiatry, Université de Montréal, PQ, Canada

3 Department of Health Sciences, Section of History of Medicine and Bioethics, University of Genoa, Italy

*Address all correspondence to: llorettu@uniss.it

Chapter 23

Plato in Contemporary Medical Ethics: Holism and Care

Tudor-Ștefan Rotaru

Abstract

There is a gap of twenty-four centuries between us and the Greek philosopher Plato. But what he had to say about illness, healing and the human being can be unexpectedly relevant in contemporary medical ethics. We argue that the contemporary principles of autonomy and beneficence can be revisited by means of platonic philosophy. We present an old and consistent idea of care which includes the empowerment of people in choosing the good by means of a virtuous character. We connect this idea to the contemporary notion of autonomy. We also show how a holistic approach was present for a long time in our cultural history. We argue that, despite its almost definitive loss, holism can and should be present in revisiting the principle of beneficence. For both holism and care, we provide samples of philosophical history. We conclude that an autonomous choice should be revisited as a wise choice and that medical beneficence should be reconsidered as holistic.

Keywords: Plato, autonomy, beneficence, care, holism

1. Introduction

Connecting the ancient philosopher Plato with contemporary medical ethics seems a bold endeavor. However, such a connection is relevant and useful. Many of the contemporary medical ethics' problems stem from how we define life, choices, beneficence and similar difficult notions. And, throughout history, the various implicit definitions of such notions have changed. Therefore, how human life was philosophically understood from one era to another had an impact on the practice of medicine. From the magical medicine of the XVIIIth century B.C. until the contemporary evidence-based medicine, the way of acting with respect to an illness, to the body and to the human being has changed [1, 2]. There is, however, one especially relevant nexus of ideas in history, with respect to how we understand life, healing and what it is to be human. This work gathers older and important ideas from its past but also serves as an influential philosophy for the whole subsequent way of Western thinking. These ideas are important now, because tracing back some crucial elements of our cultural DNA can provide valuable insights to the contemporary struggles of medical ethics. Such philosophical nexus is Plato. His Dialogs are ingrained with two key implicit notions that are essential to medicine: holism and care. We try to trace back these two implicit notions and to connect them to the contemporary principles of autonomy and beneficence. In summary, we explore what Plato can unexpectedly teach us when it comes to solving current bioethical issues.

IntechOpen

2. Contemporary autonomy and the platonic *therapeia*

The precise meaning of "autonomy" in contemporary biomedical ethics is still in dispute. A minimal definition of autonomy includes a self-governance, free from both the controlling influence of others and free from any other form of interference that would prevent a meaningful choice. Such an interference would be, for instance, an incomplete understanding or some type of coercion. For many contemporary bioethicists, an autonomous person is one who acts freely in line with a plan of their choosing. On the other hand, a person with diminished autonomy is, in some significant respect, controlled by other people or unable to deliberate and act based on their own plans and desires. Almost all theories of "autonomy" consider two crucial prerequisites. One of them is freedom, and the other is agency or ability to act with intent [3]. Autonomy includes both the capacity to distinguish between alternatives and the capacity to put one's plan into action [4]. There is no mention about the nature of what the patient is about to choose.

But, the way Plato understood the meaningful choice is intriguing. This is because the philosopher reveals two ways of doing something one "wants" in his Dialogs. One type is doing something that seems good for a person's opinion or in line with appearances. The other one is doing something a person genuinely wants. The difference lies in pursuing the good or pursuing the mere pleasure or appearance. The passage in Gorgias[1] is obvious: "For I say, Polus, that the orators and the despots alike have the least power in their cities, as I have stated just now; since they *do nothing that they wish* to do, practically speaking, *though they do whatever they think to be best*" [5]. An alternate translation of the original Greek fragment ἂν αὐτοῖς δόξῃ βέλτιστον εἶναι would be "though they do whatever looks better to their opinion" [6].

The distinction uses two families of words. With respect to appearance, the preferred verb is δοκέω (to expect, suppose, imagine, seem). Its family of words includes δόξα (expectation, mere opinion, conjecture). On the other hand, with respect to what is real, the preferred verb is βούλομαι (will, wish, be willing) [7]. This separation between two types of "doing what one wants" mirrors the thorough classifications of human endeavors[2] in the Gorgias dialog and the distinction between what is apparent and what is real. It also includes medicine among the arts of restoring what is true and not what is merely apparent.

On the side of doing what looks better for an individual's opinion (expressions using δοκέω), there are basic practices (ἐμπειρία) that target mere flattery, including apparent health. These practices are meant for pleasure, and Plato classifies them according to their aim of creating appearances (like sophistry and cosmetics) or restoring appearances (like rhetoric and cooking). All four practices pretend to deal with health, either in the soul (sophistry and rhetoric) or in the body (cooking and cosmetics). Oppositely, doing what one really wants (expressions with βούλομαι) deal with knowledge. They are the arts (τέχναι). Their quality resides in the fact that they do not target pleasure but the good. Legislation and gymnastics are generators of real health, while justice and medicine are ways of restoring real health. Finally, legislation and justice deal with the soul, while gymnastics and medicine deal with the body. A previously published table can offer more details on this classification [8].

In many parts of Plato's Dialogs, they make mention about θεραπεία, a term which holds meanings like service, attendance, treatment, cure and care [7]. But, this term holds strong connections with the whole platonic philosophy about the

[1] Gorgias, 466d-e

[2] Gorgias, 462c-469

betterment of the human being[3]. "The approach of the ultimately real, in Plato's thought, is properly made by way of his interpretation of man, man whose rational existence is in jeopardy, because he is divorced from his ground of Being despite de telltale signs of his essential kinship with it" [9].

I argue that platonic θεραπεία is meant to restore wisdom, which can be considered as an authentic form of autonomy, based on doing the things a person genuinely wants (βούλομαι) and not on things that one finds best in appearance (δόξῃ βέλτιστον εἶναι). This authentic form of autonomy includes the basic attitude of care. Plato's cave is a symbol of the inverted life of man[4], who cheerfully exchanges shadows for reality, ignorant of himself and of his own bondage. For this very condition Plato tries to devise a θεραπεία, a scheme of educating the man in adequately coping with it [9].

Now, returning to the field of contemporary medical ethics, we can explore new meanings about what it is to care about one's patient and what their autonomy might mean. For a patient, to act in accordance with what they genuinely want, the patient should act in line with what is good for them. In a platonic understanding, this would be true, because true τέχναι deal with the real health and with authentic volition[5]. If we are to understand autonomy in a platonic way, being autonomous implies knowledge about what is genuinely good for oneself and not the freedom to do what merely appears to be good.

Did Plato speak about the patient's freedom to choose? We argue that he did but with the appropriate vocabulary of his time and of his philosophy. For Plato, freedom to choose is genuinely exerted in the realm of good. The good, as the patients understands it at a given moment in time, might not be properly grasped. There is always the danger of substitution the true good for the apparent one. And one fragment of Alcibiades[6] is illustrative in this sense: "Socrates: For if a man, my dear Alcibiades, is at liberty to do what he pleases, but is lacking in mind, what is the probable result to him personally, or to the state as well? For instance, if he is sick and at liberty to do what he pleases, without a medical mind, but with a despot's power, which prevents anyone from even reproving him, what will the result be? Will not his health, likely, be shattered?" [10].

From these platonic fragments we draw some valuable information. Not only can we map a superposition over the contemporary definition of autonomy (capacity and information), but there is also a nuance which deals with what a patient should really know and do. So, θεραπεία as care should prepare one to be the type of person who is able to genuinely want something that is good for themselves. In short, θεραπεία makes one wiser, when and if it is possible. The battle between what a patient genuinely wants and what merely seems appropriate is externalized in another example of Gorgias[7]. By a values inversion, those who pursued the patient's good get blamed: "servants you tell me of, and caterers to appetites, fellows who have no proper and respectable knowledge of them, and who peradventure will first stuff and fatten men's bodies to the tune of their praises, and then cause them to lose even the flesh they had to start with; and these in their turn will be too ignorant to cast the blame of their maladies and their loss of original weight upon their regalers, but any people who chance to be by at the time and offer them some advice—just when the previous stuffing has brought, after the lapse of some time, its train of disease, since it was done without regard to what is wholesome, these are the people they will accuse and

---

3 p. xvi

4 p. 47

5 Gorgias, 466d-e

6 Alcibiades, 135a

7 518c-d

chide and harm as far as they can, while they will sing the praises of that former crew who caused the mischief" [5].

Does platonic θεραπεία mean caring for the person? I argue that it does. Since[8] "the tragedy of human existence and, therewith, the problem of philosophia as a method of education, are signalized in men's contentment with living an unreal and alien life [...] the art is not one of conveying truth to man in the form of propositions, but rather one of conducting men, by exacting scrutiny of opinions, into the presence of reality" [9]. What Plato desired for people is, in this sense, unexpectedly actual.

Plato's type of care consisted of enabling people to seize the good and make right decisions about it. However, this type of θεραπεία was not the only way of understanding the relationship among the patient, illness and choices. Even before Plato, the complete care for the person started to be conceptually dismantled. This was due, in part, to the Hippocratic tradition, which focused more on the specifics of the disease and on the interaction between signs of illness and individual particularities in showing these signs[9]. "The physician must be able to tell the antecedents, who know the present and foretell the future—must meditate these things and have two special objects in view with regard to diseases, namely, to do good or to do no harm. The art consists of three things—the disease, the patient and the physician. The physician is the servant of the art, and the patient must combat the disease along with the physician" [11]. We can notice that the good this section refers to does not seem to have the same meaning as Plato's good in Gorgias. While Plato's good seems to have more to do with how we should understand autonomy, Hippocrates's good sound like the current-day principle of beneficence. One of them seems more preoccupied with servicing the person; the other one is more inclined in servicing the profession.

However, in history, medicine was on the brink of losing care altogether. Beginning with the spectacular discoveries of the XVIIIth century, the disease, once an element of the Hippocratic triad, almost became the sole center of attention. And Foucault[10] unmasked it: "In the rational space of disease, doctors and patients do not occupy a place as of right; they are tolerated as disturbances that can hardly be avoided: the paradoxical role of medicine consists, above all, in neutralizing them, in maintaining the maximum difference between them, so that, in the void that appears between them, the ideal configuration of the disease becomes a concrete, free form, totalized at last in a motionless, simultaneous picture, lacking both density and secrecy, where recognition opens of itself onto the order of essences" [12]. Care, as a patient's guidance towards the truth, is completely absent, and the only truth being sought is the scientific model of the illness itself.

In modern times, some philosophers were able to offer indirect explanations about the way in which care got lost on the way. For instance, we can draw some insight from the works of Emmanuel Lévinas. Phenomenologically, the philosopher shows that the relationship between "I" and "Other" cannot be a mere representation. A simple perception turns the Other into a mental object. But for a genuine understanding of the relationship between I and Other, "I" am forced to accept that the phenomenological distance between I and Other is infinite. In short, I cannot access their phenomena and experiences. All assertions using "we" are unable to circumvent totalization and lose the Other among all the objects of my mind. The only process at our disposal is to look towards the Other, to acknowledge the infinite distance. If infinity is acknowledged, the relationship with the Other ceases to be a

[8] p. 45

[9] Epidemics, Book 1, Sect II, 5. p. 360

[10] p. 9

mere relationship like the one I have with the objects of my mind. Looking towards the Other becomes an ethical relationship, and its only honest feature becomes care [13]. For current-day medicine, Lévinas' conclusion is valuable: it shows us that care is embedded in the very core of the human relationship. Since Hippocrates made the physician the servant of the profession (ὁ ἰητρὸς ὑπηρέτης τῆς τέχνης) and not of the Other, the understanding of care gradually changed.

Today, we understand autonomy in terms of agency and ability to choose. But what is chosen by patients and communities is in dispute. This is because, in the patient-physician relationship, free choice has become an honored process but with no moral content. It is a procedural morality, based on the principle of permission, but often lacking in content [14]. We hold dear values that allow us to choose, but medicine and care have no true insight in what should be chosen as good and for whom. Therefore, the process itself of choosing is doubtful since appearance and truth can easily switch places. Plato held a view that the patient (on a physical and spiritual level) must be accompanied towards the truth and enabled to choose the appropriate path; this included healing situations and medicine. Getting back to some of these incredibly old insights might trigger a certain degree of reconsideration of the actual tenets in medical ethics: autonomy, not separated of care, but enhanced by care itself might be a provocation worth exploring.

3. Contemporary beneficence and Plato's holism

Beneficence is another celebrated principle of medical ethics, from the time of the Belmont report itself [15]. This principle demands doctors and researchers to make an active contribution to the welfare of patients. In the common morality, it includes obligations[11] of persons like protecting the rights of others, preventing harm from occurring to others, remove conditions that cause harm to others, help persons with disabilities and rescue those in danger [3]. In contemporary medical practice, many efforts have been directed towards what we call Evidence Based Medicine, a methodic approach that aims to validate or to invalidate separate interventions for separate pathologies. The process is meant to guarantee the beneficial effect of medical interventions. However, many professionals feel that this type of partitioning in medical knowledge misses the very art of healing and loses touch with the human patient [2].

This is where the holistic approach comes into discussion. The word's etymology is self-explanatory. It comes from the Ancient Greek term ὅλος, -η, -ον, meaning whole, entire, utter [7]. A contemporary definition of "holism" states that it is "the theory that certain wholes are greater than the sum of their parts, the opposite of atomism. In medicine, it is the treating of the whole person, rather than just the symptoms of a disease" [16]. We can notice in the above definition two elements. The first one deals with the whole as superior to a mere sum of parts. The second one speaks specifically about the entirety of the human being in medical thinking. On the opposite side, the term "atomism" means "a theoretical approach that regards something as interpretable through analysis into distinct, separable, and independent elementary components, the opposite of holism" [16].

Although contemporary medicine seems to lean towards atomist thinking, Plato held a more holistic view about what was beneficial for a person or a community. This view was still popular in his time and culture. Setting aside the enormous differences in medical scientific knowledge between current day practices and the medicine in the 5th and 4th centuries, the philosophy Plato held about the human

[11] p. 204

being was intriguing. We argue that many passages in the Dialogs demonstrate a holistic approach. For instance, in Charmides[12], Plato shows us the important relationship between the part and the whole. In order to cure Charmides' headache, Socrates states that one cannot look for a cure destined to the part without a cure for the whole: "This Thracian said that the Greeks were right in advising as I told you just now: "but Zalmoxis," he said, "our king, who is a god, says that as you ought not to attempt to cure eyes without a head, or a head without body, so you should not treat a body without a soul"; and this was the reason why most maladies evaded the physicians of Greece—that they neglected the whole, on which they ought to spend their pains, for if this were out of order, it would have been impossible for the part to be in order" [10].

In the Republic[13] Plato mentions a connecting order with respect to parts of the soul and parts of the body. This is illustrative for holism in platonic thinking: "But, to produce health is to establish the elements in a body in the natural relation of dominating and being dominated by one another, while to cause disease is to bring it about that one rules or is ruled by the other, contrary to nature. - Yes, that is so. - And is it not likewise the production of justice in the soul to establish its principles in the natural relation of controlling and being controlled by one another, while injustice is to cause the one to rule or be ruled by the other, contrary to nature? - Exactly so, he said. - Virtue, then, as it seems, would be a kind of health" [17] .

In the Laws[14], Plato teaches about different global lifestyles, suggesting that disease and lack of virtue are somehow connected: "The lives of us men must all be regarded as naturally bound up in these feelings, and what kinds of lives we naturally desire is what we must distinguish, but if we assert that we desire anything else, we only say so through ignorance and inexperience of the lives as they really are. What, then, and how many are the lives in which a man—when he has chosen the desirable and voluntary in preference to the undesirable and the involuntary, and has made it into a private law for himself, by choosing what is at once both congenial and pleasant and most good and noble—may live as happily as man can? Let us pronounce that one of them is the temperate life, one the wise, one the brave, and let us class the healthy life as one; and to these let us oppose four others—the foolish, the cowardly, the licentious and the diseased" [18].

We can see that, in Plato's view, healing was almost never meant to be some isolated intervention. We can also notice that, partially, Plato did not hold the views of ancient religious medicine which connected all health states with maleficent spirits [1]. Plato shows that one could not heal the part without healing the whole. This was the case for body parts, soul parts or city parts. We argue elsewhere that Plato's obvious holistic approach can entirely change the semantics of what we understand by "patient" in the Dialogs [19]. On the other hand, what constitutes health and justice is an appropriate order or hierarchy of those parts that constitute the whole to be healed. Las but not least, health and virtue go hand in hand: one cannot expect to restore health without restoring virtue in the entire individual.

This cultural DNA string did not begin with Plato and did not disappear entirely in the following centuries. We could say it "traveled" in parallel with a more atomist approach, for a long time until it almost got lost in the XVIIIth century. We start our argument by mentioning the millennium that passed between the documentary attesting of doctors in Egypt and the doctors' attestation in Greece. The Egyptians are the first to speak about surgery around 1550 B.C. [1]. We can consider this moment as the first known separation between a more "rational" and atomist

[12] 156d-e

[13] 444d

[14] 733d-e

approach of the human being and the magic-religious medicine that held exclusive ground until then. This first rupture marks different attitudes towards man as a whole and as a sum of reparable parts.

The Greeks took on this task of separation. The reputation of Hippocrates as the father of medicine comes, especially, from the majority who currently practices a "scientific" medicine. This type of approach systematically studies components of the human body. For instance, we speak today about gastroenterology, cardiology, nephrology and so on. Indeed, there are numerous physical explanations in Hippocrates's works. However, he and his followers were not alienated from the whole-part relationship. They did not totally abandon the global explanations either. For instance, there are examples[15] where astronomical explanations leaned towards some kind of religious or magic effect: "One ought to also be guarded about the rising of the stars, especially of the Dogstar, then of Arcturus, and then the setting of the Pleiades; for diseases are especially apt to prove critical in those days, and some prove fatal, some pass off, and all others change to another form and another constitution. So it is with regard to them" [11].

However, the Hippocratic doctrine gives a secondary place when it comes to holistic approaches, frequently preferring an atomist explanation. The most important feature of these teachings makes room for the doctrine of the four humors: phlegm, yellow bile, black bile and blood. Many diagnostics[16] turn to this way of thinking: "That vomiting is of most service which consists of phlegm and bile mixed together, and neither very thick nor in great quantity; but those vomitings, which are more unmixed, are worse. But if that which is vomited be of the color of leeks, or livid, or black, whatever of these colors it be, it is to be reckoned bad; but if the same man vomits all these colors, a very fatal symptom is to be reckoned. But of all the vomitings, the livid indicates the most imminent danger of death, provided it is of a fetid smell. But all the smells, which are somewhat putrid and fetid, are bad in all vomitings" [11]. This direction of medicine, seeking physical or "naturalistic" explanations for disease will make history and will lead, in time, to present-day scientific medicine.

Later, Avicenna will still use holistic notions. The humors doctrine will continue to hold an important place but will be frequently connected to the person's entire demeanor[17]: "One must not get the idea that every temperament gives rise to its like and never its opposite. A temperament often gives rise to its exact opposite, indirectly (of course); it cannot do so directly. A cold and dry temperament may give rise to visible moisture, though this would not be beneficial but would indicate that the digestion is feeble. A person with such a temperament would be thin, with supple joints, and hairless skin, cold to the touch, the surface veins narrow, and he would be gentle and timid in nature" [20].

The contemporary philosophy will succeed in demasking the total transformation of an individual in an object of observation. This phenomenon started in the 18th century. Michel Foucault explains[18] that the language of things started to be authorized with respect to humans as well: "The task lay with this language of things, and perhaps with it alone, to authorize knowledge of the individual that was not simply of a historic or esthetic order. That the definition of the individual should be an endless labor was no longer an obstacle to an experience, which, by accepting its own limits, extended its task into the infinite. By acquiring the status of object, its particular quality, its impalpable color, and its unique, transitory form

---

[15] Air, Waters and Places, 11 (p. 205)

[16] Prognostics, 13 (p. 245)

[17] 107, III (pp. 90-91)

[18] p. xiv

took on weight and solidity. No light could now dissolve them in ideal truths, but the gaze directed upon them would, in turn, awaken them and make them stand out against a background of objectivity. The gaze is no longer reductive; it is, rather, that which establishes the individual in his irreducible quality. And thus, it becomes possible to organize a rational language around it" [12]. The almost complete deterioration of the holistic attitude came to be as a natural consequence of another limiting principle: the separation of sciences and the separation of disciplines of study. Inside a certain discipline, the rules got more constraining. The production of discourses to cover global or holistic attitudes became an almost impossible task [21].

It is Gadamer[19] that links us back with Plato with respect to medicine. He shows that Plato tried to find the connections that tie the spheres of the soul, the city and the Universe as a whole. This type of awareness must be regarded as a superior type of wisdom compared with the arrogance inspired by our ever-expanding technical skills. The German philosopher denounces a crisis of humanity: we developed the technical aptitudes in such an extent, so they became an all-encompassing attitude [22]. Gadamer shows that doctors are, by virtue of their profession, involved with two key aspects of our life: life and death. Plato has shown us that is impossible to cure the body without knowing something about the soul. Or, more precisely, one cannot heal without knowing something about the nature of the whole. The notion of the "whole", here, does not mean a mere methodological concept. It speaks about the unity of being itself[20]. "It is the whole in the sense of the movement of the stars above and the changes of weather below, the rise and fall of the oceans and living nature of the woods and fields. It is what surrounds and encompasses the nature of human beings that determines whether they find themselves in a condition of safe health or exposed to dangerous threats. Medicine seems to be a genuinely universal science, especially if this whole of nature is extended to include the whole that is our social world" [22].

Nowadays, we understand medical beneficence in terms of functionality, mobility, absence of suffering, alleviation of symptoms and removal of organic causes. But it is obvious that many medical acts fail to restore true health in an individual. The doubt in the physician-patient relationship, the psychiatric comorbidities, the side effects of various treatments, the lack of compliance in diet and lifestyle and the changes in a patient's social network often make healing impossible. Philosophy teaches us that holism has survived for a long period of time in medicine, despite less scientific medical knowledge in the past centuries. But its presence was not a mere artifact of a primitive world. It was strongly connected with the intent of a genuine beneficence. Plato believed that healing occurs in the whole of an individual and even in the whole of society itself.

4. Concrete ideas for contemporary medical ethics

We argue that Plato may help us reconsider contemporary medical ethics in two ways at least. First, useful input might come by revisiting the respect for autonomy. Nowadays, respecting a patient's autonomy includes making sure there is decisional capacity, agency, reasonable information, lack of coercion and all conditions for a meaningful deliberation between options. Plato taught us that "doing what seems best" and "doing what one really wants" are two different ways of acting. And what separates them is a will being directed towards the good, in contrast with a false

[19] pp. 84-85

[20] p. 115

sense of choice in doing what looks best to one's opinion (δόξα). However, the will is directed towards the good when and only when one's character allows it to.

In a platonic reading, an autonomous choice is a *wise* choice. To the extreme, we might argue that some implicit definition of beneficence is present in the very substance of the wise choice. Doing what seems best and not what is good for oneself is not an exercise of wisdom, thus not an autonomous choice in the new semantics of this term. Even when all reasonable information is provided, a lack of virtuous character might prevent a potential patient from seeing what is genuinely good for oneself, thus invalidating what should have been a wise choice. Therefore, the question arises if respect for autonomy as we understand it nowadays is, indeed, sufficient for a competent (i.e. wise) decision.

The closest medical practice to this ideal would be the deliberative model of the patient-physician relationship. In contrast with paternalistic and informative models, in this type of relationship, the objective of the physician-patient interaction is to help the patient determine and choose the best health-related values that can be realized in that situation. The physician helps identify the values included in the available options but also suggests why certain health-related values are worthier of consideration. However, in this model, the physician discusses only health-related values and considers that many other values are unrelated to health and disease. The deliberative model allows the doctor to act as a teacher or friend who sees respect for autonomy as moral self-development. Objections to this model include the fact that physicians would not possess privileged knowledge of those values which should have priority in health situations. This objection is intimately linked to the pluralistic moral reality of modern societies. Other objections emphasize the fact that the physician should never engage in moral deliberation or that this type of endeavor might easily turn into unintended paternalism [23].

However, Plato offers us more. And it is not mandatory to adopt his theory of Forms or his ideas about learning as remembering to figure out that people can be accompanied towards more truth and better choices. Cushman explains to us what Plato means by his theory of αναμνέσις[21]: "however valid true opinion may be, επιστήμη requires a community of kindred minds, wherein truth is jointly acknowledged and so, is removed from the closet of merely private surmise" [9]. In leading the other towards the truth, it is never a matter of coercion. Opinions arise in the individual soul, and it is by friction with other minds, that these become converted into matters of knowledge[22] [9].

The deliberative model allows the patient and doctor to discuss the worthiness of different health-related values. It also aspires to a certain moral self-development in a patient. These features match the message in the platonic dialogs. However, critics of this model argue that physicians do not possess privileged knowledge of values which should have priority in health situations. Inspired by Plato, I argue that they *should* possess this privileged knowledge. A certain type of wisdom *should* be a part of their build as doctors. This is especially true because, as Cushman explains[23], one is unable to share the perspective of his own virtue unless they are in possession of true knowledge as an integrating part of their virtuous character [9]. In short, a physician would do his job properly when they are able to help a patient become wiser. In a platonic reading, we equate true autonomy to wisdom. Therefore, to help patients become wiser, the physicians themselves should have a virtuous character. *The only way a doctor can respect patient's true autonomy is for this doctor to be wise himself/herself.*

---

[21] p. 93

[22] p. 103

[23] p. 93

Critics might argue that the definition of *good* is, often, a political one. They are right. But the platonic reconsideration of the doctor-patient relationship in terms of wisdom is not meant to violate the principle of permission. This principle remains the *sine qua non* condition of a peaceable secular community [14]. I argue that doctors have a duty to try, to the best of their ability, to stir a moral (not religious) development in their patients, as the only way to enable true autonomy. To do that, doctors have a duty to stir moral development in themselves. They cannot escape the essential universality of medical science Gadamer[24] speaks about [22]. If we are to believe Plato, failure to accomplish either of the two duties is, in fact, failure to respect the principle of autonomy. It becomes a "blind leading the blind" situation.

Secondly, Plato helps us rethink beneficence in more holistic terms. What is beneficial in a medical act should be beneficial for the entire human being but also for the entire community of beings. The over-specialization of medicine has helped scientists to expand knowledge and discover new and revolutionary treatments. However, I argue that, when it comes to medical practice, this partitioning of medical interventions represents a high risk of doing more harm than good. Patients often lack treatment compliance. They are, often, unable to change their lifestyle to make a treatment work or to avoid complications. What is mended by a medical specialty often gets broken by behaviors outside the medical area of expertise. It is the case with smoking, drug abuse or generally unhealthy behaviors. Multiple comorbidities get treated by separate specialists, with different protocols. They often result in a high number of drugs getting ingested daily. Similarly, depression is usually accompanied by lack of interest in one's health and self-care behaviors. Psychosocial problems and difficulties are both causes and effects of depression and anxiety. Moreover, poverty, lack of education, different forms of discrimination and abuse towards minorities, all diminish or altogether block access to healthcare.

A more holistic approach should be able to put together the pieces of this medical, social, behavioral and spiritual puzzle. The nexus of information about disease, lifestyle, behaviors, genetics, social status, economic status and so on might be in the hands of the general practitioner in their role as the family doctor. They might be the most suitable physician to take on the difficult task of accompanying patients on their way to true, better lives. Besides technical skills, the family doctor might be particularly trained in deliberating values and shared decision making. They should be able to have a good understanding of spiritual, social, economic and educational aspects of life. This will enable them to direct the use medical and non-medical resources to empower patients to make beneficial decisions in the web of all medical specialties and non-medical realities. This would also give a special place to psychotherapies as a resource that might favor life-changing personal leaps. In short, multidisciplinary approaches cannot and should not become the patient's task. The reason for it is that the patient is not able to build the holistic approach needed for their care on their own. The task of approaching things holistically should be in the hands of a doctor. This doctor is not required to be over-specialized in some areas of the human body. They are expected to be a wise mentor in the health of the human being.

5. Conclusions

The advancements in medical ethics and bioethics allow for more precise useful frameworks to judge different ethical dilemmas and questions. These frameworks are also easier to include in codes and regulations pertaining to the medical

[24] p. 115

profession. However, it is still not clear what exactly an autonomous patient does when making a health-related choice. Nor it is clear what the moral content of beneficence is, since the patient is a whole in themselves and is part of a larger system of relationships and interactions. I have shown that Plato, twenty-four centuries ago, spoke about a type of choice that we can construe, today, as genuine autonomy. This type of choice has been cultivated inside a relationship of care between the doctor and the patient. The philosophy of care about one's genuine want of something good for oneself has traveled the history. It almost got lost once disease (and not the ill human) became the central focus of the medical profession. Plato also harbored a holistic approach about people and their illnesses. This holistic approach was present before him and survived after him, in parallel with an atomist view that began with the first Egyptian surgical attempts. Holism was almost completely lost in the 18th century. Recovering both holism and care is an endeavor that might dramatically change the way medicine is practiced but also the moral choices it implies. Empowering a patient for an autonomous choice means caring about them making a wise choice. Healing is not due to parts but to the whole of patients, communities and to the environment. It seems that Plato had this type of wisdom. We might want to recover something our Western culture had back at its roots. This content was lost once the scientific revolution promoted partitioned technical skills and reductive formulas of autonomy.

Author details

Tudor-Ștefan Rotaru[1,2]

1 University of Medicine and Pharmacy "Gr. T. Popa", Iași, Romania

2 "Alexandru Ioan Cuza" University of Iași, Iași, Romania

*Address all correspondence to: tudor.rotaru@umfiasi.ro

References

[1] Jouanna J. Médecine egyptienne et médecine grecque. In: *La médecine grecque antique. Actes du 14eme colloque de la Villa Keyros a Beaulieu-sur-Mer les 10 et 11 octobre 2003.* Paris: Académie des Inscriptions et Belles-Lettres, 2004, pp. 1-21.

[2] Welch CJ. Evidence based medicine misses the "art" of helping humans. *BMJ* 2017; 358: j3426.

[3] Beauchamp TL, Chidlress JF. *Principles of biomedical ethics*. 7th ed. New York ; Oxford: Oxford University Press, 2013.

[4] Shannon TA, Kockler NJ. *An Introduction to Bioethics, fourth edition revised and updated*. New York/Mahwah, NJ: Paulist Press, 2009.

[5] Plato. *PLato in Twelve Volumes. Vol. 3, translated by W.R.M. Lamb.* London: William Heinemann Ltd., 1955.

[6] Plato. *Platonis Opera, ed. John Burnet*. Oxford University Press, 1903.

[7] Liddell, H.G., Scott R. *Greek-English Lexicon: with a revised supplement*. Oxford: Claredon Press, 1996.

[8] Rotaru T-Ștefan. How Can Plato Be Relevant for Contemporary Medicine? *J Intercult Manag Ethics* 2020; 59-65.

[9] Cushman R. *Therapeia: Plato's conception of philosophy*. Chapel Hill, North Carolina: The University of North Carolina Press, 1958.

[10] Plato. *PLato in Twelve Volumes: Charmides. Alcibiades I and II. Hipparchus. The lovers. Theages. Minos. Epinomis. with an English translation by W.R.M. Lamb, vol. XII*. Cambridge: Harvard University Press, 1979.

[11] Hippocrates. *The Genuine Works of Hippocrates. Translated from the Greek with a Preliminary Discourse and Annotations by Francis Adams. Vol. 1.* London: Printed For Sydenham Society, 1899.

[12] Foucault M. *The Birth of the Clinic. Translated from the French by A.M. Sheridan*. London: Taylor & Francis e-Library, 2003.

[13] Lévinas E. *Totalité et infini. Essai sur l'exteriorité*. Dordrecht, New York, London: Kluwer Academic, 1971.

[14] Engelhardt JR HT. *The Foundations of Bioethics, second edition*. Oxford; New York: Oxford University Press, 1996.

[15] National Commission for the Protection of Human Subjects of Biomedical and Behavioral Research. *Ethical Principles and Guidelines for the Protection of Human Subjects of Research*, https://www.hhs.gov/ohrp/regulations-and-policy/belmont-report/read-the-belmont-report/index.html (1979).

[16] Oxford Dictionaries. *The Concise Oxford Dictionary*. Tenth, rev. New York, NY: Oxford University Press, 2001.

[17] Plato. *Plato in twelve volumes: V The Republic translated by Paul Shorey*. Cambridge: Harvard University Press, 1982.

[18] Plato. *Plato in twelve volumes: Volume 10, The Laws, books I-VI translated by R.G. Bury*. London: William Heinemann Ltd., 1967.

[19] Rotaru T-Ş. Who is the Patient in Plato's Dialogues? *Hermeneia J Hermeneut Art Theory Crit* 2018; 227-233.

[20] Avicenna. *The Canon of Medicine of Avicenna*. New York, NY: AMS Press, 1973.

[21] Foucault M. The order of Discourse. In: Young R (ed) *Untying the text: a*

post-structuralist reader. Edited and introduced by Robert Young. Boston, London and Henley: Routledge and Kegan Paul, 1981, pp. 48-78.

[22] Gadamer H-G. *The Enigma of Health: The Art of Healing in a Scientific Age. Translated by Jason Gaiger and Nicholas Walker*. Stanford, CA, US: Standford University Press, 1996.

[23] Emanuel EJ, Emanuel LL. Four Models of the Physician-Patient Relationship. *JAMA* 1992; 267: 2221-2226.

Chapter 24

Evolution of Catholic Marriage Morality in the Twentieth Century from a Baby-Making Contract to a Love-Making Covenant - Part II: Code of Cannon Law to Vatican II

Edward Collins Vacek S.J.

Abstract

Sexual ethics in the West has been evolving, in practice and in theory, over the last century. The official Catholic Church teaching was challenged by many Christian churches and by the changing culture of the West. The Vatican insisted that no change could be made in its timeless truths. Nevertheless, each challenge required ever more sophisticated and convoluted arguments. The impetus for change came through the Western shift from seeing sexual activity as a procreative act toward viewing it as a way for husbands and wives (and gradually also any consenting adult) to express and deepen love. The Second Vatican Council accepted this new view, but subsequently the official teaching became more strict, insisting that both procreation and marital love-making must be present. The teaching of Pope Paul VI prohibiting contraception was the proverbial straw that broke the camel's back for many Catholics. They abandoned the official teaching, recognizing that it was the new personalist view itself that complicated the meaning of marriage. Subsequently, the Canon Law tried reestablish the validity of loveless sex in marriage–the dominant view through the centuries. That move was rejected.

Keywords: sexual ethics, marriage, birth control, covenant, Canon law, Pius XI, Pius XII, Paul VI, Vatican II

1. Background introduction

The official teaching of the Catholic Church on sexuality evolved significantly during the twentieth century. The Church encountered rapidly changing understandings of sexuality among both secular and religious groups. Church teachings, Charles Taylor observes, had been "connected with the denigration of sexuality, horror at the Dionysian, fixed gender roles" and involved an "unfortunate fusion of Christian sexual ethics with certain models of the 'natural.'"[1] From 1917 to 2000, the Church monumentally changed the foundations of its sexual and marital ethic.

[1] Charles Taylor, "Sex and Christianity: How has the Moral Landscape Changed?" *Commonweal* 134, no. 16 [September 18, 2007], 16. *Notes for Church documents at end.*

IntechOpen

likewise is twofold, namely, 4a] "mutual help" and 4b] "allaying of concupiscence." Lastly, marriage has two "essential properties," which are 5] "unity" and 6] "indissolubility."

In general, a contract is an agreement of wills between two or more parties. The nature of this consent is laid out by the Code in exact form: "Matrimonial consent is an act of the will by which each party gives and accepts a perpetual and exclusive right over the body, for acts which are themselves suitable for the generation of children" [1917: 1081.2].[7] The Code presupposes that the marriage contract is not subject to negotiation. People are free to enter or not into marriage, but they are not free to alter the rights and obligations of this institution. This contract is not structured primarily for the individual needs of the spouses but for producing children for the species.

According to the Code, spouses consent to give and accept a right to use the other's body for purposes of sexual activity. In this, it follows St. Paul: "the wife does not have authority over her own body, but the husband does; likewise the husband does not have authority over his own body, but the wife does" [1 Cor. 7:4]. Once married, spouses must, if asked, pay the marital "debt" of sexual intercourse. Spouses do not have the authority to say, "No." In this sense, marital rape is not possible.

Contracts typically focus chiefly on behaviors and not on the interior attitudes that became so important in personalist philosophies later in the century. In the Code it is not necessary, for validity of the contract, that the spouses have any affection for one another. Indeed, spouses can get married even though they live in different countries and have not previously met.[8] This is not to say that the Church encouraged loveless marriage. For example, Pope Leo XIII earlier wrote that spouses "are bound, namely, to have such feelings for one another as to cherish always very great mutual love" [1880: 11; L:6] [1]. But such love is not necessary for the validity of the contract.

One major reason why mutual love is not necessary for marriage is due to the influence of the Pauline writings. While there is a precedent for a connection between marriage and love in Ephesians [5: 25–32], the love urged there is not mutual.[9] More importantly, few if any biblical texts have shaped Christian sexual ethics as much as 1 Corinthians 6–7. This text tends to make sex and love incompatible bedfellows. Paul said that a husband should relate to his wife as if he had no wife [1 Cor. 7: 29]. Augustine reinforced this attitude: "Thus it is characteristic of a good Christian to love in one woman the creature of God whom he desires to be transformed and renewed, but to hate corruptible and mortal intimacy and

[7] Throughout this essay, references to Church documents will be made in the text itself. The online sites where these documents can be found are listed in the bibliography. Usually, in references to Church documents, I will add an "L." The "L" in the citation refers to translations made by Odile M. Liebard, *Love and Sexuality: Official Catholic Teachings* [Wilmington, NC: McGrath Publishing, 1978]. Liebard's numbering often makes it easier to locate exact citations, since Liebard numbers each paragraph. Unfortunately, he does so consecutively in a way that makes later documents begin their numbering with the next number after the last number in the previous document.

[8] This is still true; see the current *Code of Canon Law*, [1983: 1104.1]; Ladislas Örsi, S.J., *Marriage in Canon Law: Texts and Comments, Reflections and Questions* [Wilmington, DL: Michael Glazier, 1986], 52–53.

[9] Ephesians' recommendation [5:21–33] of mutuality, when spelled out, holds only that husbands should love their wives, in imitation of Christ's love for the Church and, curiously, as a form of loving their own bodies, but that wives should obey and respect their husbands.

copulation–that is, to love the human being in her creaturehood but to hate that which makes her a wife."[10]

The Code makes a specifically theological claim that "Christ our Lord elevated the very contract of marriage between baptized persons to the dignity of a sacrament" [1917: 1012.1]. It is now widely recognized that this is not a historically true statement.[11] Still, after Trent, this claim is asserted to be theologically true, with the hazard that theological truth and historical truth follow separate paths. A further divergence appears in that the biblical Jesus and St. Paul recommend celibacy in sacramental terms, but the Church has chosen not to make vowed celibacy a sacrament.[12]

The Code draws a not-obvious conclusion from Christ's elevation of marriage: "Therefore it is impossible for a valid contract of marriage between baptized persons to exist without being by that fact a sacrament" [1917: 1012.2].[13] This contention leads to some severe problems, which make church law foreign to the intuitions of most people. First, although most Protestant Churches deny that marriage is a sacrament, the Catholic Church teaches that Protestants who marry in fact receive the sacrament in spite of their sincere intention or adamant determination not to receive a sacrament. Second, the Church teaches that Catholics who are baptized but no longer believe are simply unable to get married, even though everyone has a right to get married. On the one hand, they will not and should not ask the Church to marry them, since they no longer believe. On the other hand, their attempt to marry civilly outside of the Church is invalid. Such people, as well as all those around them, likely think they are married, but the Church says they are in fact fornicating. Well-known canonist Ladislaus Orsy, S.J., describes these results as "absurd."[14]

Central to the Code is its natural law view of marriage's purposes. Activities are distinguished by the ends or goals they pursue. "The primary end of marriage is the procreation and education of children; its secondary end is mutual help and the allaying of concupiscence" [1917: 1013.1]. It should be noted that neither the flourishing of the individual spouses nor their personal communion are ends of marriage. Behind the Code's teaching on the primacy of procreation are the theologies of Augustine and Aquinas. For them, sexual activity was directed to the continuance of the species, not to the good of the spouses. In fact, Augustine thought that sexual activity was usually immoral, though excused by the good of procreation. Thomas developed a rather complete theology of marriage out of the nature of sperm. For him, unlike other bodily fluids, sperm is not directed to the man's good.

[10] Augustine, *The Lord's Sermon on the Mount* in *Ancient Christian Writers* [Westminster, MD: Newman, 1948], bk. 1, ch. 15, #41.

[11] Örsi, Marriage in Canon Law, 53.

[12] St Paul [1 Cor. 7] recommended celibacy instead of marriage. When Paul comments that a wife will be anxious to please her husband, he does not see this desire to please as an expression of love. Rather, he interprets it as an occasion for her to turn away from the Lord. The unmarried are described as holy in both body and spirit, while the married are those concerned about the things of the world. In other words, Paul's advice in First Corinthians does not present marriage as a central relationship where love of God and marital love of neighbor unite.

[13] The "therefore" seems to be a reverse reading of history. For much of history prior to Trent, Christians got married without thinking of marriage as one of Christ's sacraments. When then the Church decided that marriage was a sacrament, it became necessary to say that all those previous marriages had been sacraments even if people were not aware of receiving a sacrament. See Joseph Martos, "Marriage: A Historical Survey," *Perspectives on Marriage: A Reader*, Kieran Scott & Michael Warren, eds., 3rd ed. [New York: Oxford, 2007], 60.

[14] Örsi, *Marriage in Canon Law*, 56.

Rather, sperm is designed by God to continue the species. Thomas then justifies the long-term bond of marriage because the education of children takes many years and women are not naturally capable of doing that task alone.[15]

Marriage, according to the Code, has a two-part secondary end. This end is to remedy the spouses' insufficiencies and evil tendencies. The first part is "mutual help," which refers to tasks that need to be done in the ordinary course of living, e.g., laundry. The focus is on deeds, not on sharing personal life with the spouse. The second part says that people get married in order to remedy concupiscence. For St. Paul [1 Cor 7:2, 5, 8, 36], marriage is a solution to the problem of lust. Similarly, Aquinas held that marriage is a sacrament because it is a remedy against sin.[16] Luther memorably opined, "The temptation of the flesh has become so strong and consuming that marriage may be likened to a hospital for incurables which prevents inmates from falling into graver sins."[17]

While the primary and secondary ends of marriage focused respectively on the child and on the limitations and problems of spouses, the essential properties of marriage, which are "unity" and "indissolubility," name characteristics of marriage as an institution [1917: 1013.2]. "Unity," in the mind of the Code, like *fides* in Augustine,[18] is not the same as love. Rather, it is a negative term, meaning exclusivity. It forbids sexual activity with anyone other than one's lawful spouse. "Indissolubility" likewise is negative: it forbids divorce. It is said to acquire "a peculiar firmness in Christian marriage by reason of its sacramental character." The Code had to add a qualification like "peculiar firmness" because from its very beginning the Church has dissolved indissoluble marriages. The Church–due to the pressure of real life difficulties, the Pauline privilege, the Matthean exception for *porneia*, the distinction between *ratum* and *consummatum*, and the Petrine privilege–altered any absoluteness deriving from Jesus's prohibition of divorce. Because of these exceptions, the vast majority of all indissoluble marriages in the world are, in principle, dissolvable. Throughout much of the twentieth century, pressure within the Church for further exceptions increased, often masquerading under the rubric of annulments.

It can be noted in passing, though the point is significant, that procreation is given as the purpose of marriage, not the purpose of sexual intercourse. Subsequently, there arose a focus on the specific nature of sexual act and of how it itself might be violated. Thus birth control even outside marriage eventually became an intrinsically evil act. Subsequently, this allowed the Church to teach that while, for good reasons, it was no violation of marriage to be infertile, it was a violation of sex to prevent it from being fertile. Similarly, there is no assertion of any inseparability between procreation and love since love was not necessary for the validity of a marriage.

Christians should not live by legal codes alone, and so it was important for theologians and the papacy to develop theologies of marriage during the rest of the twentieth century. More sensitive to communal reception and pastoral practice, such theologies progressively modified official Church teaching.

---

[15] Thomas Aquinas, *Summa Contra Gentiles*, trans. Vernon J. Bourke [Notre Dame: University of Notre Dame: 1975], bk. 3. ch. 122. Nevertheless it should be noted that Aquinas also describes a sweet friendship that grows between the spouses.

[16] For Aquinas, friendship and mutual help that are part of marriage belong not to its pre-lapsarian essence nor to its sacramental quality, but to its institution in civil law. ST 3:42.1–2.

[17] Martin Luther, *Luther's Works*, vol. 44 [Philadelphia: Fortress, 1966], 9.

[18] Augustine: *Against Julian* in *Fathers of the Church* [New York: Fathers of the Church, 1957], bk. 3, ch. 16, # 30.

DOI: http://dx.doi.org/10.5772/intechopen.95101

3. Pius XI

The winds of the twentieth century were pushing against the wall of tradition. In response, Pius XI devoted an important encyclical, *Casti connubii*, to the topic of marriage. He resists some changes but welcomes others. Along with his predecessor Leo XIII, he still holds that, in things like marriage and sex, it is "more useful and salutary" that they "remain in their natural state, unimpaired and unchanged." God knows best, and only the wickedness of men would try to change this natural order [1930: 95; 1880: 25] [2]. Crucially, the pope holds that this order "is entirely independent of the free will of man" [1930: 6]. In this view, marriage is not an institution that humans through "trial and error" devised to meet certain needs and that might change when those needs change. The underlying image of marriage is that of entering an institution that has established rules and purposes. One cannot change these rules and purposes. And, once inside the institution, one cannot choose to leave.

Pius XI upholds the absolute sexual prohibitions that have been the hard core of Church teachings in the area of sexuality: "Since, therefore, the conjugal act is destined primarily by nature for the begetting of children, those who in exercising it deliberately frustrate its natural power and purpose sin against nature and commit a deed which is shameful and intrinsically vicious" [1930: 54]. The probable background for these claims is, on the one hand, "sanctity" concerns surrounding sexuality and, on the other, the historical connection often made between contraception and murder, since both are understood to be against "life." The Church does not explain why sexual aberrances are "intrinsically vicious" or "intrinsically evil." In Church teaching few acts other than genital acts are placed in this category. For example, inspite of Genesis 3:16, there has been no prohibition of anesthesia during childbirth, Caesarian sections, or subsequent wet-nursing. To say that some acts are intrinsically evil is to say there are no exceptions. It renders needless any consideration of the particulars of real situations. That is, some acts are wrong, no matter how much good they might bring about or how much evil, for example, the death of a wife, they might prevent [1930: 61].

Nevertheless, contrary to this absolutism, the pope makes two strange concessions. Pastorally, he proposes that when one spouse is practicing contraception, the other spouse is guiltless as long as that spouse does not formally consent to the sin [1930: 59].[19] For moral theologians, this should be an astounding claim. In no other area of life is such immediate and indispensable cooperating in serious sin allowed. In allowing this exception, which goes back to Augustine,[20] the rights and duties of marriage override the strictness of moral theory.

Pius XI makes a second adaptation that had implications that occupied theologians through much of the rest of the twentieth century. Augustine held that sexual intercourse when procreation could not happen was sinful.[21] Instead, Pius XI wrote that, although the "conjugal act is destined primarily by nature" for begetting children, it is not against this nature to engage in sexual intercourse when, due to "natural reasons either of time or certain defects," no children can be begotten [1930: 59]. Around the time of this encyclical, the menstrual cycle of women was being better understood. That new understanding laid the biological basis for the rhythm method and later for natural family planning. Pius proposed that, even

[19] The concession is repeated in the Pastoral Council for Families document, "Vademecum for Confessors," [1997: III.13].

[20] Augustine, *Letters*, vol. 5 [New York: Fathers of the Church, 1956], #262.

[21] Augustine, *Catholic and Manichaean Ways of Life* [Washington: Catholic University of America, 1966], 2.18.65.

though sexual activity cannot achieve its reason for existing, it is permissible. Thus, an opening was made that there is no necessary moral connection between marriage, sex, and procreation. As a result of this concession, countless trees have been felled for books and articles debating the issues that opened up. More importantly, countless lives have been thrown into religious and moral turmoil.

Pius XI points to the secondary ends of sexual intercourse as justification for engaging in sex when it cannot achieve its primary end. When he does so, however, he adds a new secondary end: the "cultivating of mutual love," which, he says, goes beyond "mutual help." He then draws a revolutionary implication: the "mutual inward molding of husband and wife. .. can in a very real sense. .. be said to be the chief reason and purpose of marriage" [1930: 23–24]. He says that the traditional view that marriage is primarily for procreation is itself only a restricted sense of marriage. In its fuller sense, marriage is for "mutual interchange and sharing" [1930: 24]. Then, perhaps for the first time in official Catholic teaching, he describes sexual intercourse as "the cultivating of mutual love" [1930: 59].

When describing this love, however, Pius XI reverts back to that strand of the tradition that insists that such love includes no seeking of one's own advantage but only the good of the other, much as Christ loves the Church [1930: 23]. Throughout much of the century, Church descriptions of love tend to use terms such as self-sacrifice or self-gift. That people get married and engage in sexual activity also as a form of self-love or to receive love has only gradually been admitted. Fulfilling a basic human drive has usually been described negatively as concupiscence and lust. Church teaching during the early part of the century offered little affirmation that seeking pleasure can be healthy and normal. More importantly, there was little awareness that sexual intercourse is a "pleasure-bond" that contributes greatly to holding loving marriages together.[22] Western culture has come to affirm openly that marriage is a central locus of *eros* and *philia* or mutual love.[23] Gradually, without denying the place of self-sacrifice and the importance of *agape*, the Church has accepted this view of marriage.

4. Pius XII

The enormous cultural changes of the twentieth century in the Western world made more urgent the question why people should simply follow nature. Developments in biology and medicine changed the so-called natural order and enhanced human life. The control made possible by science abetted the Enlightenment's emphasis on human freedom. A growing sense of personal dignity made it unseemly to describe marriage as an agreement to use another's body as a means to make babies and to temper lust. Similarly, it seemed dualistic to speak of "using" one's own body. Reacting against these and other challenges, Pius XII reasserted a natural law foundation for marriage and devised new characteristics to proscribe recent medical possibilities. According to the pope,

"In forming man, God regulated each of his functions, assigning them to the various organs.. .. God fixed, prescribed, and limited the use of each organ. He cannot allow man now to arrange his life and the functions of his organs according to his own taste, in a manner contrary to the intrinsic and immanent function assigned them. Man, in truth, is not the owner of his body nor its absolute lord,

[22] William H. Masters and Virginia E. Johnson, *The Pleasure Bond* [New York: Bantam, 1974].

[23] For the development of these three distinct types of love, see Edward C. Vacek, S.J., *Love, Human and Divine: The Heart of Christian Ethics* [Washington, DC: Georgetown University, 1994].

but only its user. A whole series of principles and norms derives from this fact" [1944; L: 204] [3].

This text again sets down biological nature as the basis for the Church's teaching in sexual morality. By identifying the work of God and the work of nature, the Pope implied that any direct interference with or alteration of our bodily nature is in fact a direct rejection of God's work. But, since the Church allowed such interference in all of the rest of creation, even though God is also the "owner" of such creatures, it became clear that the only area off-limits to human intervention was human nature. Again, since the Church came to allow considerable interference in all the organs of the human body, except the sexual organs, it became clear that the ban on changing nature was restricted solely to human sexual organs. For example, while artificial insemination is permissible in other animals, for human beings "In the case of artificial insemination one should not only keep a very cautious reserve," but also one "must exclude it altogether." This is so, even though it would enable the couple to achieve the primary end of their sexual organs and of marriage [1949; L: 256; 1951: 318] [4, 5].

Pius XII uses the same biologistic argument against contraception. He repeats his predecessor's claim that, no matter how grave the consequences, it is wrong to "deprive this [marital] act of its inherent force or to impede the procreation of new life" [1951; L: 288, 291]. Thus, direct sterilization, even when the removal of a woman's ovaries might protect her from a life-killing pregnancy, was completely forbidden. Her sexual organs do not exist for her good but "for the conservation of the human race" [1951; L: 300]. She is not allowed to change the function of her organs "in a manner contrary to the intrinsic and immanent function assigned them." Still, indirect sterilization, such as in the removal of cancerous ovaries, was permitted to save her life [1951: 45]. Here compassion wins out.

Like his predecessor, Pius XII held that it is permissible for spouses to restrict sexual activity to infertile periods. Remarkably, he added that for good reasons, such as eugenics or health, spouses may choose to avoid procreation for the entire duration of their marriage, even though, again, this practice seems to undermine the primary purpose of marriage. He holds that persons can make the decision not to procreate even in advance of getting married [1951; L: 296, 298]. One could surmise that it was becoming clearer that marriage was a great good, quite apart from procreation. Catholic authors explain this seeming exception by insisting on an "in-principle" view of marriage and sexual activity.[24] Both are essentially procreative, even when they actually cannot or morally ought not be procreative.

Other authors, embracing personalist philosophy, pressed the Church to include a greater emphasis on the person in its sexual teaching. They argued that sexual intimacy, "the expression and actuation of the personal and affection union," is equal to procreation or independent of it [1951; L: 310–314].[25] Against that view, Pius XII insists that marriage "is not ordered by the will of the Creator towards the personal perfection of the husband and wife as its primary end, but to the procreation and education of a new life.. .. This principle holds good for all marriages, even if they are unfruitful" [1951; L: 312]. He holds that all that is profound in married love should be at the service of the children, such that complete self-sacrifice of their own needs is demanded of the spouses [1951; L: 314, 316]. The Pope cryptically

[24] Dennis Hollinger, "Good Sex: Its Meaning and Morals," in *Moral Issues and Christian Reponses*, 8th edition, ed. Patricia Beattie Jung and L. Shannon Jung [Minneapolis: Fortress Press, 2013], 122–26.

[25] In 1944, the Roman Rota had taken up a new theological challenge from personalism, namely, that "the evolution and perfection" of the husband and wife is not secondary but a primary end of matrimony." In response, it said: "These newcomers to matrimonial matters stray from true and certain doctrine" *Acta Apostolicae Sedis* 36 [1944], 103.

added that if God had wanted sex to be primarily about mutual love, God would not have designed the sexual act the way God created it [1951; L: 328].

Nevertheless, Pius XII introduced somewhat of a personalist argument in response to those who were arguing that artificial insemination would in fact enable some spouses to fulfill the procreative purpose of marriage. Pius argued that if all God desired was the "union of two life-germs," God would not have devised nature so that procreation requires the "personal cooperation" of the husband and wife [1951; L: 318]. Indeed, Pius's theology gradually shifted to allow personalist concerns to play a greater role in the Church's theology of marriage. For example, he describes a child as "the true and complete expression of [spouses'] reciprocal love" [1956; L: 500]. Then, making a novel addition to the tradition, he set out the precursor to what became for Paul VI the inseparability principle: "Never is it permitted to separate these various aspects to the positive exclusion either of the procreative intention or of the conjugal relationship" [1956; L: 503] [6]. It is this "never" that increasingly split the official Church teaching from the intuitions and practice of so many of the people of God.

5. Vatican II

The Second Vatican Council is justly famous for its shift in style. It made no normative changes to official teaching on particular sexual practices. Rather it offered a strongly positive affirmation of both marriage and sexuality. Whereas, in the popular Catholic mind, the Church taught that sexuality was the locus of sin, now it became a locus of grace. The Council demonstrated this change of attitude when it wrote that sexual actions within marriage are "noble and worthy ones.. .. These actions signify and promote that mutual self-giving by which spouses enrich each other." Shifting the emphasis from procreation, *Gaudium et spes* describes marriage as "the primary form of interpersonal communion," a "community of love," a "conjugal covenant," and an "intimate partnership of married life and love" [1965: 12, 47, 48, 49] [7]. This married sexual love is affirmed as eminently human, involves the good of the whole person, and, most remarkably, merges the human with the divine. In this context, children are described as the "ultimate crown" of this married love and the "supreme gift of marriage" [1965: 48–50].

For most of its history, the Church did not understand sexual activity in terms of love. Now, without giving any explanation for the change, the Church insisted that sexual activity both expresses and perfects love [1965: 49]. To use colloquial language, spouses "make love." Even more, this love is now described as a "total love" [1965: 49]. One might best appreciate the exuberance of these descriptions by contrasting them with that of Augustine, who encouraged spouses to give up sexual activity as soon as they were able, because sexual activity usually involves sin.[26] That advice makes little sense if sexual intercourse is "love-making." In contrast, the Council said that sexual activity signifies and promotes mutual self-giving [1965: 49].

It is well known that Vatican Council II did not reaffirm the past teaching of primary and secondary ends. The Council fathers assert that the "other purposes of matrimony" should not be made "of less account." Like previous popes, the Council insisted that both marriage and marital love are ordained toward children. It now elevates this role by proclaiming that through procreation parents participate in

[26] Augustine, *The Good of Marriage*, in *Fathers of the Church* [New York: Fathers of the Church, 1955], 3.3.

God's own creative work; they cooperate with God's own love in enlarging God's family [1965: 50].

The Council fathers then took up the contentious question of "harmonizing conjugal love with the responsible transmission of life." The underlying challenge was that of birth control. The phrase "transmission of life," referring to the life of the species, reflects a biological perspective. The challenge was that many people had the experience that methods of birth control seemed to help spouses grow in conjugal love. As is well known, Paul VI decided that all the Council fathers, meeting in solemn assembly, along with their expert theologians, were not qualified to make a decision on this topic. So constrained, the Council members simply recognized that there are difficulties in the present era that make this harmony difficult. In an irenic, if naïve, fashion, they asserted that there cannot be any contradiction between these two natural inclinations, since God created both [1965: 51]. Breaking with a past attitude that encouraged large families, they recognized that there may be strong reasons to limit family size. They encouraged prospective parents to take into account a host of personal, social and historical factors in deciding the number of children [1965: 50]. The Council importantly affirms that it is the right of the parents to make this decision. However, these "responsible parenthood" decisions must include more than good intentions and should refer objectively to the "nature of the human person and his acts," in the context of love [1965: 50]. Then, in a restrictive clause, it tells "sons of the Church"--thus not making a natural law claim--to avoid what the teaching authority might eventually determine to be blameworthy [1965:51].

6. Paul VI

The encyclical *Humanae vitae* by Paul VI summarized, solidified, and somewhat extended points that had been made earlier in more informal "addresses" by Pius XII and by Vatican II. In that sense, the teaching was not particularly new. Paul VI tried to preserve the same norms, while also honoring the more positive theology of Vatican II. Thus, he affirms, "husband and wife tend toward the mutual communion of their beings in view of personal perfection, to collaborate with God in the generation and education of new lives" [1968a: 8] [8]. Paul VI points out that, even if not fertile, sexual acts always remain "ordained to expressing and consolidating their union" [1968a: 11].

The pope then addressed the central question of the encyclical: whether there is a moral difference between natural family planning methods and other methods such as the "pill." *Humanae vitae* masterfully acknowledges almost all the arguments made in favor of these other methods of birth control and against the old restrictions. Then he rejects these arguments on two grounds: the intrinsic evil criteria and the inseparability principle. The first subtly appeals to a sacrosanct character of sex, whose violation Aquinas thought was worse than sacrilege.[27] The second appeals to the laudable goal of a fully integrated life. Both grounds deserve consideration.

First, it bears repeating that an intrinsically evil act is said to be wrong, no matter how much evil would be avoided or how much good would be achieved [1968a: 14]. Thus, on the one hand, any positive reason for using contraception is immediately cast into the framework of doing evil (sin) to bring about good. On the other hand, any suggestion that using contraception might avoid evil is shunted off and directed to spirituality considerations: it is better to suffer evil or death (with

[27] *The Summa Theologica of Saint Thomas Aquinas* [Westminster, MD: Christian Classics, 1948], I-II.154.12.

Christ) than to do something wrong. In short, although Paul VI blesses human intelligence, which makes humans similar to God, he holds that this intelligence cannot be used to intervene to alter the biological sexual order established by God [1968a: 16]. Thus, even though "God has wisely disposed natural laws. .. which, of themselves, cause a separation" between sexual activity and procreation, that is, during most of the menstrual cycle, humans may not use their intelligence to do the same. Rather, "each and every marriage act must remain open to the transmission of life" [1968a: 11]. As is well known, such a claim was not widely persuasive. Taking a pill did not seem intrinsically evil to most people. In fact, it seemed prudent and loving. They agreed with Vatican II that sexual activity is holy in the sense that it cooperates with God, but they did not agree that it is sacred in the sense that it is off limits to human intervention.

Second, Paul VI reasserted the inseparability principle [1968a: 12]. Shifting from the teleological language of "purposes" or "ends," he uses the word "meanings" (or "significations"), which point to an essential "nature." The point is that sexual activity, abstractly, is the "kind" of activity that has a procreative "meaning." This reference to kind of activity allows the procreative "meaning" to be honored even when no procreation is possible, that is, when the body itself is not "open to the transmission of life." Contrary to at least one strand in John Paul II's theology of the body, Paul VI does not demand that spouses be psychologically open to children. For Paul VI, there is no sin when spouses agree "in the positive will of avoiding children for plausible reasons, seeking the certainty that offspring will not arrive" [1968a: 16]. Rather "inseparability" forbids only actively doing something to cause infertility. Again, what God does (namely, separate fertility and union), humans may take advantage of; but humans may not actively do what God does [1968a: 11].

Paul VI nicely finessed the traditional teaching that husbands and wives have the right to use the body of their partners. He says that having rights over the body of another for sexual acts does not mean that one can impose a conjugal act on one's partner. The Pope does not argue that marital rape would be a violation of rights (marriage remained a contract that gives to another rights over one's own body); rather it is a violation of love, which is the new personalist criterion shaping the Church's sexual and marital ethics [1968a: 13].

The Pontiff extended the Council's teaching on love by describing it as a "special form of personal friendship, in which husband and wife generously share everything without undue reservation" [1968a: 8]. This is an important addition, since previous teaching tended to treat love in a generic way, sometimes describing it simply as "spiritual and disinterested." Because marriage is a special form of friendship which is both bodily and interested, spousal love can allow or expect certain expressions of this love that are not permissible in other forms of friendship. Thus, the Church now offered a personalist argument for sexual exclusivity.

A week after *Humanae vitae*, Paul VI further embraced the "personalist conception" of married love [1968b; L: 1238]. Such love is "preeminent" in the subjective dimensions of marriage [1968b; L: 1238]. Paul describes it somewhat more carefully than his predecessors. It is "not a total fusion. Each personality remains distinct" [1970; L: 1235] [9]. This point balances out the frequent Church assertion, citing Genesis, that the two become one flesh. The latter assertion had been used as a basis for denying the independence of women within marriage. More expansively, Paul VI writes, "There is no married love that is not, in its exultation, an impulse toward the infinite, and that does not wish to be, in the impulse, total, faithful, exclusive and fecund" [1970; L: 1345; 1968a: 9]. As will be seen below, John Paul II will drop the qualifiers of "in its exultation," "wish to be," and "in the impulse." For him, married love must be total, or else it is wrong. For Paul VI, conjugal chastity is a step by step process. He rebuffs the commonplace assumption that, after the marriage ceremony, any sexual desire or practice is chaste. Rather, marital chastity is a life-long process of integrating "manifold tendencies" [1970; L: 1362]. John Paul II will insist, with the idea of total love, that love must always be complete.

Not surprisingly, the new theology of marriage as a personal covenant created problems with the contract notion enshrined in the 1917 Code. Paul VI himself underscored the disjunction: "conjugal love" plays a "lofty and necessary role in marriage," but it plays no role in the canonical law about marriage [1976; L: 1609] [10]. He spoke against those newcomers who make the "validity" of marriage dependent on the presence of love [1976; L: 1603]. Any lack of love among married persons affects not in the least the traditional teaching of the absolute impossibility of divorce. Hence, Paul VI insists, all that is needed for a valid marriage is the "indivisible moment" of consent [1976; L: 1606]. After saying "I do," no other decision of the will or any absence of love can make the slightest difference in the validity of marriage [1976: L: 1606]. This legal logic hardly matched the messiness and the narrative character of married life. The cognitive dissonance between official Church teaching and people's experience of loveless marriages as well as life-giving second marriages led many people to abandon the Church.

The ecclesial consequences of *Humanae vitae* were enormous. They spring from two of the concerns that Haidt argues, as we saw, distinguish political and religious conservatives, namely, authority and loyalty. Coming at the time of an array of cultural world revolutions, *Humanae vitae* fractured the back of Church authority and divided Catholics. While Paul VI reaffirmed the tradition in order not to undermine Church authority, the result was the opposite. Many of the liberated Catholics insisted that they had to "follow their own conscience," a teaching that had received some support in Vatican II but that also fed upon an expanding individualism. Mostly in vain, the defenders of Paul's document pointed out that Vatican II's affirmation of conscience was accompanied by restrictions such as the demand to be "submissive to Church's teaching office," the assertion of "objective standards," the rejection of relying on sincere intentions, and the proscription of any "methods of regulating procreation which are found blameworthy by the teaching authority of the Church" [1965: 51]. Disagreement was often characterized as "dissent," a term that unfortunately recast any debate from being a disagreement over truth to being disobedience to authority and a lack of loyalty to the Church. For many, the disjunction was cataclysmic.

7. Congregation for the doctrine of the faith

During Paul VI's tenure, the Congregation for the Doctrine of the Faith (CDF), in its succinct "Declaration on Sexual Ethics," charted new territory in understanding the fundamental topic of sexuality itself, even as it resisted any changes in the norms for specific sexual behaviors that might flow from that understanding. Following developments in psychology, the Vatican asserts that sexuality "so profoundly" affects the human person that it is one of "the principal traits that distinguish" an individual's life [1975: 1] [11]. Where the tradition had primarily focused sexuality on prolonging the human race, here sexuality is considered to be central to a person's identity.

This acknowledgment brought new challenges because many people were engaging in expressing their sexual identity in ways that the Church disapproved. Hence, the CDF decried "erroneous opinions" and a "growing permissiveness" that contradicted the traditional norms. In spite of evidence of widespread uncertainty and disagreement throughout much of the West, the Church declared that it "knows with certainty" that its own norms "are in complete harmony with the

Divine order of creation and with the spirit of Christ, and therefore also with human dignity" [1975: 13].

The CDF was quite explicit: its goal is "to repeat the Church's doctrine" [1975: 6]. To understand the CDF's response, it is helpful to recall what Haidt says about the human tendencies of both sides of an issue. "In moral matters, we. .. deploy our reasoning skills to support our team, and to demonstrate commitment to our team." Haidt adds: "Conscious reasoning functions like a press secretary who automatically justifies any position taken by the president." Haidt notes that all humans practice "confirmation bias," which is "the tendency to seek out and interpret new evidence in ways that confirm what you already think." At the same time, reason works hard to dismiss contrary evidence. In short, Haidt observes, "Moral matrices bind people together and blind them to the coherence, or even existence, of other matrices."[28]

Thus, on the topic of masturbation, the CDF dismissively writes that "facts do not constitute a criterion for judging the moral value of human acts"[1975: 9]. It adds, "Whatever the force of certain arguments of a biological and philosophical nature,. .. both the Magisterium. .. and the moral sense of the faithful have declared without hesitation" that it is "an intrinsically and seriously disordered act" [1975: 9]. In other words, contrary evidence does not affect the Church's position. Taking loyalty for granted, the CDF presumes that the faithful's attitude is in agreement, so no further evidence of that support is needed. Put another way, if any of the faithful were to hesitate or disagree, then they are not among the "faithful." Finally, the CDF, after admitting that no text of scripture condemns masturbation, asserts that "the tradition of the Church has rightly understood it to be condemned in the New Testament." Even if it is not there, it is there.

The CDF acknowledged that many psychologists were arguing that masturbation "is a normal phenomenon of sexual development, especially among the young" [1975: 9]. This appeal to development presented a new kind of challenge, namely, an activity might be appropriate at an early stage of life, even if it is not appropriate for adults. For example, some taught that it was a matter of indifference that young children fondle their own genitals. But in Church teaching this would objectively be one of the worst sexual sins possible (although subjectively innocent). That is, these act would be objectively evil since in children there is neither any possibility of procreation nor any sense of unity with another person. The response of the CDF was not to undertake the task of inserting a developmental understanding of sexuality. Rather, the CDF shifted the argument. It notes that many people who masturbate do not have the freedom to be fully responsible for sexually sinning [1975: 9–10]. This approach preserves the norm that the child's act is objectively a grave matter, but it offers the relief that in children there is no personal guilt.

Not surprisingly, the CDF also rejected premarital sex. It upholds the traditional standard, saying, "every genital act must be within the framework of marriage" [1975: 7]. It explains that nonmarried persons "cannot ensure, in sincerity and fidelity, the interpersonal relationship between a man and a woman" [1975: 7]. It is noteworthy that the Congregation appeals to personalist values. Unfortunately, this approach overlooks that marriage itself cannot guarantee interpersonal stability, as divorce statistics indicate. Only a contract theory, claimed to be from God and both unalterable and unbreakable, keeps marriage technically stable even when in all other ways the interpersonal relationship has died.

The CDF next took up the topic of homosexuality, which had scarcely been present in earlier twentieth century Church teaching but was now being positively assessed in the psychological sciences. Social acceptance of homosexuality was gradually increasing since sexuality was now considered part of one's very identity.

[28] Haidt, *The Righteous Mind*, 79–80, 91, 110.

The CDF, at this point in history, generally understood homosexuality as a pathology due to some incurable instinct or condition. It insists that "homosexual acts are intrinsically disordered and can in no case be approved" [1975: 8]. Even if homosexuals are incurable, they ought not act with their limitations.

Finally, the CDF had to deal with a new challenge raised by theologians who said that many people who engage in prohibited sexual actions may be acting wrongly, but they may not be drastically separated from God. It seemed highly implausible to these critics that, say, an adolescent boy who on one occasion freely enjoyed sexual fantasies was fit for damnation. To this, the CDF repeated the seventeenth century declaration that all sexual sins, no matter how slight they may seem, are objectively serious sins [1975: 10]. The plausibility of Church teaching grew thinner.

Author details

Edward Collins Vacek S.J.
Loyola University New Orleans, Louisiana, United States of America

*Address all correspondence to: evacek@loyno.edu